The Biochemistry of the
Polypeptide Hormones

The Biochemistry of the Polypeptide Hormones

M. Wallis
School of Biological Sciences
University of Sussex

S. L. Howell
Department of Physiology
Queen Elizabeth College
University of London

K. W. Taylor
Department of Biochemistry
The London Hospital Medical College
University of London

A Wiley–Interscience Publication

JOHN WILEY & SONS
Chichester · New York · Brisbane · Toronto · Singapore

Library of Congress Cataloging in Publication Data:
Wallis, M.
   The biochemistry of the polypeptide hormones.

'A Wiley–Interscience publication.'
Includes index.
1. Peptide hormones.   I. Howell, S. L.   II. Taylor, K. W.
III. Title. [DNLM: 1. Hormones.
2. Peptides. WK 185 W214b]
QP572. P4W35   1984   599'.01927   84—5248

ISBN 0 471 90484 8
ISBN 0 471 90897 5 (paperback)

British Library Cataloguing in Publication Data:
Wallis, M.
   The biochemistry of the polypeptide hormones.
   1. Peptide hormones   2. Biological chemistry
   I. Title   II. Howell, S. L.   III. Taylor, K. W.
   574.19'27   QD431

ISBN 0 471 90484 8 (cloth)
ISBN 0 471 90897 5 (paperback)

Printed and bound in Great Britain.

This book is dedicated to the late Professor Asher
Korner, the first Professor of Biochemistry at the
University of Sussex.

Contents

Abbreviations

*Note that for representation of amino acids in Figures and Tables the standard 3-letter code is usually used (J. Biol. Chem. **243**, 3557–3559 (1968)). Nucleic acid bases are represented by the standard one-letter code (A, T, G, C, U).*

ACTH	Corticotropin (= adrenocorticotropic hormone)
ADH	Antidiuretic hormone (= vasopressin)
ADP	Adenosine 5'-diphosphate
AMP (= 5'AMP)	Adenosine 5'-monophosphate
APP	Avian pancreatic polypeptide
APUD-theory	Amine precursor uptake and decarboxylation theory
ATP	Adenosine 5'-triphosphate
BPP	Bovine pancreatic polypeptide
BPTH	Bovine parathyroid hormone
Ca^{2+}	Calcium ion
CCK	Cholecystokinin (= pancreozymin-CCK)
cDNA	Complementray DNA
CLIP	Corticotropin-like intermediate lobe peptide
CNS	Central nervous system
CoA	Coenzyme A
CRH/CRF	Corticotropin releasing hormone/Corticotropin releasing factor
Cyclic AMP	Cyclic 3',5'-adenosine monophosphate
Cyclic GMP	Cyclic 3',5'-guanosine monophosphate
DA	Dopamine
D form	Dependent form (of glycogen synthetase)
DNA	Deoxyribonucleic acid
EC cells	Enterochromaffin cells
EDTA	Ethylenediaminetetraacetic acid
EGF	Epidermal growth factor
EIA	Enzyme immunoassay
ELISA	Enzyme-linked immunoabsorbent assay
END	Endorphin
ER	Endoplasmic reticulum

FDNB	Fluorodinitrobenzene
FGF	Fibroblast growth factor
FSH	Follicle stimulating hormone
FSH–RH/FSH–RF	FSH releasing hormone/FSH releasing factor
GABA	γ-Aminobutyric acid
GDP	Guanosine 5'-diphosphate
GH	Growth hormone
GHRH	Growth hormone releasing hormone
GIP	Gastric inhibitory peptide
GI tract	Gastrointestinal tract
GLI	Glucagon-like immunoactivity
GMP	Guanosine 5'-monophosphate
GnRH	Gonadotropin releasing hormone
G protein	Guanine nucleotide regulatory protein
GTP	Guanosine 5'-triphosphate
hCG	Human chorionic gonadotropin
hGF	Hyperglycaemic factor
hGH	Human growth hormone
hPL	Human placental lactogen
5-HT	5-Hydroxytryptamine (serotonin)
ICSH	Interstitial cell stimulating hormone ($= $ LH)
I form	Independent-form (glycogen synthetase)
IGF	Insulin-like growth factor
K	Kilo (e.g. Molecular Weight 14 K : Molecular Weight 14,000)
K^+	Potassium ion
Ka	Association constant
Kd	Dissociation constant
Km	Michaelis constant
[Leu]enkephalin	Leucine enkephalin
LH	Luteinizing hormone
LHRH	Luteinizing hormone releasing hormone
LPH	Lipotropin
[Met]enkephalin	Methionine enkephalin
Mg^{2+}	Magnesium ion
MIF	Melanotropin inhibiting factor
M_r	Molecular weight
MRF	Melanotropin releasing factor
mRNA	Messenger RNA
MSA	Multiplication stimulating activity
MSH	Melanotropin (Melanocyte stimulating hormone)
Na^+	Sodium ion
NAD(H)	Nicotinamide adenine dinucleotide (reduced form)

NADP(H)	Nicotinamide adenine dinucleotide phosphate (reduced form)
NEFA	Non-esterified fatty acids
NGF	Nerve growth factor
NH_2 group	Amino group
NSILA	Non-suppressible insulin-like activity
OT	Oxytocin
Pancreozymin-CCK	Pancreozymin cholecystokinin
PDGF	Platelet-derived growth factor
PIF	Prolactin inhibiting factor
PIP_2	Phosphatidyl inositol bisphosphate
PP	Pancreatic polypeptide
Pr	Prolactin
PTH	Parathyroid hormone
pyro-Glu	Pyroglutamic acid
PZ-CCK	Pancreozymin-cholecystokinin
RIA	Radioimmunoassay
RNA	Ribonucleic acid
SDS	Sodium dodecyl sulphate
SH-group	Sulphydryl group
SRIF	Somatostatin (somatotropin release inhibiting factor)
S-S linkage	Disulphide linkage
STH	Somatotropic hormone (= growth hormone)
$t_{\frac{1}{2}}$	Half-life
T_3	Triiodothyronine
T_4	Thyroxine
TRH	Thyrotropin releasing hormone
TSH	Thyrotropin (= thyroid stimulating hormone)
U	Units (μU, Microunits; mU, Milliunits)
VIP	Vasoactive intestinal peptide
VP	Vasopressin

Preface

Some years ago, while teaching endocrine biochemistry to third-year science students, we realized that there is a need for a textbook which deals exclusively with the polypeptide hormones. Some of what is written in this book has been the subject matter of lectures given to science students at the University of Sussex, the University of Sydney, Australia, and Queen Elizabeth College, London, as well as to preclinical students at the London Hospital and Charing Cross Medical Schools. The book is directed primarily towards advanced undergraduate science and preclinical students, but we hope it may also be useful as an introduction to the subject for some postgraduates.

In view of the enormous growth in the literature of the subject over the past few years, we have had to be highly selective about what material was included. Inevitably in an area that is moving so rapidly there will be gaps; we hope these have been kept to a minimum. Some of the topics that have been developing most rapidly, such as applications of recombinant DNA technology and hormone receptors, have been dealt with in separate chapters. With regard to references it has been necessary to exercise great selectivity, and we have chosen where possible suitable reviews for further reading.

Finally, but by no means least, our wives, secretaries, colleagues and students must be thanked for their help and tolerance during the lengthy period during which this manuscript was being prepared.

M. Wallis
S. L. Howell
K. W. Taylor

January, 1984

Chapter 1

Ultrastructure of endocrine glands and their secretory mechanisms

1. INTRODUCTION

Secretion—the process by which a substance elaborated within a cell is eliminated from it for use elsewhere in the organism—is a widespread phenomenon which is not restricted only to the traditional gland cell types which occur in discrete tissues and organs. Thus, collagen production by fibroblasts, histamine release from mast cells, albumin release from liver and antibody production by plasma cells, all provide examples of the secretion of biologically important molecules by cells not normally regarded as glandular in function. Furthermore,

1

the actual mechanism of elaboration and secretion of product may be quite similar in different cell types regardless of whether they are individual cells in the circulation (lymphocytes, etc.), exocrine secretory tissues which secrete their products into ductal systems for local use (salivary gland, exocrine pancreas) or endocrine cells which secrete their products directly into the extracellular fluid and thence the blood stream. By definition, the products of endocrine cells are hormones, and the major advantage of their mode of secretion directly into the blood stream is that it enables the hormones to have simultaneous multiple sites of action on different target tissues, which may be far removed from their cells of origin.

2.   CLASSIFICATION OF SECRETORY CELL TYPES

From the point of view of ultrastructure of the secretory cells, a far more important distinction than that between endocrine and exocrine lies in the nature of the secretory product, and here endocrine cells are separable into three distinct cell types.

(1) Cells which secrete their product without a significant period of storage—particularly steroid-producing cells.

(2) Cells which store their product in an extracellular pool before it is secreted.

(3) Cells which store their product within the cell of origin in membrane-bound vesicles. This type of mechanism in particular is common to both endocrine and exocrine cells.

Since the majority of protein and polypeptide hormones come from this last type of cell in which the product is stored within the cells before secretion, these will be considered in this section for the most part. For completeness, however, the principal features of all three cell types are discussed very briefly. A full discussion of the ultrastructure of endocrine cells is given by Fawcett.[1]

2.1   Cells which do not store their products

These are mainly steroid-producing cells in which the starting material, usually cholesterol, is stored and converted enzymatically to the hormonal product required, at the time when secretion is stimulated. Thus, the adrenal cortex contains sufficient corticosteroids to maintain secretion for 3 minutes if hormone synthesis were prevented, the ovary sufficient oestrogens for 10 minutes and the testis sufficient androgen for 18 minutes of normal secretion in the absence of further hormone production. These contrast with storage times of the order of 24 hours for many polypeptide hormones which are stored within the cells. Steroid-producing cells are characterized by a very extensive smooth endoplasmic reticulum, relatively poorly developed rough surfaced endoplasmic reticulum, prominent Golgi complex and the presence of abundant lipid

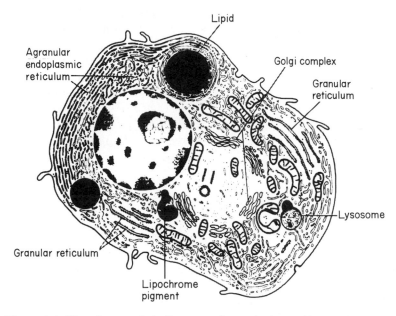

Figure 1.1 The characteristic features of a typical steroid-secreting cell.
(Reproduced with permission from Fawcett, D. W., Long, J. A. and Jones,
A. L. (1969). *Recent Progr. Hormone Res.* **25**, 315–380. Copyright (1979)
Academic Press).

droplets 'liposomes' and lysosomes (Figure 1.1). The liposomes represent the
storage pools of cholesterol which act as precursor of the steroids. The cells thus
store the hormonal precursor rather than the product. The enzymatic pathways
which are involved in the production of the various steroids are summarized in
Figure 1.2. The pathways are complex, involving both smooth-microsomal and
mitochondrial enzymes.

2.2   Extracellular storage of product

The thyroid provides a unique example of extracellular storage of its products:
in this case thyroxine and triiodothyronine. Detailed discussion of the mechanism of production of the thyroid hormones is beyond the scope of this book:
suffice it to say that the hormones are produced in the thyroid follicular cells as
part of a large glycoprotein molecule (thyroglobulin, MW 680 000) which is
then stored extracellularly in the lumen of the follicles and colloid. On initiation
of thyroid hormone secretion the colloid is reabsorbed into the follicular
cells by pinocytosis and the thyroglobulin is degraded by lysosomal enzymes
with release of the native thyroid hormones (Figure 1.3). The regulation of
these pathways of production of thyroxine and triiodothyronine by thyroid-
stimulating hormone is discussed in Chapter 6.

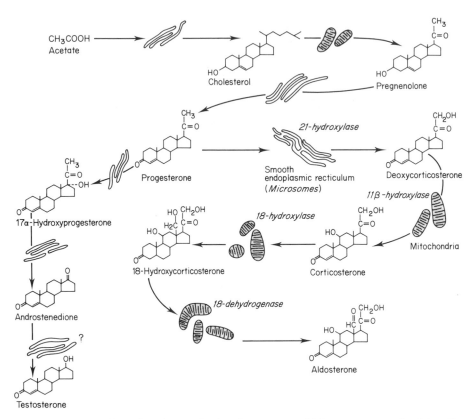

Figure 1.2 Pathways for the synthesis of some steroid hormones emphasizing the endoplasmic reticulum and mitochondrial localizations of the enzymes involved. (Reproduced with permission from Fawcett, D. W., Long, J. A. and Jones, A. L. (1969). *Recent Progr. Hormone Res.* **25**, 315–380. Copyright Alan R. Liss, Inc).

2.3   Intracellular storage of product

Endocrine cells which store their product in membrane-limited vesicles include, in addition to most polypeptide and protein-secreting cells (growth hormone, insulin, etc), those in which the secretory product, although of low molecular weight, can be aggregated in a suitable way to enable it to be stored. Examples of this are the catecholamines (e.g. adrenaline and noradrenaline) which are stored along with specific binding proteins (chromogranins) and adenosine triphosphate, and oxytocin and vasopressin, the octapeptides of the posterior pituitary, which are stored in association with binding proteins called neurophysins. It seems likely that this type of mechanism may represent a way to overcome the problem of aggregation and storage of relatively small molecules by themselves, so that they may be classified on ultrastructural grounds as polypeptide-type storage mechanisms involving membrane-limited granules.[2]

Before a detailed assessment of the ultrastructure of the polypeptide-secreting

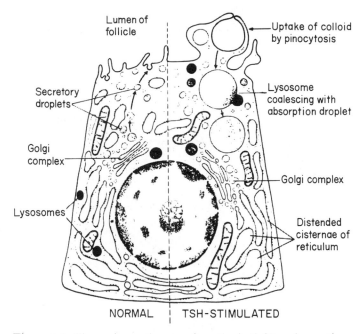

Figure 1.3 The major pathways of synthesis (left) and secretion (right) of thyroxine in the thyroid follicular cell. (Reproduced with permission from Fawcett, D. W., Long, J. A. and Jones, A. L. (1969). *Recent Progr. Hormone Res.* **25**, 315–380. Copyright (1979) Academic Press).

endocrine glands is attempted, it is necessary first to look in outline at the methods which have been used, and are still used, to study their structure.

3.   METHODS OF STUDY

The 'classical' approach to the study of endocrine glands involves extirpation of the gland with subsequent observation of the results, or alternatively/ additionally extraction of the active principle from the gland and injection into a recipient. The observed effects were attributable to the actions of the extracted hormone. The most serious drawback to this approach came in the case of glands which produce more than one hormone, for instance the anterior pituitary (Chapter 3), or where there was accidental removal of more than one gland, e.g. parathyroid when in association with thyroid (see Chapter 13).

3.1   Histology

By the turn of the century histological techniques were becoming established and were used extensively in the characterization of endocrine tissue. Available

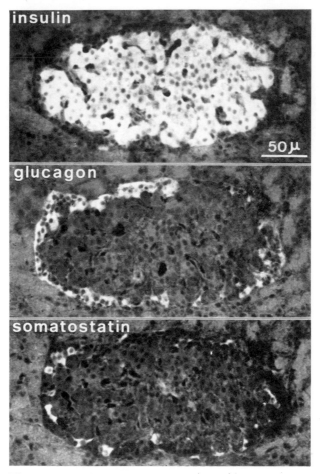

Figure 1.4 Series of three consecutive sections of a rat pancreatic islet stained respectively with anti-insulin, anti-glucagon and anti-somatostatin fluorescent antisera. The pattern is characteristic of islets situated in the body, and at the periphery of the pancreas. (Reproduced from Orci, L., *Metabolism*, **25**, no. 11, suppl. 1, 1303–1313 (1976). Copyright (1976) Grune and Stratton Inc.).

stains concentrated on identifying the acidophilia (affinity for acidic stains) and basophilia (affinity for basic stains) of the cells, resulting in a classification (acidophils, basophils and neutrophils) which survives to this day. Subsequently more specific stains allowed for identification of glycoprotein hormones (Periodic acid-Schiffs) and disulphide groups (aldehyde fuchsin) which proved particularly useful in the identification of cells producing, for example, TSH (a glycoprotein) or insulin (a disulphide-containing protein), respectively. Full specificity of histological staining was only achieved by adaptation of staining methods which allowed visualization of antibodies to hormones at the sites where they were found in the tissue. This is commonly achieved by coupling a specific antiserum to a fluorescent dye, such as fluorescein, which is then incubated with the tissue sections. The antiserum binds to the hormone-containing cells and marks them out by fluorescence. This method has allowed positive indentification of the cells of origin of many of the polypeptide hormones to which specific antisera can be raised and has allowed positive identification of each cell type in, for instance, anterior pituitary[3] or islets of Langerhans (Figure 1.4).

3.2   Electron microscopy

Electron microscopy was first applied to endocrine glands in the late 1950s. It immediately revealed details of the structure of the cells and, in particular, of the nature of the hormone storage granules, whose existence had been inferred from light microscopy, and within a short time allowed identification of the mechanism of their secretion (by exocytosis). It was subsequently possible to perform immunochemical identification of cells and organelles producing specific hormones by coupling of the antisera (or of a 'second antibody') to peroxidase which could, by a series of reactions, produce an electron-opaque deposit visible in the electron microscope.

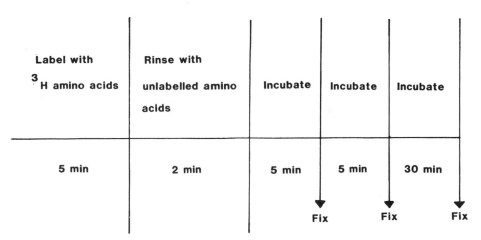

Label with ^{3}H amino acids	Rinse with unlabelled amino acids	Incubate	Incubate	Incubate
5 min	2 min	5 min	5 min	30 min
		Fix	Fix	Fix

Figure 1.5 Sequence of incubations involved in the labelling and incubation of tissue for autoradiography. After fixation the tissue is processed for electron microscopy by standard procedures

A further refinement of electron microscopic technique which has been widely applied to polypeptide hormone-producing cells is that of auto-radiography.[4] The newly-synthesized hormone is labelled biosynthetically for a short period by incorporation of ^{3}H-labelled amino acids (pulse-labelling),

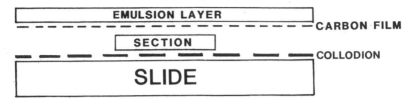

Figure 1.6 Sections of tissue are mounted on collodion-coated glass slides before exposure to the photographic emulsion. The carbon film prevents chemographic exposure of the film by materials in the section, but permits passage of radioactivity

and then tissue is fixed at various times during a subsequent incubation (chase incubation) (Figure 1.5). The tissue is then embedded and sectioned in the usual way, the fixation and washing procedures serving to wash away the free labelled amino acids, while those incorporated into the protein are fixed at their final position in the cells. The sections are then coated with a thin layer of a photographic emulsion and incubated in the dark for 4–8 weeks (Figure 1.6). The emulsion is then developed in the same way as for photographic film, the resultant product being a section overlaid with emulsion. The developed grains appear as black dots in the light microscope and as 'worms' in the electron microscope (Figure 1.7). By examining tissue fixed at various times after such a

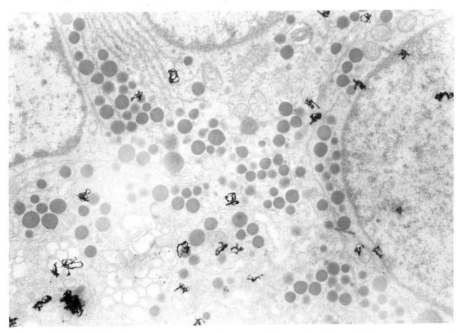

Figure 1.7 Typical electron microscopic autoradiograph. The cell is a somatotroph in the anterior pituitary, fixed 30 minutes after exposure to ^{3}H labelled amino acids. The disposition of the grains over organelles is clearly seen

pulse-labelling with labelled amino acids, it is possible to determine the actual sites of biosynthesis of a polypeptide hormone. This technique allows determination of the intracellular pathway of transport of newly-synthesized protein through the cells prior to its secretion. Its drawback is that it is impossible to determine the nature of the labelled protein whose location is studied: it is inferred but not proven that it is the exportable proteins which are predominantly labelled.

3.3 Subcellular fractionation

Subcellular fractionation involves breakage of the cells of the gland by homogenization in cold isotonic medium and subsequent separation into fractions on the basis of their sedimentation by centrifugation (differential centrifugation). This allows only partial separation of the organelles. A refinement involves further separation of the organelles on the basis of their density (density gradient centrifugation). The traditional material for construction of such gradients is sucrose, although more recently the less osmotically damaging dextran derivative Percoll has been used. The gradients can be prepared in two ways. In the first method, the density of the sucrose used varies continuously from bottom to top of the gradient; in the second, termed 'discontinuous

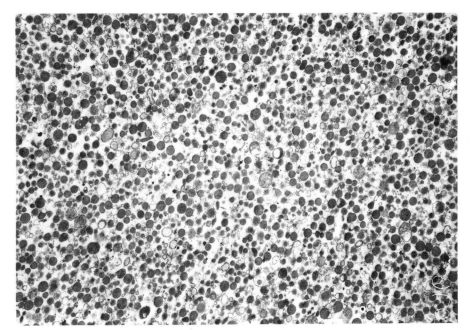

Figure 1.8 Isolated insulin storage granule fraction obtained after differential and subsequent density gradient centrifugations in conditions designed to purify these organelles

gradients', solutions of decreasing density are overlaid one on the other to provide a series of layers. The material then tends to separate out in discrete bands
at the individual interfaces. The sample can be layered at either the top or the
bottom of the tube, which is then spun at very high speed (commonly 105,000 g)
until the organelles reach their equilibrium positions in the gradient (60–120
minutes). An example of a storage granule fraction prepared by this technique is
shown in Figure 1.8. Similar techniques, often using CsCl as the material for
manufacture of the gradients, are used to separate polyribosomes and characterize them on the basis of their sedimentation characteristics, which are often
expressed as Svedberg (S) units (e.g. 4.5 S). This type of procedure permits isolation of relatively pure fractions for use in studies of, for instance, properties of
the isolated hormone storage granules or transfer of material through the cells
in pulse chase experiments using ^{3}H amino acids as precursors. This method, in
conjunction with electron microscopic autoradiography, can permit a rather
complete study of the intracellular pathways of hormone production.

4.   THE INTRACELLULAR PATHWAY OF SYNTHESIS, STORAGE AND SECRETION OF POLYPEPTIDE HORMONES

4.1   Biosynthesis

In those polypeptide-hormone-secreting tissues in which extensive studies have
been made, it has been found that hormone biosynthesis occurs in the rough-
surfaced endoplasmic reticulum, specifically by membrane-attached polysomes.
Certainly in many secretory cell types the initial synthetic product is a pre-
hormone, a precursor of very short half-life (30 seconds) which is elongated, by
means of a 15–30 amino-acid, predominantly non-polar, N-terminal extension
to the main polypeptide sequence. This facilitates transfer of the newly-synthesized protein into the cisternal space of the endoplasmic reticulum, thereby

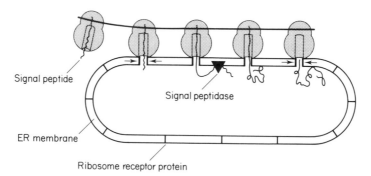

Figure 1.9 Role of signal peptides in the translocation of newly-
synthesized secretory proteins through the membrane of the endoplasmic reticulum. After Blobel, G. In: *International Cell Biology*.
Rockefeller University Press, 1977

inducing the binding of active (poly)ribosomes (Figure 1.9). This additional sequence of amino acids is called the signal sequence and the concept that it acts to facilitate transport of the polypeptide chain is generally called the signal hypothesis.[4]

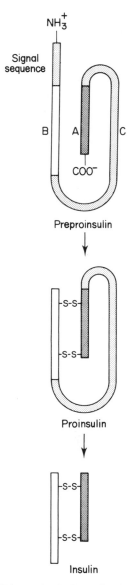

Figure 1.10 Steps in the enzymatic conversion of preproinsulin into proinsulin and then into insulin

The signal sequence is probably removed immediately the main body of the protein enters the cisternae of the endoplasmic reticulum. Such a pre-sequence has been identified and characterized for insulin (pre-proinsulin) (Figure 1.10), parathyroid hormone (pre-proparathyroid hormone) and placental lactogen (pre-placental lactogen) and many other protein hormones.[5]

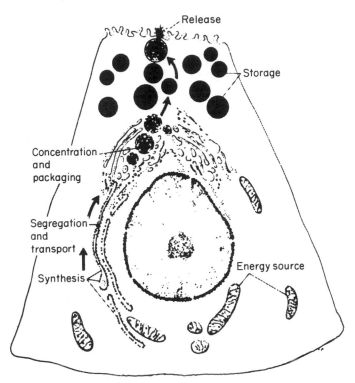

Figure 1.11 The organelles and intracellular pathway involved in the synthesis, storage in granules and secretion in a typical protein secreting cell. The overall pathway is likely to be similar for endocrine and exocrine cell types. (Reproduced with permission from Fawcett, D. W., Long, J. A. and Jones, A. L. (1979). *Recent Progr. Hormone Res.* **25**, 315–380. Copyright (1969) Academic Press).

The newly-synthesized protein which remains can, at least in some glands (e.g. pancreatic B cells and parathyroid), still be a precursor of the hormonal product itself, the so-called prohormone (proinsulin or proparathyroid hormone). The conversion of precursor to hormone occurs in these two well-characterized examples in the Golgi complex, so that the major storage product of the cells is the hormone itself—insulin or parathyroid hormone, respectively.

The glycosylation of those hormones which are glycoprotein in nature, e.g. thyroid-stimulating hormone, follicle-stimulating hormone, luteinizing hormone, occurs within the endoplasmic reticulum (in the case of mannose residues), or in the Golgi complex (in the case of galactose).[6] There was a suggestion that all exportable proteins must be glycosylated in the course of their intracellular processing prior to secretion, but this has not so far been demonstrated to be the case in many types of secretory cell. It seems likely that the production of many other polypeptide hormones follows similar pathways in other cell types and in particular that the overall pathway, rough endoplasmic reticulum → Golgi complex → storage granule, is nearly always applicable. The time course of this transfer has also been identified in several cell types and is summarized for growth hormone,[6] insulin[7] and parathyroid hormone[8] in Table 1.1. The cellular events involved are shown in Figure 1.11.

Table 1.1   Time course of intracellular transport of newly-synthesized hormones

	Time (minutes) after labelling		
	Insulin	Growth hormone	Parathyroid hormone
Endoplasmic reticulum	10–20	10–20	15–20
Golgi complex	20–45	60	30
Storage granules	30–120	120	50
Secreted hormones	60 onwards	120 onwards	50 onwards

4.2. Storage granules

The hormone storage granules are membrane-bound vesicles which enclose hormonal products, together with other material which in many cases has been only poorly characterized. The form, and in particular the diameter of the granule, is a characteristic of the hormone which is stored. Thus, anterior pituitary cells can be identified by the diameter of their granules, which in the rat are: ACTH 100–120 mμ, TSH 80–100 mμ, GH 340–360 mμ and prolactin 800–900 mμ. There may be many of these in each cell, for instance approximately 13 000 in the case of the pancreatic B cell. Up to 10% of these granules may be secreted per hour by a process involving their transfer from the cytoplasmic granule pool to the plasma membrane where fusion of the granule membrane and plasma membrane occurs, allowing release of the granule contents which may rapidly dissolve (Figure 1.11). This process is termed 'exocytosis'. The incorporation of the granule membrane into the plasma membrane results in some initial enlargement in the volume of the plasma membrane and this is counteracted by the reabsorption of areas of membrane back into the cell by a pinocytotic mechanism. These reabsorbed membrane vesicles are recycled back into the Golgi complex where they may be reutilized.[9]

4.3   Secretory mechanisms

The mechanism of secretion by exocytosis is a common property for almost all of the polypeptide secretory cells which have been studied to date. In some cases there is quantitative or semiquantitative evidence of correlation between the rate of hormone secretion (in biochemical terms) and the number of exocytotic events which can be observed by electron microscopy. How the frequency of exocytosis is accelerated by stimulation of secretion is far from clear. In many cell types the final mediator of hormone secretory rates appears to be the concentration of intracellular free calcium[10,11] and this can be changed by alteration of rates of influx or efflux of calcium across the cell membrane or by changing the intracellular distribution of calcium. This last is achieved by altering the calcium accumulation by organelles, in particular the mitochondria, endoplasmic reticulum and secretory granules.

This rise in calcium concentration is then translated in some way into granule movement to the cell membrane and exocytosis. The movement is achieved in some cell types by microtubules and/or microfilaments which may play a role as cytoskeletal elements in granule transport. Evidence for this is derived principally from the fact that drugs which interfere with microtubule function (colchicine, vinblastine) or microfilament function (cytochalasin B) alter rates of hormone secretion; an example of this is in the regulation of insulin secretion.[12] In some other cells, e.g. somatotrophs of the anterior pituitary, microtubules seem to be involved at a rather earlier stage, at the point where the newly-synthesized hormone is transferred from the rough surfaced endoplasmic reticulum to the Golgi complex. These intracellular movements may involve both microtubules as rigid cytoskeletal elements and microfilaments (composed of actin) which can provide a contractile element. As indicated in Chapter 9, an elevation of intracellular calcium could lead to activation of this intracellular transport by inducing actin (or possibly actomyosin) contraction.

The fusion of granules with plasma membrane may be separate from intracellular transport, although again the factors involved are not completely clear. It is possible that calcium could act simply to reduce the net negative charge on the granule membrane, thereby allowing the granule to approach the plasma membrane more closely, the probability of their coming into contact and fusing being correspondingly increased.

Detailed information which is available about secretion mechanisms and their regulation for particular polypeptide hormone cell types is given in the appropriate chapters.

4.4   Crinophagy

Crinophagy is a term used to describe the intracellular destruction of hormone storage granules as a result of their fusion with lysosomes, which leads to

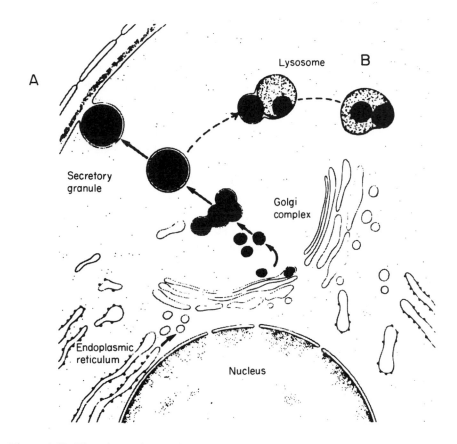

Figure 1.12 The alternative pathways for the processing of storage granules in the prolactin secreting cell of the anterior pituitary. The normal process of secretion is shown at (A), while intracellular destruction of the granules by crinophagy is represented by pathway B. (Reproduced with permission from Fawcett, D. W., Long, J. A. and Jones, A. L. (1969). *Recent Progr. Hormone Res.* **25**, 315–380. Copyright (1969) Academic Press).

degradation of their hormone content by proteolytic action (Figure 1.12). This does not appear to be a fate of storage granules in normal conditions and has in fact been demonstrated in two sets of circumstances in which, because of changed circumstances, the cells are greatly over-producing hormone. The first and best-documented case is that of the prolactin-secreting cells of the anterior pituitary of lactating rats.[13] If the young are removed from the mother, a considerable excess of prolactin production results, and this excess is apparently destroyed by crinophagy. A similar situation may arise in the glucagon-producing A cells of the islets of Langerhans when hyperglycaemia (diabetes) is induced. Again a temporary over-production of hormone is apparently counter-balanced by its intracellular destruction by crinophagy.

In addition to this mechanism, the cells may be able to destroy hormone molecules intracellularly before they are packaged into granules for storage, and this could provide an alternative mechanism to prevent over-production of hormone in the short term.[14]

REFERENCES

1. Fawcett, D. W., Long, J. A. and Jones, A. L. (1969) *Recent Progr. Hormone Research,* **25**, 315–380.
2. Smith, A. D. (1972) *Scientific Basis of Medicine Annual Reviews,* **1972**, 74–102.
3. Pelletier, G., Robert, F. and Hardy, J. (1978) *J. Clin. Endoc. Metab.,* **46**, 534–542.
4. Blobel, G., Walter, P., Chang, G. N., Goldman, B. M., Erickson, A. H. and Lingappa, V. R. (1979) *Symp. Soc. Exp. Biol.,* **33**, 9–36.
5. Tager, H. S., Patzelt, C., Assorian, R. K., Chan, S. J., Duguid, J. R. and Steiner, D. F. (1980) *Ann. N.Y. Acad. Sci.,* **343**, 133–147.
6. Howell, S. L. and Whitfield, M. (1973) *J. Cell Sci.,* **12**, 1–21.
7. Howell, S. L., Kostianovsky, M. and Lacy, P. E. (1969) *J. Cell. Biol.,* **42**, 695–705.
8. Habener, J. F. and Potts, J. T. (1976) In: *Handbook of Physiology,* Section 7, Vol. VII. (Ed. Aurbach, G. D.) American Physiological Society, Washington D.C. pp. 313–342.
9. Orci, L. (1974) *Diabetologia,* **10**, 163–187.
10. Rasmussen, H. and Goodman, D. B. P. (1977) *Physiol. Rev.,* **57**, 421–509.
11. Ruben, R. P. (1982) *Calcium and Cellular Secretion.* Plenum Press, N.Y.
12. Howell, S. L. and Tyhurst, M. (1982) *Diabetologia,* **22**, 301–308.
13. Smith, R. E. and Farquhar, M. G. (1966) *J. Cell Biol.,* **31**, 319–347.
14. Bienkowski, R. S. (1983) *Biochem. J.,* **214**, 1–10.

Chapter 2

Hormone assay

---

1. INTRODUCTION[1-3]

Most advances in endocrinology have been dependent on the ability to measure the amount of a specific hormone present in a given sample. Thus, the original isolation and purification of a hormone requires an assay, usually based on a recognizable biological effect, which allows the purification process to be followed. Once the pure hormone is available, investigation of the relationship between its structure and biological activity also requires assays which can measure such activity. Studies on the synthesis and secretion of a hormone and its distribution in the tissues of the body require assays of high sensitivity and specificity, as do studies requiring estimates of hormone concentration in blood (including clinical studies of oversecretion or undersecretion).

The wide application of hormone assays has led to a vast effort being expended on the development of suitable methods. This effort has been dominated by two overriding factors. First, where biological test systems form the basis of an assay, the inherent variability of such systems has to be accommodated. Secondly, hormone levels often have to be measured at very low concentrations (down to 10^{-10} M or less) and in the presence of many potentially interfering substances—giving a need for sensitivity combined with specificity.

The type of hormone assay chosen to deal with any individual problem will depend on that particular problem. Four main types of hormone assay can be recognized.

(1) *Chemical assays,* which depend on the specific physicochemical properties of the hormone.

(2) *Biological assays* (bioassays), which exploit the specific biological activity of the hormone.

(3) *Receptor binding assays,* which exploit the ability of the hormone to bind to a specific receptor molecule.

(4) *Immunological assays* (immunoassays), in which binding of the hormone to a specific antibody forms the basis of the assay.

Receptor binding and immunoassays are the main examples of a wider class known as competitive-binding assays. All of these forms of assay, except for bioassays, require the availability of a reasonably pure preparation of the hormone. Bioassays are usually therefore necessary for the original purification of a hormone, as well as for studies on the relationship of structure to biological activity and for pharmacological standardization of hormone preparations. However, for studies of hormone secretion and for measurement of hormone levels in blood, immunoassay or receptor binding assay is usually the method of choice.

2. CHEMICAL ASSAY

The general principle of chemical assay of a hormone is straightforward, and similar to that of chemical assay of other organic molecules. The hormone is detected and measured, often after application of a chromatographic or other separation method, either by virtue of an intrinsic physicochemical property (such as UV absorbance) or after reaction with a reagent which produces a specific colour reaction, fluorescence, etc. Such methods are frequently used for assay of steroids, catecholamines and other small hormones for which they may be extremely sensitive.

In the past few years a type of chemical assay has been applied to peptide and protein hormones. The hormone is resolved from other peptide and protein material by a high resolution electrophoretic method, such as electrophoresis in polyacrylamide gel, and its concentration in a particular electrophoretic band is measured by densitometry after staining with a general protein stain. This type of approach is rather insensitive, but has been usefully applied, for example, to the measurement of the prolactin content of pituitary glands from a range of different species.[4]

A potentially important, relatively recent development along similar lines uses high performance liquid chromatography—a rapid separation technique of very high resolving power which can be combined with very sensitive measurement of optical absorbance to give a sensitive technique for the measurement of minute amounts of peptides and proteins. This technique is likely to have an increasing impact on the assay of small peptides during the next few years.

3. BIOLOGICAL ASSAY (BIOASSAY)[5,6]

3.1 General principles

The basis of all bioassays is that they measure a biological response, the amount of which can be related to the quantity of hormone being studied. All hormones can be estimated by bioassay, and in cases where the pure hormone has not been isolated this is usually the only type of assay that can be applied. Bioassay techniques are also used for many substances other than hormones, and are employed extensively in the pharmaceutical industry for standardizing drugs, including, for example, antibiotics and vitamins.

It is instructive to compare hormone bioassay and enzyme assay. In an enzyme assay, the enzyme is used to modify a substrate under conditions where enzyme concentration is limiting. The response comprises a rate of reaction (either appearance of products or disappearance of substrate; initial rate in each case). Given careful experimental work, for the majority of enzymes the rate of reaction should vary directly, linearly, with the enzyme concentration, and experimental points should fall very close to the straight line (Figure 2.1a). In hormone bioassay, the hormone is used to induce a response in a biological system—either in an intact animal or in a cell or tissue preparation *in vitro*. However, the response is rarely related to the amount (dose) of hormone applied in a linear fashion and, however careful the experimental technique, there will be a good deal of variation of data about the dose–response line (Figure 2.1b).

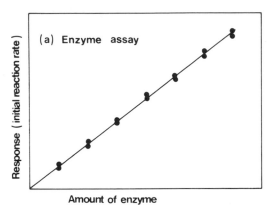

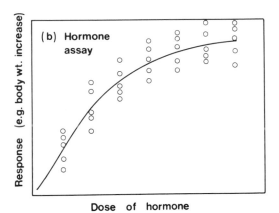

Figure 2.1 A comparison of enzyme assay and hormone bioassay. Hypothetical dose–response curves are shown to illustrate (a) the linearity and reproducibility encountered in enzyme assay and (b) the non-linearity and variability encountered in hormone assay

The non-linearity and variability associated with the dose–response curve for a bioassay are intrinsic features of the biological test systems used. A great deal of the effort which has been, and still is, devoted to biological assay is concerned with developing methods to cope with these features. The problem of non-linearity can be overcome by a log dose transformation of the data. In most cases if the response is plotted against the log of the dose, a part (but only part) of the resultant log dose–response curve is linear, or sufficiently so for practical purposes (Figure 2.2). The vast majority of bioassay methods use such a transformation. The resultant log dose–response curve can then be used as a 'standard curve' for the assay—the response observed for an 'unknown' preparation is related to a standard curve obtained using a standard of known potency. It is normal practice to run both standards and unknowns in the same assay, so that the responses can be compared directly. Attempts to establish a 'standard response' to a given dose of hormone, which is reproducible from assay to assay, have now largely been abandoned.

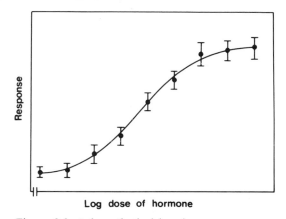

Figure 2.2 A hypothetical log dose–response curve for a hormone bioassay. The central part of the curve approaches linearity. The response to each dose of hormone is shown as a mean value ± SEM (standard error of the mean)

The problem of variability of test material encountered in bioassays is dealt with in two ways. First, such variability is reduced to a minimum by standardizing and simplifying the test system as far as possible. Use of inbred animals and maintenance of standard conditions is an important aspect of this. Bioassays which use tissue or cell preparations rather than whole animals will often show much reduced variability. Secondly, the problem of the remaining biological variability is minimized by good experimental design and statistical treatment of results. Good design may allow some errors due to variability to be eliminated, but more importantly ensures that intrinsic variability does not lead to biased results. Appropriate statistical procedures (usually including analysis of variance) allow the validity of the experimental design to be tested and enable a potency estimate of the unknown samples to be derived and the confidence limits of this estimate to be assessed.

3.2   Types of bioassay: graded and quantal responses[5,6]

Responses observed in bioassays may be of two main types: graded and quantal. In the case of a graded response, the degree of response in each test animal or other system will vary according to the dose of hormone applied. For example, a standard bioassay for pituitary growth hormone involves injecting the hormone into hypophysectomized rats and determining the increase in weight that results. For any one rat, the weight increase will depend, within limits and modulated by the usual variability, on the dose of hormone injected. In a quantal-response assay each animal or other test system used shows only a presence or absence of the response. For example, a standard assay for insulin involves injecting different doses of the hormone into batches of mice and observing the convulsions caused by the subsequent lowering of blood sugar levels. Any one mouse either undergoes convulsions (after a fixed time) or does not—no attempt is made to measure the 'extent of convulsion'. The overall response to any one dose of hormone will be the percentage of animals giving a positive response which should, of course, increase with increasing dose.

Clearly, each individual animal will give more information in a graded-response assay than in a quantal-response assay, and the former type is therefore usually preferable. However, in some cases a graded response cannot be obtained, and in others it requires a good deal of experimental manipulation, so use of quantal-response assays is by no means unusual. The statistical analysis of results from quantal-response assays is rather different from that for those from graded-response assays.

3.3   Statistical treatment of bioassay results[5–9]

The analysis of bioassay data has become a branch of statistics in its own right, and cannot be treated in any detail here. In a well designed assay, in which for example randomization of test animals has ensured uniformity between groups receiving all treatments, the statistical analysis allows tests of the validity of the assay and determination of the potency of test samples (relative to a standard) and of the confidence limits of this potency.

For the standard preparation and test preparations (if these have been assayed at more than one dose) linear regression allows the dose–response relationship to be established. This is illustrated in Figure 2.3 for a '6-point assay'—one in which standard and test preparation have both been assayed at three dose levels. Analysis of variance allows the testing of the linearity of the dose–response curves and parallelism of standard and test lines. The horizontal distance between the two lines is a measure of the log potency ratio and the variability of observations encountered in the assay is used to determine the reliability of this estimate. Potency of the test material will usually be expressed in the form:

$$x \text{ units/unit wt (95\% fiducial limits } z_1 - z_2).$$

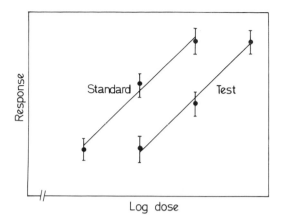

Figure 2.3 A '6-point' bioassay. Standard and test sample have each been assayed at three different doses, and log dose–response curves for each constructed by linear regression. The potency of the test sample relative to the standard can be determined by comparing the doses of the two that give an identical response

The 95% fiducial limits indicate the range within which the potency of the sample lies, with a 95% confidence.

Bioassays can be assessed using the following criteria.

(1) *Accuracy*—a measure of the extent to which the results obtained represent a 'true' result. It is tested by determining the potency of the same sample by different assay methods, in several different laboratories.

(2) *Precision*—a measure of the amount of deviation between assays carried out under exactly the same conditions, i.e. the intrinsic error associated with a particular assay method. Precision can be estimated by determining the *index of precision* (λ):

$$\lambda = \frac{\text{standard deviation of responses}}{\text{slope of the dose–response curve}}$$

The smaller λ, the more precise the assay. A workable assay normally requires a value of < 0.4; a value < 0.2 is preferable.

(3) *Specificity*—a measure of the extent to which other hormones or impurities interfere with the assay response by synergism (potentiation of the action of the hormone), inhibition or simply intrinsic activity. The less such interference, the more specific (and better) the assay. Bioassays completely specific for a single hormone are unfortunately rare.

(4) *Sensitivity*—a measure of the amount of material needed to get a significant response in the assay. The less material, the more sensitive the assay.

Sensitivity can be a very important criterion when material is scarce or concentrations are low. Bioassays are frequently too insensitive to deal with the range of concentrations met in biological fluids such as blood.

3.4   Standards[10]

The importance of standard hormone preparations in bioassay has already been emphasized. The ideal is a standard preparation which is available to all laboratories as a reference—thus enabling comparison of results between laboratories. Such 'International Standards' are now available for many protein and polypeptide hormones, and have been widely distributed. A standard preparation does not have to be very pure (although purity is obviously desirable)—the main requirements are uniformity and stability. The first International Standard preparation for a given hormone is normally designed (arbitrarily) as having a certain number of International Units per unit weight, and subsequent International (or local) standards are calibrated relative to this.

3.5   Biological test systems; examples of bioassay

Assay systems obviously differ widely according to the hormone that is being measured. The modern trend is towards use of tissue, cellular or tissue-culture preparations rather than whole animals, since this often increases sensitivity and precision. However, use of whole animals is unlikely to be eliminated completely. This trend will be exemplified here. It should be noted that test systems of the kind discussed here are not only used for bioassay but also for many studies on the biochemical mode of action of hormones.

3.5.1   In vivo assays

Assays using whole animals have already been mentioned—the mouse convulsion test for insulin and growth-promotion assays for growth hormone. In most such *in vivo* systems, where the hormone often reaches its target tissue via the circulation, the response will be determined by several factors, including the half-life of the hormone in the circulation and the rate of absorption from the site of administration, as well as intrinsic activity at the target tissue. The effects of degradation in the circulation are usually reduced in assays using tissue or cell preparations, and results obtained with such *in vitro* systems may therefore differ considerably from those obtained *in vivo*.

3.5.2   Tissue preparations

The use of tissue preparations for bioassay is illustrated by an assay for insulin utilizing rat fat pads.[11] Insulin stimulates uptake and oxidation of glucose by adipose tissue *in vitro,* and its conversion into lipid, and each of these effects has been adapted for use as a bioassay, using ^{14}C-labelled glucose. Other tissue preparations which have been used *in vitro* include quartered rat adrenal glands (used for assay of corticotropin), explants of mouse mammary gland (used to assay prolactin) and contracting uterus preparations (used to assay oxytocin). More information about some of these assays is given in later chapters.

3.5.3   Dispersed cells

The use of tissue preparations will often increase sensitivity and precision, but variation between animals remains important because any one animal can usually provide only a few preparations. This limitation can often be overcome by using isolated (dispersed) cell preparations. Cells in tissues are of course normally linked firmly to one another by a variety of junctions or an extracellular matrix or both. The links between the cells can be disrupted, often without seriously damaging the cells themselves, by treatment with enzymes and/or chelating agents. Enzymes frequently used include trypsin and other general proteases, collagenase (which degrades collagen, but also often contains other important enzymes as contaminants) and hyaluronidase. Chelating agents such as ethylenediaminetetraacetic acid (EDTA) sequester calcium and other multivalent cations, which appear to play an important part in cell–cell adhesion.

If these methods are used, dispersed cell preparations can be obtained from most tissues.[12] If necessary, tissue from several different animals can be pooled. The dispersed, suspended cells will often then respond to hormones, and these responses can be used as the basis for assay systems. Assays of this kind have been described using adrenal cells (which respond to ACTH by giving dramatically enhanced corticosteroidogenesis) and fat cells (which respond to various hormones; enhanced glucose oxidation in response to insulin has been made the basis of one assay). Most of the problems encountered in bioassay due to variation between animals are avoided by using such cell preparations, since a single cell dispersion is used for production of large numbers of assay tubes.

3.5.4   Cell culture[13]

One disadvantage of the use of freshly dispersed cells is the possibility that enzymic dispersion may cause some damage to the cells. In particular, hormone receptors on the outside of the cells may be enzymically modified, leading to altered ability to bind (and respond to) hormones. This problem can often be overcome by putting the cells into short-term culture, either in suspension or with attachment to a solid surface such as a petri dish. Many dispersed cells can be maintained in an active and viable state for 1–2 weeks (or often more) by such an approach, and even 1–2 days in culture is often sufficient to allow receptors to be resynthesized and hormonal responses restored.

In some cases hormones can be assayed using cells maintained in more permanent tissue culture. This involves culture of permanent cell lines, often lines derived from tumours. Such an approach eliminates the need for using whole animals completely (except of course when the line is first being established), but the hormonal responses of transformed cell lines are often altered or completely lost when compared with those of the tissue from which they originated. Use of permanent cell lines for hormone assay has been relatively little developed, but it seems likely that they will become increasingly important, especially in the assay of hormonal factors that regulate cell growth.

3.5.5   Cytochemical bioassay[14,15]

Cytochemical bioassay has been developed for assay of a fairly small number of polypeptide hormones, but may prove to be quite widely applicable. Its application to the assay of corticotropin will be described here.

Fragments of guinea pig adrenal glands are cultured for a few hours in a medium containing ascorbic acid, and then rapidly frozen and sectioned on a cryostat. Sections are freeze-dried, incubated briefly in solutions containing standard or unknown quantities of corticotropin and then stained with a modified Chèvremont–Frederic stain (containing ferric chloride and potassium ferricyanide). The stain measures the 'reducing potency' of the tissue, and is largely a measure of ascorbic acid concentration under these conditions. Corticotropin causes a depletion of ascorbic acid concentration in adrenal cortical tissue, and the intensity of stain is thus inversely proportional to the log of the corticotropin concentration. The intensity of staining is measured by microdensitometry, which can integrate absorbance over an area as small as a single cell. The method is extremely sensitive (about 100 times more sensitive than the radioimmunoassay for corticotropin) and gives a precise measurement of a biological response, but is time consuming and expensive.

The very high sensitivity and good precision obtainable with the cytochemical bioassay are stimulating efforts to apply it to many hormones, but the sophisticated technology needed for the microdensitometry, together with the associated high cost, are serious disadvantages.

4.   RECEPTOR ASSAY[16,17]

4.1   The principle of competitive binding assays

Receptor assays and immunoassays are the most important examples of competitive binding assays. Binding of the hormone to a specific macromolecule (usually a protein) is used as the basis of the method, competition for binding between the hormone sample under study and a labelled form of the hormone (usually, but not always, radioactively labelled) being followed. The approach is best illustrated with reference to immunoassay, where it will be described in more detail.

The nature of the specific binding protein can vary considerably. In the case of immunoassays it is obviously a specific antibody. For some hormones specific binding proteins occur in the circulation where they are concerned with transport of the hormone, and in such cases these binding proteins can be used for competitive binding assays; such binding-protein assays have been described for thyroid hormones and many steroids, but most peptide hormones are not associated with a binding protein in the circulation (the somatomedins provide an exception).

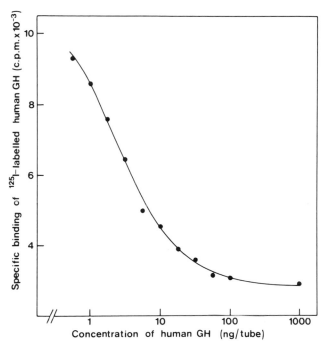

Figure 2.4 A log dose–response curve for a radioreceptor assay of growth hormone (GH). Binding of ^{125}I-labelled human growth hormone to a preparation of rabbit liver membranes is followed. Increasing concentrations of unlabelled growth hormone compete with the labelled hormone and decrease the amount bound. Potency of test samples may be observed by relating their ability to compete with labelled hormone for binding, using a standard curve of the kind shown here

All hormones bind to specific receptors in their target cells, and these receptors also provide suitable binding proteins for competitive binding assays. Such receptor assays have been described for most hormones. In most cases binding of a radioactively labelled hormone (usually radioiodinated—see the next section) is studied; binding of this to the receptor is reduced competitively by increasing concentrations of unlabelled hormone (sample or standard). An example is shown in Figure 2.4. As in the case of radioimmunoassay, the problem then becomes one of separating the receptor-bound and free hormone, determining the proportion of labelled hormone bound, and relating this to the amount of unlabelled hormone present using a standard curve.

4.2   Types of receptor-binding assay

4.2.1   Membrane-bound receptors

Most receptors for peptide and protein hormones are associated with the plasma membrane (see Chapter 17). Preparations of plasma membranes show specific binding properties for appropriate hormones, and can be used directly in radioreceptor assays. Such membranes will bind labelled hormone, competitively, as discussed above, and receptor-bound hormone can then be rapidly separated from unbound hormone by centrifugation, often after alteration of the medium to facilitate association and precipitation of the membrane fragments. The proportion of labelled hormone bound is then easily determined by radioactivity counting of the membrane-precipitate.

4.2.2   Solubilized receptors

The heterogeneous nature of assay mixtures containing membrane fragments may prove inconvenient and lead to some loss of precision, and various methods have been developed for solubilization of membrane-bound receptors. Many of these use detergents to disrupt the structure of the membranes, although in some cases the detergents may cause inactivation of the receptor proteins. Solubilized receptors have formed the basis of many receptor assays. Separation of soluble hormone–receptor complexes from unbound hormone is more difficult than for membrane-bound receptors, although the fact that the receptors are usually very much larger than the hormone usually allows selective precipitation of the complex, for example by polyethylene glycol. Intracellular receptors for steroid hormones are not membrane-bound, and do not require detergents for solubilization.

4.2.3   Receptors on intact cells

In the intact cell, a polypeptide hormone interacts with a receptor on the outside of the plasma membrane. Binding of hormone directly to cells can therefore often be observed. Receptor assays have been described which exploit this feature. Incubation of the labelled hormone with the dispersed target cells allows the binding of hormone to receptors on the cell surface, and the addition of unlabelled hormone is followed by competition for these binding sites and displacement of the labelled hormone in proportion to the amount of unlabelled material added. Bound hormone associated with the cells is then very easily separated from unbound by low speed centrifugation or filtration. Such assays have been described for several hormones, including growth hormone (using a lymphoma-derived cell line—IM9).[16]

A problem which may be encountered when intact cells are used for receptor-binding assays is that hormone binding may be followed by internalization and degradation of the bound hormone. Such processes can often be selectively slowed by studying binding at 4°C, but may not be eliminated completely.

5. RADIOIMMUNOASSAY

5.1 Introduction and historical

The term 'radioimmunoassay' will be used to describe assays which are based on the reaction between a labelled hormone and an antibody ligand. They are now very widely used in endocrinology and endocrine biochemistry, having in many instances completely superseded earlier bioassay techniques. In fact, such techniques now have a place in many other aspects of biochemistry where small quantities of substances may be measured with considerable specificity, and precision.

The earliest successful attempts to assay hormones by radioimmunoassay methods were due to Yalow and Berson.[18] Prior to that, there had been some assays based on haemagglutination procedures which involved reaction between antibody and a protein hormone antigen, although these are no longer used.

Yalow and Berson were the first to observe that radioactively labelled insulin could bind to antibodies in the sera of insulin treated diabetics. Since commercial insulins used in the treatment of such patients are derived from pig and beef pancreas, antibodies to these species insulins are readily generated. Relatively large quantities of added insulin may be bound in this way, though generally the binding of insulin to antibody does not present a therapeutic problem. Yalow and Berson, however, realized that this highly specific combination of insulin with its antibody might be used as a sensitive assay procedure. In the original methodology, insulin bound to antibody was separated from unbound insulin by paper electrophoresis. Unbound insulin was attached to the paper, whereas insulin bound to the immunoglobulins readily moved away, because of the electrical charge on the proteins. Although this method is no longer widely used, the theory of the method was worked out using this system, and the first reliable assays for insulin, which could be used clinically, employed this technique. The method was soon also applied to a number of other protein hormones, such as growth hormone, glucagon and ACTH.

5.2   Theoretical considerations

For this purpose it is assumed that a protein hormone reacts with only one
species of antibody, at a single binding site. The reaction leading to binding may
be designated as follows:

$$\text{hormone} + \text{AB} \rightleftharpoons \text{hormone antibody complex.}$$

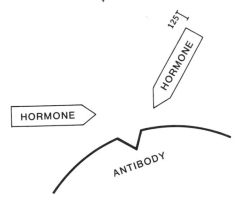

Figure 2.5   Reaction between labelled and
unlabelled antigen and antibody

Assuming a sufficiency of antibody, the hormone bound to it will bear a relationship to its original concentration. If labelled hormone is added to such a
system, then both labelled and unlabelled hormone will compete for the same
binding site (Figure 2.5). In the presence of a given quantity of labelled hormone,
therefore, the greater the concentration of unlabelled hormone, the less radioactivity will be bound to a limited quantity of antibody. The labelled hormone
associated with the antibody will thus bear a relationship to the quantity of
unlabelled hormone in the mixture. It necessarily follows that there must be an

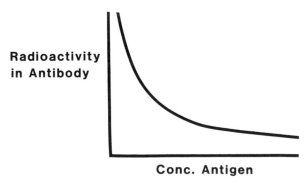

Figure 2.6   Radioactivity associated with antibody
plotted against hormone concentration

adequate method for separating the free hormone from that which is antibody bound. Techniques which can achieve this are discussed below.

If the radioactivity which is associated with the antibody is plotted against hormone concentration, a hyperbolic curve is obtained (Figure 2.6). From this the concentration of unlabelled hormone may be calculated. There are in practice many transformations of such curves. Thus, by plotting the free/bound ratio against concentration, a straight line may be obtained (Figure 2.7). These are discussed elsewhere in detail.[19,19a]

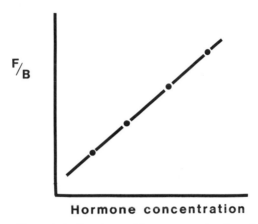

Hormone concentration

Figure 2.7 Free/bound hormone ratio plotted against hormone concentration

5.3 Methods of separating free from bound hormone

Earlier techniques were by paper electrophoresis, gel filtration and salt precipitation of globulins.[20] Very many other methods are now in use.

A frequently used variant of this type of procedure is the precipitation of immunoglobulins by a second antibody. Such methods are usually termed 'double antibody precipitation' techniques.[21,22] In these techniques, the antibody is generated against the globulins of one species in another species and the anti-gammaglobulins so made are used to precipitate immunoglobulins, including those which bind hormones. This is illustrated in Figure 2.8. Such a precipitate may be filtered off, or alternatively centrifuged down. One of the most interesting variants of this system is the use of solid phase antibodies. In such systems antibody may for example be coupled with Sepharose using cyanogen bromide activation. Again labelled and unlabelled hormones will compete for antibody attached to the Sepharose. After incubation, the Sepharose is centrifuged off and counted. Such methods, although involving a rather tedious preparation of reagents, may give an excellent separation of free and bound hormone, with high precision.

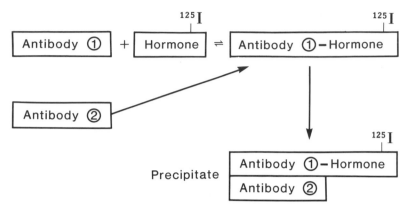

Figure 2.8 Diagrammatic representation of double antibody immunoassay technique

5.4 Other variants of radioimmunoassay

An interesting variant of these procedures employs the use of iodine-labelled antibody. Such methods, which were originally pioneered by Miles and Hales, have been called immunoradiometric assays. Their principle is illustrated in Figure 2.9.

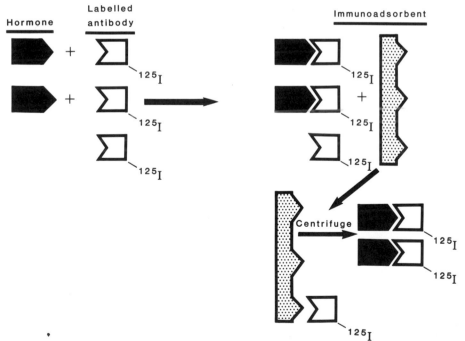

Figure 2.9 Principles of immunoradiometric assay

Such methods provide a straight line relationship between radioactivity in antibody and concentration of hormone. To avoid rather high blanks, 'sandwich' assays in which unlabelled antibody is coupled to a solid phase have been used.

5.5 Specificity of assays

Early in the development of immunoassay systems it was realized that cross-reactivity with closely related substances might be a major problem.

One instance in which this difficulty will arise is when hormones and their precursor forms are assayed together. For example, the same antigenic determinants are present in insulin and proinsulin, and there are similar problems when glucagon is assayed with its precursor forms; in addition, very closely related hormones such as growth hormone and placental lactogen may interfere with the assay of one another. Several examples of this will appear later in this book.

Another major problem arises from the species difference between hormones. This is shown in Figure 2.10, which compares the standard curves for rat and beef insulins using a beef insulin antibody and beef insulin standards.[24] The curves do not coincide, and indeed if the structural differences are very great there may be large errors in the assay.

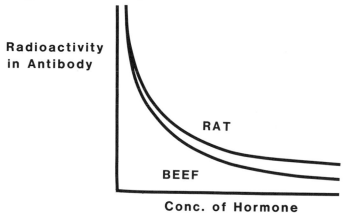

Figure 2.10 Standard curves for beef and rat insulins, assayed using a beef insulin antibody

5.6 The generation of antibodies

While many polypeptide hormones of molecular weight greater than 5,000 are antigenic in suitable animals, this is not so for a number of smaller molecular weight hormones. A notable example of this is glucagon, of molecular weight

3,500, against which antibody can only be raised with difficulty. Nevertheless, some quite small polypeptides may produce antibodies if injected into animals under appropriate conditions. The antigenic potency of such molecules is a function of their structure. To overcome the difficulty, smaller hormones may be conjugated with larger protein molecules such as albumin. The complex so formed may then be much more antigenic. The standard method for conjugation involves reaction of free NH_2 groups of a lysine residue on the albumin with a free carboxyl on the hormone molecule.

It is in any case customary to mix the antigen with Freund's adjuvent (killed tubercle bacilli, mineral oil and emulsifier) to ensure better antibody production.[19a]

A further problem with antibody production by classical techniques is that a mixed population of antibodies may be induced. This may react against various portions of the antigen molecule. The development of monoclonal antibodies will greatly assist in the production of large amounts of a more specific antibody which may then be available in high titre (see also Chapter 15).

5.7   Iodination of proteins

5.7.1   *Iodination techniques*

Most labelling techniques employ attachment of one or other of the radio-isotopes of iodine to the antigen. Originally ^{131}I was universally used, although more recently ^{125}I, which has a longer half-life, is generally employed.

Iodine is customarily associated with tyrosine residues, though frequently these will not be uniformly labelled within a protein. An ideal situation is one when only a single tyrosine molecule per antigen molecule or subunit is labelled. Over-iodination may result in two special kinds of difficulty. In the first place radiation damage may be produced in some proteins. This particular problem is very evident on storage, leading to the production of fragments of damaged hormones. In addition, the introduction of an atom of large size such as iodine may influence the hormone's immunoreactivity.

The method most commonly used is based on the technique of Hunter and Greenwood,[25] in which chloramine T—a powerful oxidizing agent—is used to liberate free iodine from iodide. In this method care must be taken not to over-expose proteins to the effect of chloramine T. Excess iodine is removed by the addition of sodium metabisulphite. Subsequently, the iodinated protein is purified on Sephadex. With some iodinated proteins it is advisable to purify the hormones before each assay on account of radiation damage.

Iodination may also be carried out, using lactoperoxidase, in the presence of hydrogen peroxide. This technique avoids the alteration of proteins that follows exposure to chloramine T.

5.7.2   Use of antigens lacking a tyrosine residue

Some polypeptides (e.g. secretin) lack a tyrosine residue so that conventional iodination techniques are not possible when using them. A method[26] which avoids the problem is the use of a tyrosine derivative of succinic acid which can subsequently be conjugated with the NH_2 group of the N-terminal amino acid or with the ε-amino group of lysine. The sequence of reactions is shown in Figure 2.11.

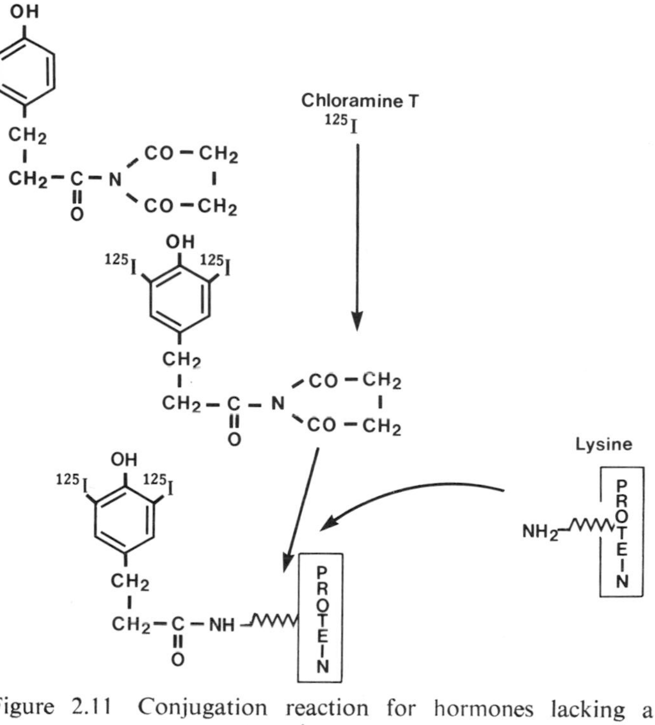

Figure 2.11   Conjugation reaction for hormones lacking a tyrosine group

6.   OTHER TYPES OF IMMUNOASSAY

Quite recently, methods for trace labelling of proteins have been used, other than with radioactive agents.

Thus, fluorescent substances (e.g. fluorescein derivatives) may be directly linked with protein hormones to serve as labels in an immunoassay. Among hormones which may be assayed by these methods are insulin and placental lactogen. Fluorescence is measured with standard fluorimeters.

Enzymes may also be used in a similar way. These are coupled to a hormone by chemical methods. The enzyme-labelled material is then used in the assay as if it were isotopically labelled. An enzyme immunoassay (EIA) is frequently employed with a solid phase type of assay (called enzyme-linked immunoabsorbent assay, or ELISA). The enzyme is detected by adding a suitable substrate and then measuring its conversion to some other chemically detectable material. Among enzymes which have been used in this way as labels are peroxidase, glucose oxidase and alkaline phosphatase. Since only very small quantities of enzyme are needed to produce measurable chemical effects, the method may be made as sensitive as radioimmunoassay. This method has been applied to the estimation of TSH.

7.   RELATIONSHIP BETWEEN BIOASSAY AND OTHER FORMS OF ASSAY

7.1   Receptor assays

Unlike other competitive binding assays, receptor-binding assays may give a direct measure of biological activity, since the receptors they utilize may be mediators of the hormonal actions. However, it is possible that some hormone 'receptors' mediate degradation of the hormone rather than biological actions. Although results obtained with receptor assays generally agree well with those obtained using bioassays, agreement is rarely perfect, and caution must always be used in interpreting the results of a receptor assay in terms of the biological actions of the hormone.

7.2   Immunoassays

It is less easy to compare the results obtained by bioassay methods with those of immunoassay. It will be evident from what has been said that immunoassay techniques measure only immunoreactivity, which does not always equate with biological activity. It is thus possible for a molecule to display considerable immunoreactivity without any biological activity. An example of this is seen in the endorphins. The N-acetyl derivative of β endorphin is biologically inactive even though it has the same immunoreactivity as the non-acetylated molecule.[27] Similarly, some species of immunoactive glucagon may be devoid of the glycogenolytic activity of pancreatic glucagon. Such differences generally arise because those parts of the molecule responsible, say, for receptor combination, are not the same as those which combine with antibodies. Immunoreactivity may neither closely parallel biological activity nor receptor-binding reactivity.

8.   PREPARATION OF SAMPLES FOR ASSAY

Ideally, hormones should be capable of being assayed directly in biological fluids, and in blood, but even with assays such as the immunoassay, which is highly discriminatory, this may not always be possible. Preliminary extraction will sometimes be necessary and this applies also when the hormone content of solid glandular tissues is being assayed. Such methods were formerly much used when bioassay techniques were the rule, in order to avoid interference from the many other substances which could interfere with an assay. As used at present, they frequently involve extraction of a hormone by organic solvents and the use of preliminary chromatography or gel filtration techniques. It is probably rather simpler now to produce an antibody of a more specific kind (if immunoassay is employed) to avoid preliminary extractions.

REFERENCES

1. Loraine, J. A. and Bell, E. T. (Eds.) (1976) *Hormone Assays and their Clinical Application,* 4th edn. Churchill Livingstone, Edinburgh.
2. Gray, C. H. and James, V. H. T. (Eds.) (1979) *Hormones in Blood,* 3rd edn, 3 vols. Academic Press, New York.
3. Emmens, C. W., Berson, S. A., Yalow, R. S., Samols, E., Catt, K. J., Weintraub, B. D., Rosen, S. W. and Tashjian, A. H. (1973) In: *Methods in Investigative and Diagnostic Endocrinology,* Vol. 2A (Eds. Berson, S. A. and Yalow, R. S.), pp. 61–153.
4. Jones, A. E., Fisher, J. N., Lewis, U. J. and VanderLaan, W. P. (1965). *Endocrinology,* **76**, 578–583.
5. Dorfman, R. I. (Ed.) (1969) *Methods in Hormone Research,* 2nd edn. Vol. IIA. Academic Press, New York.
6. Emmens, C. W. (1948) *Principles of Biological Assay.* Chapman and Hall, London.
7. Bliss, C. I. (1952) *The Statistics of Bioassay.* Academic Press, New York.
8. Finney, D. J. (1978) *Statistical Method in Biological Assay,* 3rd. edn. Griffin & Co., London.
9. Finney, D. J. (1952) *Probit Analysis: A Statistical Treatment of the Sigmoid Response Curve,* 2nd edn. Cambridge University Press, London.
10. Bangham, D. R. (1979) In: *Hormones in Blood,* 3rd edn, Vol. 1 (Eds. Gray, C. H. and James, V. H. T.). Academic Press, New York, pp. ix–xiv.
11. Froesch, E. R., Bürgi, H., Ramseier, E. B., Bally, P. and Labhart, A. (1963). *J. Clin. Invest.,* **42**, 1816–1834.
12. Bashor, M. M. (1979) *Methods in Enzymology,* **58**, 119–131.
13. O'Hare, M. J., Ellison, M. L. and Neville, A. M. (1978) *Current Topics in Exptl. Endocrinology,* **3**, 1–56.
14. Chayen, J. (1980) *The Cytochemical Bioassay of Polypeptide Hormones.* Springer-Verlag, Berlin.
15. Chayen, J. and Bitensky, L. (1982). In: *Recent Advances in Endocrinology and Metabolism,* Vol. 2 (Ed. O'Riordan, J. L. H.) Churchill Livingstone, London, pp. 261–285.
16. Roth, J. (1975) *Methods in Enzymology,* **37**, 66–81.
17. Odell, W. D. and Daughaday, W. H. (Eds.) (1971) *Principles of Competitive Protein-Binding Assays.* Lippincott Co., Philadelphia.
18. Yalow, R. S. and Berson, S. A. (1960) *J. Clin. Invest.,* **39**, 1157–1175.
19. Ekins, R. P. (1976) In: *Hormone Assays and their Clinical Application,* 4th edn (Eds. Lorraine, J. and Bell, E. T.). Churchill Livingstone, Edinburgh, pp. 1–72.
19a. Chard, T. (1982) *An Introduction to Radioimmunoassay and Related Techniques.* Elsevier, Biomedical Press, Amsterdam.

20. Ratcliffe, J. G. (1974) In: *Radio-immunoassay and Saturation Analysis* (Ed. Sönksen, P. H.), *Brit. Med. Bull.,* **30**, 32–37.
21. Hales, C. N. and Randle, P. J. (1963) *Biochem. J.,* **88**, 137–146.
22. Morgan, C. R. and Lazarow, A. (1963) *Diabetes,* **12**, 115–126.
23. Binoux, M. A. and Odell, W. D. (1973) *J. Clin. Endocrin. Metabolism,* **36**, 303–310.
24. Taylor, K. W., Howell, S. L., Montague, W. and Edwards, J. C. (1968) *Clin. Chim. Acta,* **22**, 71.
25. Hunter, W. M. and Greenwood, F. C. (1964) *Biochem. J.,* **91**, 43–56.
26. Bolton, A. E. and Hunter, W. M. (1973) *Biochem. J.,* **133**, 529–539.
27. Zakarian, S. and Smyth, D. G. (1982) *Biochem. J.,* **202**, 561–571.

Chapter 3

The pituitary gland and hypothalamus

1.   INTRODUCTION

The pituitary gland, or hypophysis, is a small organ (about the size of a large pea in man) situated in a bony cavity, the *sella turcica,* at the base of the skull. It is attached to the hypothalamus (the floor of the mid-brain) by a stalk. The

39

proximity of the hypothalamus is of great functional importance, for this organ controls the secretion of most pituitary hormones and is the site of biosynthesis of some of them.

The pituitary has been known since antiquity. For many centuries it was thought to control the secretion of phlegm (whence the name—*pituita* is Latin for mucus). Its role as an endocrine organ was first suspected at the end of the last century, when various medical disorders were found to be associated with disease of the pituitary. Most notably, gigantism and acromegaly were frequently accompanied by pituitary tumours.[1] The gland is now recognized as being of prime importance in regulating the activity of other endocrine organs, including the adrenal cortex, the gonads and the thyroid, as well as in affecting various metabolic functions including growth.

2.   THE ANATOMY, DEVELOPMENT AND CYTOLOGY OF THE PITUITARY GLAND[2]

A longitudinal and diagrammatic section through the pituitary gland is shown in Figure 3.1. The gland is readily divisible into two main parts, which have quite separate embryological origins, and which could well be considered as separate organs. One part, the neurohypophysis, originates from the brain, while other, the adenohypophysis, is derived from epithelial tissue. The separate

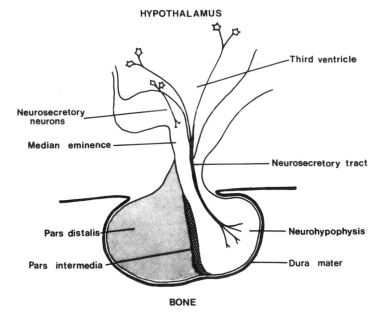

Figure 3.1   A diagrammatic, longitudinal section through the pituitary gland. The bulk of the gland is encased in a pocket in the skull; it is attached by a 'stalk' to the base of the brain

origins of these two regions are seen quite clearly if the embryological development of the gland is followed.[2] A process growing down from the floor of the
brain associates with a pouch formed from ectodermal tissue in the roof of the
mouth (Rathke's pouch). Both ectodermal and neural tissue proliferate to form
the pituitary gland. The connection between the epithelial tissue and the roof of
the mouth is lost completely, but that between the neurohypophysis and the
brain is retained and forms part of the pituitary stalk (Figure 3.2).

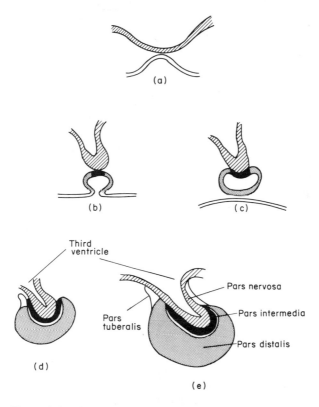

Figure 3.2 Diagram showing stages (a–e) in the development of the pituitary gland from the brain (, giving
the pars nervosa/neurohypophysis) and epithelial tissue
(Rathke's pouch; giving the adenohypophysis-pars distalis
 , pars intermedia , and pars tuberalis)

The neurohypophysis is rather simple in structure and is similar in most of the
vertebrate groups. The adenohypophysis is more complex and varies from one

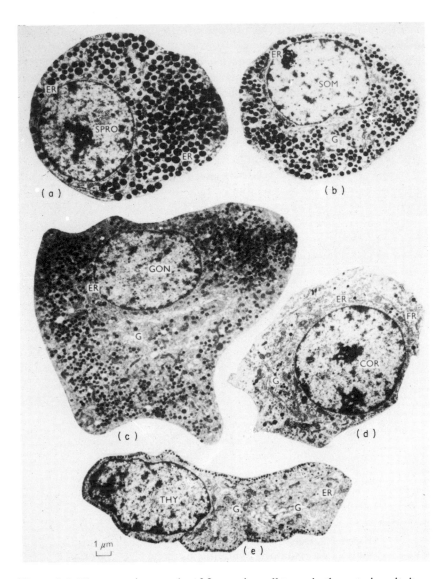

Figure 3.3 Electron micrographs of five major cell types in the anterior pituitary gland. (a) Lactotroph (prolactin secreting); (b) somatotroph (growth hormone secreting); (c) gonadotroph (FSH and LH secreting); (d) corticotroph (ACTH secreting); (e) thyrotroph (TSH secreting). ER, endoplasmic reticulum; G, Golgi body. Note the different size, form and distribution of the secretory granules in each cell type. (Reprinted with permission from Foster, C. L. (1971) *Memoirs of the Society for Endocrinology* **19**, 125–146. Copyright (1971) Journal of Endocrinology)

group to another. In mammals it is often divided into the *pars distalis* (the bulk of the secretory tissue), the *pars tuberalis* (which grows up around the pituitary stalk) and the *pars intermedia* (Figure 3.1). The last of these is often not easily distinguishable, and sometimes becomes partly fused with the neurohypophysis. In some cases, such as the human adult, it may be missing completely. Alternative names often used for parts of the pituitary are anterior lobe *(pars distalis)*, intermediate lobe *(pars intermedia)* and posterior lobe (neurohypophysis; also called *pars nervosa)*. Yet further schemes of nomenclature are sometimes encountered, but will not be used in this book.

The pituitary occurs in all vertebrates, although the anatomy of the adenohypophysis in lower vertebrates is rather different from that of mammals.[2] It is also possible to find an organ which may be homologous with the pituitary in some non-vertebrate chordates.[3] Extensive studies on the anatomy and development of the pituitary gland in lower vertebrates have been carried out by comparative endocrinologists, but cannot be considered in detail here.

Within the adenohypophysis a range of cell types has been recognized. Most of these are typical polypeptide-hormone-producing cells, with granules and a well-developed secretory apparatus. The different cell types can be distinguished by their overall form, the size and shape of their secretory granules and their response to cytochemical stains. Different pituitary hormones have been assigned to different cell types on the basis of a range of evidence, including behaviour in certain physiological conditions (e.g. thyroidectomy induces proliferation of thyrotrophs; lactotrophs are particularly common in lactating mammals), staining with fluorescent antibodies[4] and immunoelectron microscopy[4a]. A diagram showing the general nature of the various pituitary cell types is shown in Figure 3.3; the nomenclature of these cells is also shown there. It is possible that further subdivision of the main cell types considered here can be made.

3.   DISCOVERY OF THE HORMONAL ACTIVITIES OF THE PITUITARY GLAND

The elucidation of the endocrine function of the pituitary followed a rather classical pattern (association of physiological disorders with disease of the gland, followed by observation of the effects of removing the gland and replacement by injection of extracts). The earliest indications came when various disease states (including dwarfism and acromegaly) were seen to be frequently accompanied by abnormalities of the pituitary.

Clear demonstration of the influence of the pituitary hormones on the body was shown by the results of experimental removal of the gland from animals. This surgical operation (hypophysectomy) proved difficult, because in the earliest experiments it was often accompanied by extensive damage to the brain, which usually led to death. However, in 1910, Crowe, Cushing and Homans[5] successfully hypophysectomized dogs, and subsequently P. E. Smith achieved success with tadpoles, and then rats.[6] Hypophysectomy has now been carried out on a wide range of vertebrates, and techniques have been developed which make hypophysectomy of rats a standard, though still difficult, operation. Hypophysectomy is used in some circumstances as a surgical procedure for treating various pathological conditions in man.

Removal of the pituitary gland is followed by a range of physiological effects, which indicates the extent of the influence of the pituitary on the body. These effects include the following.

(a) *Cessation of growth.* Growth of young rats ceases at once. A similar effect has been demonstrated in many vertebrates.

(b) *Atrophy of the thyroid.* Production and secretion of thyroxine and related hormones ceases almost immediately, and the thyroid gland loses size quite rapidly.

(c) *Atrophy of the adrenal cortex.* The secretory activity of the adrenal cortex is drastically reduced, and the gland is soon reduced in size.

(d) *Atrophy of the gonads.* In the adult, the gonads fail to produce mature germ cells, and their endocrine function degenerates. Hypophysectomy of the immature animal prevents sexual maturation.

(e) *Cessation of lactation.* Hypophysectomy of many female mammals during late pregnancy or early lactation leads to a severe reduction of mammary growth and lactation.

(f) *Prevention of skin darkening in lower vertebrates.* In many lower vertebrates (e.g. the frog) the darkness of the skin can alter in response to environmental influences. Hypophysectomy of such animals frequently leads to a permanent lightening of the skin and loss of the ability to respond to a dark background. A similar effect may be seen in some mammals, but is much less marked.

(g) *Disturbance of carbohydrate, fat and protein metabolism.* The precise nature of this effect of hypophysectomy varies from species to species but in some cases it can be quite dramatic. Hypoglycaemia (fall in blood sugar) and increased sensitivity to insulin are frequently consequences of hypophysectomy.

(h) *Disturbance of water relations and salt balance.* Again, the effect varies according to the species under study. It is now recognized that several pituitary hormones can influence salt and water relationships, especially the neuro-hypophysial hormone, vasopressin (or vasotocin in lower vertebrates) and, in some species, prolactin.

The effects of hypophysectomy can be reversed, in most cases, by some form of replacement therapy—such as the injection of pituitary extracts or implantation of pituitaries from other animals. In many cases such treatment also has effects in intact animals. A pituitary extract can cause resumption of growth in hypophysectomized rats, and in intact female rats that have reached the growth plateau. It can restore the thyroids, gonads and adrenal cortex in hypophysectomized rats and enable them to resume their endocrine and other functions. Lactation can be induced in suitably prepared animals by pituitary extracts, as can skin darkening in frogs. Extracts of the neurohypophysis show various pharmacological and physiological effects, including induction of milk ejection and uterus contraction, antidiuresis and elevation of blood pressure. Indeed, these actions of neurohypophysial hormones were recognized before the effects of hypophysectomy were established.

These observations on the effects of hypophysectomy and their reversal by pituitary extracts had mostly been made by the late 1920s. There followed a prolonged period in which the isolation of the hormonal factors responsible for these effects was pursued. The task of separating and purifying these hormones proved a difficult one, and its successful accomplishment had to await the development of suitable assays for each of the hormones involved, and of sophisticated separation techniques for proteins and polypeptides. It is now believed that most of the hormones present in the pituitary have been isolated in pure form, from at least one or two mammalian sources. It is quite possible, however, that one or two mammalian pituitary hormones remain to be isolated and characterized. Fibroblast growth factor, calcitonin, a gastrin-like peptide, and a peptide similar to the β-subunit of human chorionic gonadotropin are all peptides that have been reported recently to occur in the mammalian pituitary gland, although their hormonal status there has not yet been fully assessed. In lower vertebrates our picture of the pituitary hormones is less complete, and new ones may well remain to be discovered.

4. THE HORMONES OF THE PITUITARY GLAND

It is now recognized that the mammalian pituitary gland contains at least 10 peptides and proteins with hormonal activity. These appear to account satisfactorily for most of the actions of pituitary extracts. They are summarized in Table 3.1, where some of their main biological actions are indicated.

These hormones fall naturally into four major families, the members of which are related structurally, show some resemblance with regard to biological activities and presumably had a common evolutionary origin.[7] The families comprise:*

(a) the neurohypophysial hormones, oxytocin (OT) and vasopressin (VP), and related hormones in lower vetebrates;

(b) the polypeptides corticotropin (adrenocorticotrophic hormone, ACTH), several melanotropins (melanocyte stimulating hormones, MSH), lipotropins, (LPH) and several endorphins;

* The first names used, in each case, are those recommended by the IUPAC–IUB Commission on Biochemical Nomenclature.[8] In several cases these have not been very widely used elsewhere, and alternatives are shown; several of these older alternatives are used in this book.

Table 3.1 The hormones of the mammalian pituitary gland

Name	Abbreviation	Chemical nature	Main actions
Posterior lobe			
1. Oxytocin	OT	peptide (9 residues)	Stimulates milk ejection, uterus contraction
2. Vasopressin (antidiuretic hormone)	VP (ADH)	peptide (9 residues)	Raises blood pressure (in mammals), antidiuretic
Intermediate lobe			
3. α-Melanotropin (α-melanocyte stimulating hormone)	α-MSH	peptide (13 residues)	Causes skin darkening, especially in amphibia
Anterior lobe			
4. β-melanotropin (β-melanocyte-stimulating hormone)	β-MSH	peptide (18–22 residues)	Causes skin darkening, especially in amphibia
5. Corticotropin (adrenocorticotropic hormone)	ACTH	peptide (39 residues)	Stimulates steroidogenesis in adrenal cortex
6. Lipotropin	LPH	peptide/protein (91 residues)	Weakly lipotropic (may be primarily a precursor of β-MSH and β-endorphin)
7. β-endorphin	β-END	peptide (31 residues)	Analgesia?
8. Somatotropin (growth hormone)	STH (GH)	protein	Promotes growth, many anabolic actions, stimulates somatomedin production
9. Prolactin	(Pr)	protein	Lactogenic activity (many other actions proposed, particularly in lower vertebrates)
10. Thyrotropin (thyroid-stimulating hormone)	TSH	glycoprotein (2 subunits)	Stimulates thyroid hormone production
11. Follitropin (follicle-stimulating hormone)	FSH	glycoprotein (2 subunits)	Stimulates maturation of Graafian follicle in female, spermatogenesis in male
12. Lutropin (luteinizing hormone; interstitial cell-stimulating hormone)	LH	glycoprotein (2 subunits)	Stimulates ripening and release of ovum and early development of corpus luteum in female; steroid production by gonads in male

(c) the protein hormones somatotropin (growth hormone, GH) and prolactin (Pr); and

(d) the glycoproteins, thyrotropin (thyroid-stimulating hormone, TSH), lutropin (luteinizing hormone, LH; also sometimes referred to as an interstitial-cell-stimulating hormone, ICSH) and follitropin (follicle-stimulating hormone, FSH).

The biochemistry of each of these hormone families is considered in the next five chapters. Several of the pituitary hormones are related to hormones in the placenta, and these are considered along with the appropriate pituitary hormones.

5.   BIOSYNTHESIS AND SECRETION OF PITUITARY HORMONES

The neurohypophysial hormones are synthesized in neurones of the hypothalamus, as protein precursors. They are packaged in the form of granules while still in the cell body of the neurone in which they are made, and these granules are subsequently transported down neuronal axons to the posterior lobe of the pituitary, where they are stored. During the process of packaging (and perhaps transport) the protein precursors are cleaved to give oxytocin and vasopressin and peptide or protein 'remainders'. These last include the neurophysins, proteins which are found in the posterior lobe and which tightly bind the neurohypophysial hormones. Specific neurophysins are associated with oxytocin and vasopressin, and are derived from specific precursors of the two hormones. The process will be discussed in more detail in Chapter 4.

The hormones of the adenohypophysis are synthesized *in situ*. The adenohypophysis contains about six cell types, and most of the current evidence suggests that each adenohypophysial hormone is produced by a specific cell type,[4] except in the case of the gonadotropins and some hormones of the ACTH–MSH family. As already mentioned, these cell types can be distinguished by various cytochemical features, including size of storage granules, size of cell, staining properties, etc. The specific hormones that they give rise to can be identified by staining with fluorescent antibodies and by observing the responses of particular types of cells to certain physiological and biochemical stimuli. Thus, cells suspected of being lactotrophs (prolactin-producing) can be shown to increase in number and in secretory activity in the lactating animal; thyrotrophs respond in a similar fashion to thyroidectomy of the animal; and each cell type shows specific cytological and biochemical responses to treatment with the appropriate hypothalamic regulatory hormone (see Section 7 below).

Details of the processes of synthesis and secretion of specific adenohypophysial hormones will be given in Chapters 5–8. Here it will suffice to emphasize a few general points. The synthesis and packaging of these hormones follows the general mechanisms common to most polypeptide hormones (Chapter 1). All the hormones are synthesized as precursors (prehormones) carrying a 'signal' peptide of 15–30 amino acid residues at the N-terminal end, which 'directs' the hormone to the secretory pathway. The signal peptide is rapidly removed once the hormone has entered the lumen of the endoplasmic reticulum, giving in some cases the mature protein hormone. In the case of the peptide hormones of the ACTH–MSH family, however, there is further, complex processing of the precursor before the mature hormones are produced, and the details of this differ from one cell type to another, so that, for example, the same precursor is converted to ACTH in corticotrophs but to α-MSH in melanotrophs (see Chapter 5). The glycoproteins are modified after polypeptide synthesis by addition of carbohydrate moieties. It is also possible that some hormones (e.g. growth hormone and prolactin) are processed further, after secretion from the pituitary gland. Synthesis of each hormone occurs on polysomes attached to the endoplasmic reticulum, and from the site of synthesis the polypeptides are transported via the endoplasmic reticulum to the Golgi body, where they are packaged into granules ready for secretion (see Chapter 1).

In the main, regulation of the secretion and possibly of the synthesis of adenohypophysial hormones is effected by factors produced by the hypothalamus. There is also evidence that specific metabolites and other factors circulating in the blood, including steroid hormones, may play an important part in controlling the pituitary gland directly.

6.   THE HYPOTHALAMUS AND REGULATION OF PITUITARY FUNCTION

The hypothalamus is a small region at the base of the mid-brain, immediately above the pituitary gland, and continuous with the neurohypophysis via the pituitary stalk. It is an endocrine gland in its own right, for as well as its role in producing the hormones of the neurohypophysis, it secretes several peptides and other factors which are carried via the hypophysial portal vessels directly to the adenohypophysis, where they regulate the secretion of the pituitary hormones. The hypothalamus is also closely associated with the function of the autonomic nervous system of the body. Because of its dual function, and its close links to the rest of the brain, the hypothalamus plays a central role in integrating the nervous and endocrine systems. The main emphasis here will be on the role of the hypothalamus in regulating the pituitary.[9,10,10a]

The crucial role of the hypothalamus in this regulation was demonstrated by a series of observations and experiments, especially those of G. W. Harris and his collaborators.[11] These workers demonstrated that electrical stimulation of the hypothalamus could lead to altered secretion of several pituitary hormones. Lesions introduced into the hypothalamus also led to altered secretion, usually in the opposite direction to electrical stimulation. Removal of the pituitary gland to a site distant from the hypothalamus led to drastic reductions in the secretion rates of most adenohypophysial hormones (except prolactin and MSH). A similar effect is produced if the connection between the hypothalamus and pituitary is severed surgically, especially if regeneration is prevented by inserting an impervious plate between the two organs.

A very careful study of the circulation between the hypothalamus and pituitary showed that blood is carried by the hypophysial portal system from a capillary plexus in the hypothalamus to another in the anterior pituitary, as illustrated in Figure 3.4, providing a basis for transfer of humoral factors from one to the other. On the other hand, there are rather few nervous connections between the hypothalamus and the *pars distalis* of the adenohypophysis, and a possible basis for nervous regulation does not appear to exist (at least, in mammals), although nervous connections to, and control of, the *pars intermedia* possibly do occur.

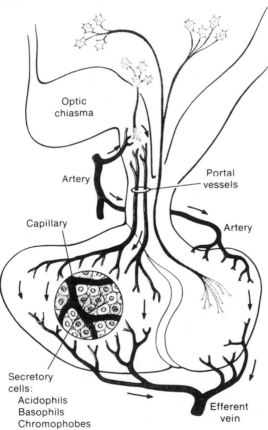

Figure 3.4 Diagram showing the blood vessels and nerves linking the hypothalamus and pituitary gland. Releasing and release-inhibiting hormones secreted by nerve endings in the median eminence are carried in the portal vessels to the adenohypophysis, where they act on hormone secretion. (Reprinted with permission from Eckert, R. and Randall, D. (1983). *Animal Physiology*, 2nd Edn. Copyright © (1983) W. H. Freeman and Co.)

Experiments with electrical stimulation of the hypothalamus and lesions within this organ have made it possible to delineate certain regions that are concerned with regulating the release of specific pituitary hormones.

The secretion of the hormones of the neurohypophysis is also controlled by the hypothalamus. Not only are these hormones synthesized in the hypothalamus, but once they have been transported down the pituitary stalk for storage in the neurohypophysis, their subsequent release is stimulated by nervous signals to this organ. These signals are thought to pass down the axons of the neurones in which the hormones are produced and transported (see Chapter 4).

7.   HYPOTHALAMIC RELEASING AND INHIBITING HORMONES[12–15]

It was established by the 1950s that specific chemical factors are produced by neurones in the hypothalamus, transported via axons to endings abutting capillaries of the hypophysial portal system (mostly in the median eminence of the hypothalamus) and carried via this blood system to the adenohypophysis where they influence the secretion (and perhaps synthesis) of the anterior pituitary hormones. These studies were followed by attempts to purify the factors. It was quite soon shown that hypothalamic extracts have the ability to regulate the secretion of various hormones *in vitro* and *in vivo*, but purification of specific factors proved much more difficult. The very low concentrations of hypothalamic hormones means that vast quantities of material have to be fractionated (one group started with 300,000 hypothalami) to obtain quite small quantities of product.

The first releasing hormone to be purified and characterized was that for thyrotropin (thyroliberin- or thyrotropin-releasing hormone, TRH), which was isolated in 1969 independently by two groups led by Schally[16] and Guillemin[17] respectively. It was shown to be a tripeptide, with blocked amino and carboxyl termini:*

$$(pyro)Glu–His–Pro–NH_2.$$

The quantities of the hormone produced from natural sources were very small, but the compound was soon synthesized chemically, so that substantial quantities became available for experimental and clinical use. TRH has been shown to stimulate release of TSH from the pituitary gland in many different experimental systems. It also stimulates release of prolactin.

---

* The (pyro)Glu residue is derived by condensation of the α-NH$_2$ group and the side chain of an N-terminal glutamic acid or glutamine residue:

$$
\begin{array}{c}
\diagup\; CO \\
NH \quad (CH_2)_2 \\
\diagdown\; CH\!-\!CO \;\cdots.
\end{array}
$$

In 1971 Schally and his co-workers[18] described the purification and structure of LH releasing factor (luliberin, LHRH) which is a decapeptide:

(pyro)Glu–His–Trp–Ser–Tyr–Gly–Leu–Arg–Pro–Gly–NH$_2$.

This too has been synthesized. It is potent in causing release of LH from the anterior pituitary, both *in vivo* and *in vitro,* and it also causes release of FSH. It has been proposed that this factor is in fact a combined releasing factor for FSH and LH (i.e. LHRH–FSHRH) and it is sometimes referred to as gonadotropin-releasing hormone (GnRH); this idea is now widely accepted. Some evidence suggests that a separate FSH–RF remains to be isolated,[19] but its existence remains controversial. It has been shown, recently, that LHRH has direct effects on the gonads, so it cannot be seen simply as a releasing factor for pituitary gonadotropins.

The influence of the hypothalamus on growth hormone secretion was thought until about 1970 to be primarily stimulatory. However, many attempts to purify and characterize a GHRH proved unsuccessful, and it is only recently that a peptide of 44 amino acids with potent growth-hormone-releasing activity has been isolated, characterized and synthesized.[20] This peptide was first isolated from a pancreatic tumour, but subsequent work strongly suggests that it is identical to the corresponding hypothalamic factor. Guillemin and his colleagues[21] earlier isolated a GH-release inhibiting hormone (somatostatin) and showed that it is a tetradecapeptide:

H–Ala–Gly–Cys–Lys–Asn–Phe–Phe–Trp–Lys–Thr–Phe–Thr–Ser–Cys–OH.

Somatostatin inhibits the release of GH both *in vivo* and *in vitro.* It seems likely that it operates together with GHRH in controlling the secretion of growth hormone. It has now been shown that somatostatin exists in several other organs, in addition to the hypothalamus, including various other parts of the brain, the islets of Langerhans of the pancreas and various regions of the gut[22] (see Section 10). It appears to inhibit the release of several pituitary hormones, insulin, glucagon and some other polypeptide hormones (and possibly other secreted proteins, including digestive enzymes) so it is clearly not a specific factor for controlling GH release. Larger, more potent forms of somatostatin have also been identified.

Evidence for a hypothalamic releasing factor (CRH) for ACTH was obtained at an early stage in the investigation of hypothalamic factors, but its purification and characterization has proved a difficult problem. Recently a peptide of 41 residues has been described, with CRH activity.[23] It seems likely that this is the main hypothalamic peptide controlling ACTH release, though other peptides may modulate its activity. ACTH is derived from the same precursor as β-MSH, lipotropins and β-endorphins, and there is evidence for co-secretion of these peptides, suggesting that CRH may stimulate secretion of several or all of them. CRH is discussed further in Chapter 5.

Release of prolactin is primarily under inhibitory control by the hypothalamus. Dopamine inhibits prolactin release *in vitro*, and a good deal of evidence suggests that it acts as a physiological inhibitor.[24] Other inhibitory factors may also exist, although evidence for these is inconclusive. TRH stimulates prolactin secretion and may do so physiologically; other peptides with prolactin-releasing activity have been described. The control of prolactin secretion is described further in Chapter 8.

Releasing and release-inhibiting factors have been described for MSH. They are structurally related to oxytocin (respectively, the 'ring' and the 'tail'). Doubts remain as to whether these are true physiological regulatory factors— the very high content of oxytocin and vasopressin in the hypothalamus means

Table 3.2 The hypothalamic hormones that regulate the adenohypophysis

Name	Abbreviation	Chemical nature	Main actions
Well-characterized factors[a]			
1. Thyroliberin (thyrotropin-releasing hormone)	TRH	peptide (3 residues)	Stimulates release of TSH and possibly prolactin[b]
2. Luliberin (LH-releasing hormone)	LHRH	peptide (10 residues)	Stimulates release of LH and FSH
3. Somatostatin (GH-release-inhibiting hormone)	SRIF	peptide (14 residues)	Inhibits release of GH (and many other secreted peptides)
4. Dopamine	DA	catecholamine	Inhibits release of prolactin[b]
5. Somatoliberin (GH-releasing factor)	GHRH	peptide (44 residues)	Stimulates release of GH
6. Corticoliberin (corticotropin-releasing factor)	CRH	peptide (41 residues)	Stimulates release of corticotropin
Incompletely characterized factors[a]			
7. MSH-releasing factor	MRF	5-residue peptide ('ring' of oxytocin)	Stimulates release of MSH
8. MSH-release-inhibiting factor	MIF	3-residue peptide ('tail' of oxytocin)	Inhibits release of MSH
9. FSH-releasing factor	FSH–RF	?	May stimulate release of FSH
10. Endorphins, enkephalins and related peptides		peptides	Postulated to play a role in regulation of release of various pituitary hormones[c]

Notes
[a]Well-characterized hypothalamic factors are often referred to as 'hormones', less well-characterized ones as 'factors'.

[b]Other factors that regulate prolactin secretion have been postulated. Vasoactive-intestinal peptide (VIP) is a stimulator of prolactin secretion and occurs in the hypothalamus.

[c]For example, β-endorphin stimulates secretion of LH in the foetus; synthetic enkephalin analogues are potent stimulators of GH secretion; enkephalin relieves the inhibition of prolactin secretion caused by dopamine.

that they can mask other factors, and several earlier suggestions that oxytocin or vasopressin may act as releasing factors for some anterior pituitary hormones proved misleading. Nevertheless, the MSH release-inhibiting factor particularly (Pro–Leu–Gly–NH$_2$) is very potent and may prove to be of physiological importance. Dopamine may also be important in the control of MSH secretion.

Table 3.2 summarizes the known hypothalamic releasing and release-inhibiting factors. Further treatment of the individual hypothalamic hormones, and their role in regulating secretion and synthesis of specific anterior pituitary hormones, will be given in Chapters 5–8. The next two sections deal with some general biochemical aspects of these factors.

8.   THE BIOCHEMICAL MODE OF ACTION OF THE RELEASING HORMONES[25–27]

Although some studies on the biochemical mechanisms whereby the hypothalamic factors produce their actions have been carried out using whole animals, the majority have used *in vitro* preparations. The rat anterior pituitary gland can be incubated *in vitro*; it divides readily into two lobes which can be conveniently used for secretion studies—one lobe being used as a control while the other is subjected to factors which may affect hormone release. Incomplete penetration of the tissue remains a problem, however, and many workers now use preparations of pituitary cells that have been dispersed by treatment with trypsin or collagenase. Such cells can be used either soon after dispersion or after culture for a few days. Pituitary cells are very heterogeneous, of course, and attempts have been made, with some success, to fractionate the different cell types and use these for biochemical studies.

A good deal of evidence gained from systems of this kind suggests that at least some of the releasing hormones operate via effects on cyclic AMP levels.[25] Thus, LHRH, GHRH and crude hypothalamic extracts, when added to *in vitro* incubations of pituitary glands or cells, all stimulate rapid and large increases in intracellular cyclic AMP levels. The effect is apparently a result of stimulation of adenylate cyclase; phosphodiesterase activity is not altered. Addition of cyclic AMP (and its derivatives) or theophylline to appropriate incubations *in vitro* stimulates release of several of the pituitary hormones, including GH, prolactin, TSH, FSH and LH, and also stimulates protein synthesis in the pituitary gland and the synthesis of (at least) prolactin and GH. Cyclic AMP in pituitary cells is thought to activate cyclic AMP-dependent protein kinases (see also Chapter 16), which in turn may cause phosphorylation and activation of various key components of the cell's synthetic and secretory machinery such as ribosomes, membrane proteins and microtubules[26] (Figure 3.5). There is some evidence that cyclic GMP levels may also be increased by some releasing factors, and may mediate their actions.

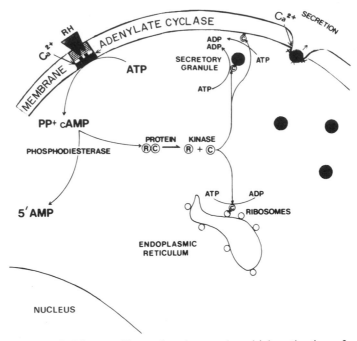

Figure 3.5 Diagram illustrating the way by which activation of adenylate cyclase by a hypothalamic releasing hormone (RH) may trigger both protein synthesis and secretory processes within the cell. (Reprinted with permission from Labrie, F. *et al.* (1978) *Recent Progr. Hormone Res.* **34**, 25–93. Copyright (1978) Academic Press Inc.)

It is not yet established whether the inhibitory hypothalamic factors, somatostatin and dopamine, act by lowering cyclic AMP levels. Some evidence suggests that they do inhibit formation of this nucleotide,[25] but other studies show that dopamine, at least, can exert its inhibitory actions on prolactin release even in the presence of high levels of cyclic AMP.[28]

Specific binding sites for TRH (and several other hypothalamic factors) have been detected in the pituitary gland.[26] These appear to be the receptor sites for the hormone. They are membrane-bound and resemble the receptors that have been described for other peptide hormones. Analogues of TRH bind to their receptors roughly in proportion to their biological activity. As in the case of most other peptide hormones, most evidence that is available suggests that the hypothalamic factors exert their actions at the cell membrane; subsequent internalization may occur, but is not essential for many of the biochemical actions. An underlying assumption of much of the work on the mode of action of the different hypothalamic hormones is that they exert their specificity by binding to specific receptors on the appropriate pituitary cell types; once they have bound, the mechanism of action may be rather similar for the various factors (at least, for those which promote release).

Although a mechanism of action of releasing hormones involving cyclic AMP is currently most favoured, evidence has been adduced for alternative or additional mechanisms. An interesting alternative is that the releasing hormones act by altering the permeability of the target cell membranes, leading to membrane depolarization, and consequent uptake of calcium. Increased intracellular calcium would then stimulate hormone release (other mechanisms for alteration of intracellular Ca^{2+} concentration may also be important). Evidence for this theory is based on the effects of altering the ionic composition when pituitary glands (or cells) are incubated *in vitro*. Thus, high extracellular potassium concentrations, which are thought to cause membrane depolarization, lead to increased release of several anterior pituitary hormones. Release of all pituitary hormones is markedly reduced when calcium is removed from the incubation medium. There have also been claims that calcium levels in the cell are raised in response to releasing hormones. The stimulation of pituitary hormone secretion by high potassium concentrations is not accompanied by elevated levels of cyclic AMP. However, it is possible that one of the effects of cyclic AMP in the cell is to alter membrane permeability and facilitate entry of calcium, which would then promote hormone release. The release of many pituitary hormones certainly involves a complex interaction between calcium and cyclic nucleotides, but precisely how it operates is not yet clear.

Recently evidence has accumulated that suggests that the products of the metabolism of the phospholipids phosphatidylinositol, phosphatidylinositol phosphate and phosphatidylinositol bisphosphate (PIP_2) may play an important role in mediating the secretion of pituitary hormones.[29] PIP_2 is hydrolysed by phospholipase C to diacylglycerol and inositol trisphosphate. The diacylglycerol may be an activator of protein kinase C,[30] while the inositol trisphosphate may play a role in mobilization of intracellular calcium stores.

Another group of compounds which may be involved in the release of pituitary hormones is the prostaglandins. These have been shown to promote release of various pituitary hormones *in vitro*. They also cause elevation of cyclic AMP levels in the pituitary (paradoxically, cyclic AMP causes elevation of prostaglandin levels). Some hypothalamic hormones cause a rise in the level of prostaglandins in the pituitary. Some prostaglandin inhibitors can block the action of LHRH, but others do not (although they completely block elevation of pituitary prostaglandin levels). It seems possible that prostaglandins are involved in some way in the mode of action of hypothalamic hormones, but their precise role remains elusive.[26]

Although the hypothalamic factors play a predominant role in regulating secretion of anterior pituitary hormones, other factors, such as steroid hormones, thyroid hormones and inhibin, may also affect the basal rate of hormone secretion and may modulate the actions of the hypothalamic factors. Interactions between the various sex steroids and LHRH are particularly intricate, for example, and may be responsible for the differential secretion of LH and FSH. LHRH stimulates secretion of both of these gonadotropins, but its actions are modulated by steroids; for example, oestrogens potentiate the action of LHRH on both LH and FSH secretion, but androgens reduce the effect on LH secretion but increase the effect on FSH secretion. The effects of steroid and thyroid hormones at the pituitary level are relatively slow, and it is likely that they produce their effects, at least partly, by actions at the nuclear level.

9.   BIOSYNTHESIS AND SECRETION OF HYPOTHALAMIC HORMONES

It is clear from the preceding sections that the hypothalamus must now be considered as an endocrine organ in its own right. The mechanisms whereby its hormones are produced, stored and secreted are as yet very incompletely understood; their elucidation presents one of the great challenges for molecular endocrinology during the next decade.

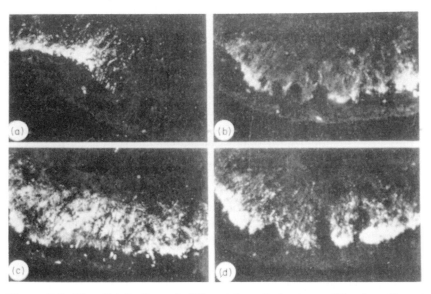

Figure 3.6 The distribution of different hypothalamic peptides in adjacent sections of the median eminence of the hypothalamus as revealed by specific immunostaining. (a) LHRH; (b) CRH; (c) vasopressin; (d) somatostatin. (Reprinted with permission from Bugnon, C., Fellmann, D., Gouget, A. and Cardot, J. (1982) *Nature* **298**, 159–161, Copyright © (1982) Macmillan Journals Ltd.)

The production of the various releasing hormones has been shown to be localized within the hypothalamus by using lesions placed within the organ as well as electrical stimulation of different regions.[11] These experiments suggest that different releasing hormones are produced by different regions. Extracts of different regions of the hypothalamus have been tested for content of different releasing hormones. More recently fluorescent antibodies have been used to localize TRH, LHRH, CRH, GHRH and somatostatin within the hypothalamus. Figure 3.6 provides examples. These various approaches have given a good deal of information about which hypothalamic region synthesizes which releasing hormone.[31-33]

Anatomical, and other, observations suggest that the main region involved in the secretion of hypothalamic releasing hormones is the median eminence of the hypothalamus. Secretion probably occurs primarily from nerve endings abutting capillaries of the hypothalamic–hypophysial portal system in this region (Figure 3.4). The median eminence is some distance from the apparent sites of origin of several of the hypothalamic hormones, and it is possible that these are transported to the region down the axons of the neurones in which they are synthesized, to nerve endings of the same neurones, from which they are eventually secreted. The process would thus be similar to that seen in the case of the neurohypophysial hormones (Chapter 4).

Very little is known about the biosynthesis of most of the hypothalamic peptides, or the mechanisms whereby they are transported to the median eminence. A biosynthetic mechanism involving a multi-enzyme system for amino acid coupling has been suggested for TRH, but the work has proved difficult to repeat and considerable doubt has been cast upon it. Somatostatin, CRH, LHRH and GHRH are now known to be synthesized as larger precursors[34] (see Chapters 5–7), and it is likely that the same applies to TRH.

There is a good deal of evidence that the production and/or release of hypothalamic hormones can be regulated by synaptic transmitters.[10a,35,36] Such a mechanism is of great interest, since it would provide a direct biochemical link between the nervous and endocrine systems. Neurones producing monoamines (dopamine, noradrenaline, serotonin) occur in many parts of the hypothalamus. A possible role for such monoamines in regulating the release of hypothalamic hormones has been demonstrated in *in vitro* experiments; dopamine incubated with hypothalamic fragments *in vitro* stimulated production of releasing hormone for FSH and LH. Similar results have been obtained *in vivo*—catecholamines injected directly into the third ventricle increased circulating levels of LHRH, LH and FSH, and reduced circulating levels of prolactin.

A considerable number of such experiments have been carried out with regard to other releasing hormones. The effects of a variety of drugs which block or stimulate synaptic transmittance also support the general hypothesis that regulation of secretion and/or production of hypothalamic hormones is at least partly under direct control of the nervous system, with neurotransmitters as the regulating agents, operating presumably via synapses formed between neuronal elements of the central nervous system and the endocrine-producing neurones of the hypothalamus.

Other factors can also control the production and release of the hypothalamic releasing hormones. Steroid hormones, for example, interact directly with receptors in the hypothalamus and play an important part in the regulation of LHRH secretion.

Once the hypothalamic hormones have been released from their nerve endings (and little is known about the details of secretory mechanisms) they diffuse into the capillaries of the hypothalamic–hypophysial portal system and are carried rapidly and directly to the pituitary gland. It seems likely that the short distance they have to travel is a vital aspect of their physiological functioning, enabling very small amounts of hormone to produce an effective concentration at the target organ. Any releasing hormones which pass through the pituitary into the rest of the circulation will probably be diluted to such an extent that they are unlikely to be very significant, at least as far as subsequent recirculation to the pituitary is concerned. The half-life of hypothalamic hormones in the circulation is very short (about 2 minutes for TRH), which further emphasizes the importance of the very rapid passage from hypothalamus to pituitary.

10.   DISTRIBUTION OF PITUITARY AND HYPOTHALAMIC HORMONES OUTSIDE THE PITUITARY AND HYPOTHALAMUS

During the past few years it has become clear that many of the hormones traditionally thought to be confined to the pituitary gland can be found elsewhere in the body, particularly in the brain.[37,37a] Furthermore, the more-recently discovered hypothalamic releasing hormones have been discovered to be distributed quite widely outside the hypothalamus.

The pituitary hormones most extensively studied in this respect are those of the ACTH–MSH–endorphin family. Since many of these hormones are produced from the same precursor molecule, discovery of one of them in a particular cell implies the potential for the cell to produce other members of the family. ACTH, α-MSH, β-lipotropin and β-endorphin have all been discovered, using radioimmunossay, immunocytochemistry, bioassay and physicochemical methods, in the hypothalamus and some other parts of the brain. Concentrations are generally much lower, by a factor of 100–1,000, than those found in the pituitary gland, although the difference for β-endorphin may be less marked. The detailed distribution of peptides of this family will be considered in Chapter 5.

Low concentrations of growth hormone, prolactin and TSH have also been discovered, using immunological methods, in the hypothalamus and some other parts of the brain.[37] Detailed physicochemical characterization has not been carried out in most cases, although in the case of growth hormone, at least, the immunoreactive material appears to have a size identical to that of the pituitary hormone. It must be noted that many other peptide hormones originally isolated from peripheral organs have now been reported in the brain, including many gut hormones, glucagon and insulin. Thus, the occurrence of pituitary hormones there is probably just one facet of a widespread distribution of polypeptide hormones in general. These recent discoveries accord with an earlier proposal by Pearse,[38] that cells producing polypeptide hormones may also produce catecholamines, and that all such cells have many features in common and belong to the so-called APUD (amine precursor uptake and decarboxylation) series.

The origin of the pituitary hormones in the brain is a matter of some controversy. Some authors consider that they originate in the pituitary gland and are transported directly to the brain via a retrograde blood flow. Anatomical evidence for some flow of blood from pituitary to brain has been provided.[39] Other authors, probably the majority, consider that the 'pituitary' hormones found in the brain are in fact synthesized there. The presence of these hormones in the brain of hypophysectomized animals, sometimes in increased concentrations, and their synthesis *in vitro* by dispersed brain cells supports this view. Of course, it may be that both sources contribute.

The hypothalamic releasing hormones are also distributed quite widely through the brain, as will be discussed more fully in Chapters 5–7. To some extent this could be a consequence of their synthesis outside the hypothalamus and subsequent transport to this region, but it seems likely that a considerable part of the TRH, LHRH and somatostatin of the brain is never associated with the hypothalamus.

The biological function of these various pituitary and hypothalamic hormones in other regions of the brain is not yet clear, but it seems possible that some of them, and other peptides, play a role as neurotransmitters or neuromodulators. It is now clear that in some cases one neurone can possess a catecholamine as well as one or more neuropeptides,[40] so the classic concept of one neurone, one transmitter may have to be modified.

In addition to the occurrence of these various pituitary and hypothalamic hormones in the brain, it is now clear that some of them are also found in various peripheral organs. Somatostatin, for example, is found in the islets of Langerhans of the pancreas and in various parts of the gut.

11. FEEDBACK MECHANISMS IN THE REGULATION OF PITUITARY AND HYPOTHALAMIC FUNCTION

As with most elements of the endocrine system, the activities of the pituitary and hypothalamus are ultimately regulated by feedback from the target tissues and organs. Feedback is partly at the level of the pituitary, where target gland hormones and other factors may modify responses to releasing hormones. Thus, thyroid hormones can suppress the actions of TRH.

However, feedback at the level of the hypothalamus is probably more important, and here feedback (by target gland hormones or metabolic factors) can interact with the neuronal regulation of hypothalamic function to give the all-important integration of the nervous and endocrine systems of the body. An example of feedback at the hypothalamic level is the regulation of LHRH production and release by oestrogens and/or progesterone in the oestrus and menstrual cycles. That such steroids operate, at least partly, at the hypothalamic level, is shown by direct effects on LHRH levels in the hypothalamus (in addition to effects on circulating FSH and LH levels) and also the observation of increased peripheral circulating LHRH in women prior to ovulation (see also Chapter 6).

Short feedback loops may also be involved in regulating secretion of hypothalamic and pituitary hormones. Thus, there is some evidence that high circulating levels of FSH may inhibit the release of FSH itself by direct action on the pituitary. In addition, FSH may inhibit FSH-RH secretion. High circulating levels of FSH-RH may inhibit production of this releasing hormone. Similar results have been obtained for some other releasing-hormone–pituitary-hormone pairs.

In accordance with these various feedback systems, receptors have been found in the hypothalamus and pituitary for steroids and many other factors likely to affect secretion.

12.   PITUITARY TUMOURS

Tumours of the pituitary gland are known in man and experimental animals. Little will be said about human pituitary tumours here (for details, textbooks of clinical endocrinology should be consulted—e.g. ref. 41). Human pituitary tumours can be derived from any of the cell types of the gland (acidophils, basophils, chromophobes, etc.) and tumours of mixed cell type are frequent. They very rarely metastasize, and clinical disorders resulting from tumours of the human pituitary are usually due to local effects, due to the large size of the tumour in a restricted region of the skull or to hormonal changes (due to overproduction or underproduction of pituitary hormones). A well-characterized disease resulting from overproduction of GH by a pituitary tumour is acromegaly (enlargement of hands, feet and facial features and many other organs). Pituitary tumours can be treated by surgery, by implanting Yttrium-90 into the gland, by X-ray irradiation, or, in some cases, by drugs.

Experimental tumours of the pituitary gland have been widely used as models for the study of the production and secretion of pituitary hormones. Tumours in rats have been most widely used experimentally. They have been induced by radiation or by prolonged hormonal imbalance. Tumours secreting prolactin, GH, ACTH, TSH and gonadotropins have been described.[42]

A common way to induce such tumours is by long-term treatment with oestrogens. Such tumours frequently secrete high levels of prolactin (and are referred to as mammotropic tumours). They can be readily transplanted to different sites within the body, and frequently grow to considerable size. The primary tumours show a fairly well preserved 'secretory cell' type of appearance (with many granules) and are usually oestrogen dependent. They often secrete GH as well as prolactin, and sometimes other hormones. During successive transplantations the nature of such tumours usually changes—dependence on oestrogen is diminished, the granular appearance is reduced, and the amounts and types of hormones secreted may change.

Several cell lines derived from pituitary tumours have been established in tissue culture *in vitro*, and maintained through many cell generations. Here they continue to secrete hormones, although during the course of growth in culture the patterns of secretion may change. In several cases such cell lines have been cloned. Such lines of pituitary tumour cells have been quite widely used for studies on secretion and synthesis of pituitary hormones, although it must be remembered that the behaviour of cells derived from pituitary tumours *in vitro* may differ markedly from the behaviour of normal pituitary cells *in vivo*. A very interesting feature which has been established concerns the way in which some cell lines can secrete more than one hormone. Thus, the cell line GH_3 derived from a rat pituitary tumour has been shown (after cloning) to produce both GH and prolactin;[43] these two hormones are thought to be produced by quite separate cells in the normal pituitary.

REFERENCES

1. Cushing, H. (1912) *The Pituitary Body and its Disorders*. Lippincott, Philadelphia.
2. Holmes, R. L. and Ball, J. N. (1974) *The Pituitary Gland*. Cambridge University Press, Cambridge.
3. Wingstrand, K. G. (1966) In: *The Pituitary Gland*, Vol. I (Eds. Harris, G. W. and Donovan, B. T.). Butterworths, London, pp. 58–126.
4. Baker, B. L. (1974) In: *Handbook of Physiology*, Section 7, Vol. IV, *The Pituitary Gland*, Part 1 (Eds. Knobil, E. and Sawyer, W. H.). American Physiological Society, Washington D.C., pp. 45–80.
4a. Pelletier, G., Robert, F. and Hardy, J. (1978) *J. Clin. Endocrinol. Metab.*, **46**, 534–542.
5. Crowe, S. J., Cushing, H. and Homans, J. (1910) *Bull. Johns Hopkins Hospital*, **21**, 127–169.
6. Smith, P. E. (1930) *Am. J. Anat.*, **45**, 205–274.
7. Wallis, M. (1975) *Biol. Rev.*, **50**, 35–98.
8. IUPAC–IUB Commission on Biochemical Nomenclature (1975) *J. Biol. Chem.*, **250**, 3215–3216.
9. Jeffcoate, S. L. and Hutchinson, J. S. M. (Eds.) (1978) *The Endocrine Hypothalamus*. Academic Press, New York.

10. Fuxe, K., Hökfelt, T and Luft, R. (Eds.) (1979) *Central Regulation of the Endocrine System*. Plenum Press, New York.
10a. Reichlin, S. (1981) In: *Textbook of Endocrinology*, 6th edn (Ed. Williams, R. H.). Saunders, Philadelphia, pp. 589 645.
11. Harris, G. W. (1972) *J. Endocrinol.* **53**, ii–xxiii.
12. Guillemin, R. (1978) *Science*, **202**, 390–402.
13. Schally, A. V., Coy, D. H., Meyers, C. A. and Kastin, A. J. (1979) In: *Hormonal Peptides and Proteins*, Vol. 7. (Ed. Li, C. H.). Academic Press, New York, pp. 1–54.
14. Vale, W., Rivier, C. and Brown, M. (1977) *Ann. Rev. Med.*, **39**, 473–527.
15. Sandow, J. and König, W. (1978) In: *The Endocrine Hypothalamus* (Eds. Jeffcoate, S. L. and Hutchinson, J. S. M.). Academic Press, New York, pp. 149–211.
16. Bøler, J., Enzmann, F., Folkers, K., Bowers, C. Y. and Schally, A. V. (1969) *Biochem. Biophys. Res. Commun.*, **37**, 705–710.
17. Burgus, R., Dunn, T. F., Desiderio, D. and Guillemin, R. (1969) *C.R. Acad. Sci. Paris*, **269**, 1870–1873.
18. Matsuo, H., Baba, Y., Nair, R. M. G., Arimura, A. and Schally, A. V. (1971) *Biochem. Biophys. Res. Commun.*, **43**, 1334–1339.
19. Bowers, C. Y., Currie, B. L., Johansson, K. N. G. and Folkers, K. (1973) *Biochem. Biophys. Res. Commun.*, **50**, 20–26.
20. Guillemin, R., Brazeau, P., Böhlen, P., Esch, F., Ling, N. and Wehrenberg, W. B. (1982) *Science*, **218**, 585–587.
21. Brazeau, P., Vale, W., Burgus, R., Ling, N., Butcher, M., Rivier, J. and Guillemin, R. (1973) *Science*, **179**, 77–79.
22. Hökfelt, T., Efendic, S., Hellerstrom, C., Johansson, O., Luft, R. and Arimura, A. (1975) *Acta Endocr.*, **80**, Suppl. 200, 1–41.
23. Vale, W., Spiess, J., Rivier, C. and Rivier, J. (1981) *Science*, **213**, 1394–1397.
24. Leong, D. A., Frawley, L. S. and Neill, J. D. (1983) *Ann. Rev. Physiol.*, **45**, 109–127.
25. Labrie, F., Godbout, M., Beaulieu, M., Borgeat, P. and Barden, N. (1979) *Trends Biochem. Sci.*, **4**, 158–160.
26. Labrie, F., Lagacé, L., Beaulieu, M., Ferland, L., De Léan, A., Drouin, J., Borgeat, P., Kelly, P. A., Cusan, L., Dupont, A., Lemay, A., Antakly, T., Pelletier, G. H. and Barden, N. (1979) In: *Hormonal Peptides and Proteins*, Vol. 7 (Ed. Li, C. H.). Academic Press, New York, pp. 205–277.
27. Tixier-Vidal, A. and Gourdji, D. (1981) *Physiol. Rev.*, **61**, 974–1011.
28. Ray, K. P. and Wallis, M. (1982) *Mol. Cell. Endocrinol.*, **27**, 139–155.
29. Berridge, M. J. (1984). *Biochem. J.*, **220**, 345–360.
30. Nishizuka, Y. (1984) *Nature*, **308**, 693–698.
31. Knigge, K. M., Hoffman, G. E., Joseph, S. A., Scott, D. E., Sladek, C. D. and Sladek, J. R. (1980) In: *Handbook of the Hypothalamus*, Vol. 2. *Physiology of the Hypothalamus*. (Eds. Morgane, P. J. and Panksepp, J.). Dekker, New York, pp. 63–164.
32. Palkovits, M. (1983). In: *Brain Peptides*. (Eds. Krieger, D. T., Brownstein, M. J. and Martin, J. B.). Wiley, New York, pp. 495–545.
33. Elde, R. and Hökfelt, T. (1978) *Frontiers in Neuroendocrinology*, **5**, 1–33.
34. Hobart, P., Crawford, R., Shen, L., Pictet, R. and Rutter, W. J. (1980) *Nature*, **288**, 137–141.
35. Müller, E. E., Nistico, G. and Scapagnini, U. (Eds.) (1977) *Neurotransmitters and Anterior Pituitary Function*. Academic Press, New York.
36. Weiner, R. I. and Ganong, W. F. (1978) *Physiol. Rev.*, **58**, 905–976.
37. Krieger, D. T. and Liotta, A. S. (1979) *Science*, **205**, 366–372.
37a. Krieger, D. T. (1983) *Science*, **222**, 975–985.

38. Pearse, A. G. E. (1978) In: *Centrally Acting Peptides* (Ed. Hughes, J.). Macmillan, London, pp. 49–57.
39. Bergland, R. M. and Page, R. B. (1979) *Science,* **204**, 18–24.
40. Hökfelt, T., Johansson, O., Ljungdahl, A., Lundberg, J. M. and Schultsberg, M. (1980) *Nature,* **284**, 515–521.
41. Hall, R., Anderson, J., Smart, G. A. and Besser, M. (1981) *Fundamentals of Clinical Endocrinology,* 3rd. edn. Pitman, London.
42. Furth, J., Ueda, G. and Clifton, K. H. (1973) *Methods in Cancer Res.,* **10**, 201–277.
43. Bancroft, F. C. and Tashjian, A. H. (1971) *Exp. Cell Res.,* **64**, 125–128.

Chapter 4

Hormones of the neurohypophysis

1. INTRODUCTION

It was recognized very early in the study of the pituitary gland that extracts of the posterior lobe contain potent pharmacological factors.[1,2] Such extracts cause contraction of smooth muscle (especially of the uterus), milk ejection, an increase in mammalian blood pressure, antidiuresis, among other effects.

When attempts were made in the 1930s and 1940s to purify the factor or factors responsible, using mild extraction conditions, an apparently homogeneous protein was isolated by van Dyke and co-workers which retained all these activities.[3] However, other workers, using acetic acid extraction, isolated two much-smaller compounds, which were shown subsequently to be peptides, which retained the pharmacological properties of neurohypophysial extracts.[4] One of these, oxytocin, had effects on uterus contraction and on milk ejection, while the other, vasopressin or antidiuretic hormone (ADH), had effects on blood pressure and diuresis. These small peptides were worked on further by du Vigneaud and his colleagues, who purified them, determined their structures and finally, in 1953, synthesized them.[5,6]

The relation between the protein of van Dyke and co-workers and the small peptides has now been elucidated. If mild conditions are used a 'pure' protein can be isolated from the neurohypophysis which possesses the activities of both oxytocin and vasopressin. It can be dissociated, with some difficulty, into oxytocin, vasopressin and a mixture of proteins (neurophysins) which are capable of binding oxytocin and vasopressin tightly. Specific neurophysins are associated with oxytocin and vasopressin, respectively.[7,8] The role of these proteins is not completely clear. They are synthesized as part of a common precursor protein (see Section 6.1 below) and may be important for transport of oxytocin and vasopressin during their passage from the hypothalamus to the posterior lobe and for their storage in that organ. Specific hormonal actions have been proposed for the neurophysins, but a physiological role of this kind has not been demonstrated convincingly.

2. CHEMISTRY AND STRUCTURE

2.1 Structure of oxytocin, vasopressin and related peptides

The structures of oxytocin and vasopressin, as elucidated by du Vigneaud and co-workers,[5] are similar. Both are nonapeptides containing a disulphide bridge and an amidated C-terminus (Figure 4.1).

It was recognized at quite an early stage that lower vertebrates must contain neurohypophysial hormones different from those of mammals. This was apparent because of the different pharmacological properties shown by neurohypophysial extracts from such lower animals. Detailed biochemical studies led

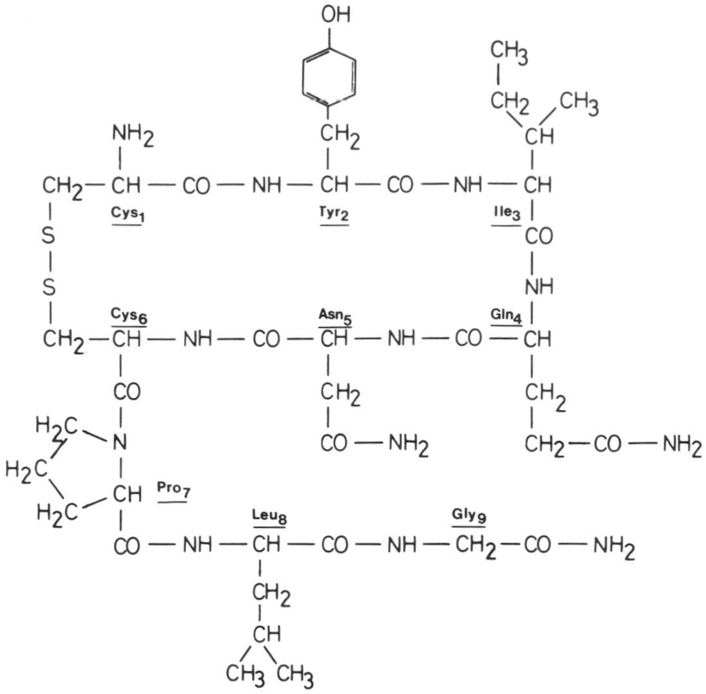

Figure 4.1 The structure of oxytocin

to the discovery of a series of new hormones from non-mammalian vertebrates,[9] as shown in Figure 4.2. Indeed, it was also shown that the domestic pig contained a different form of vasopressin (lysine vasopressin—Figure 4.2) while some other members of the pig family possessed both lysine and arginine vasopressins.[10]

The distribution among the vertebrates of these various natural analogues of oxytocin and vasopressin is indicated in Figure 4.2. They differ from each other at four positions (residues 2, 3, 4 and 8). These structural differences lead to dramatic differences in biological activity.

2.2   Molecular evolution of neurohypophysial hormones[9,11,12]

The close homology between the neurohypophysial peptides suggests that they are phylogenetically related, and various schemes have been proposed to explain their evolution. It is proposed that there was originally a single ancestral hormone in the group (perhaps arginine vasotocin, since this is found in all non-mammalian vertebrates and is the only hormone of this family found in the primitive cyclostomes). Gene duplication then allowed divergent evolution and separation of the functions concerned with reproduction and contraction of

OXYTOCIN (OT)	H-Cys-Tyr-ILE-GLN-Asn-Cys-Pro-LEU-Gly-NH$_2$	mammals, birds, some elasmobranchs; possibly some other tetrapods.
ARGININE VASOPRESSIN (AVP)	H-Cys-Tyr-PHE-GLN-Asn-Cys-Pro-ARG-Gly-NH$_2$	most mammals.
LYSINE VASOPRESSIN (LVP)	H-Cys-Tyr-PHE-GLN-Asn-Cys-Pro-LYS-Gly-NH$_2$	most Suina.
ARGININE VASOTOCIN (AVT)	H-Cys-Tyr-ILE-GLN-Asn-Cys-Pro-ARG-Gly-NH$_2$	all non-mammalian vertebrates.
MESOTOCIN (MT)	H-Cys-Tyr-ILE-GLN-Asn-Cys-Pro-ILE-Gly-NH$_2$	amphibia, reptiles, lungfish, possibly some other fish.
ISOTOCIN (IT)	H-Cys-Tyr-ILE-SER-Asn-Cys-Pro-ILE-Gly-NH$_2$	teleosts.
GLUMITOCIN (GT)	H-Cys-Tyr-ILE-SER-Asn-Cys-Pro-GLN-Gly-NH$_2$	some elasmobranchs.
ASPARTOCIN (AT)	H-Cys-Tyr-ILE-ASN-Asn-Cys-Pro-LEU-Gly-NH$_2$	some elasmobranchs.
VALITOCIN (VT)	H-Cys-Tyr-ILE-GLN-Asn-Cys-Pro-VAL-Gly-NH$_2$	some elasmobranchs.

Figure 4.2 The neurohypophysial hormones found in vertebrates. (Reprinted with permission from Wallis, M. (1975). *Biol. Revs.* **50**, 35–98. Copyright (1975) Cambridge University Press)

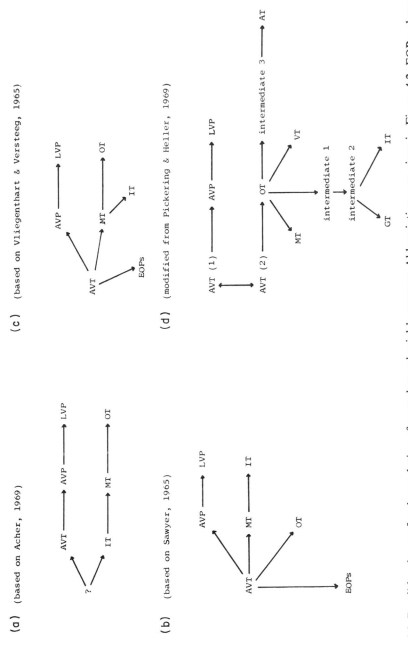

Figure 4.3 Possible schemes for the evolution of neurohypophysial hormones. Abbreviations are given in Figure 4.2. EOPs, elasmobranch oxytocin-like principles. Intermediates 1, 2 and 3 of scheme (d) have not been identified but are postulated because of the nature of the genetic code. (Reprinted with permission from Wallis, M. (1975) *Biol. Revs.* **50**, 35–98. Copyright (1975) Cambridge University Press)

smooth muscle (oxytocin-like hormones) from those concerned with osmoregulation (vasopressin-like hormones). Some of the evolutionary schemes which have been proposed are shown in Figure 4.3. More detailed schemes, showing the mutations which may have been fixed during the evolution of these hormones, have also been proposed.

The distribution of neurohypophysial hormones that has been discussed here is by no means finalized. It is likely that more natural analogues of oxytocin will be found in the elasmobranchs, and other related hormones may still be discovered among the higher vetebrates. Vasotocin, previously thought to occur only in lower vertebrates, has been reported in the mammalian pineal gland and foetal pituitary. It is also likely that peptides related to oxytocin/vasopressin occur in invertebrates, and evidence for such molecules has been reported, although detailed characterization remains to be completed.

2.3 Chemical synthesis of neurohypophysial hormones and their analogues

Since the pioneering studies of du Vigneaud, the neurohypophysial hormones have received a great deal of attention from peptide chemists. About 400 analogues have now been synthesized by chemical methods, and this group of hormones has played an important role as a model for the development of new methods of peptide synthesis. A full discussion of the synthetic work is not possible here, and just a few examples will be given of the sort of results achieved. A full account of the use of such studies in investigating structure–function relationships is given in Chapter 15.

A particularly interesting example of the use of synthetic analogues was the discovery of arginine vasotocin (Figure 4.2). This is a 'hybrid' of oxytocin and vasopressin which was synthesized chemically some time before it was detected in lower vertebrates.[13] When the biological properties of the synthetic compound were compared with those of extracts from the neurohypophyses of lower vertebrates, it was predicted that arginine vasotocin would prove to be a natural analogue occurring in these animals. Isolation and characterization of the hormones from various non-mammalian species subsequently proved that this is the case.

Analogues in which the disulphide ring is modified have proved instructive. Derivatives of oxytocin in which the ring size is increased or decreased are completely inactive, as are those in which the disulphide bridge is removed (e.g. 1/2-cystine residues are replaced by alanine[14]). However, analogues in which one of the sulphur atoms of the bridge is replaced by a methylene group retain substantial activity. The maintenance of the overall form of the ring structure, and presumably the conformation of the molecule, seems to be more important than the detailed chemical nature of the components of the ring. These observations rule out disulphide interchange as a factor in hormone–receptor interactions; such a binding interaction had been proposed previously for the initial mode of action of oxytocin.

Derivatives in which the free α-amino group of oxytocin or vasopressin is removed have increased biological activity, compared to the natural hormones;[14,15] in particular their actions are very much prolonged. It is probable that such derivatives are less readily degraded than the natural hormones (probably because the normal degradative enzymes operate from the amino terminus). Desamino derivatives of this type also lack the ability to bind to neurophysins, which suggests that these proteins play no role in the actions of the hormones after they have entered the blood.

The motivation behind the chemical synthetic work on analogues of oxytocin and vasopressin has been partly towards the preparation of compounds with interesting biological properties and partly towards the production of drugs with useful clinical applications. Several medically useful derivatives have been produced, the most valuable being those with prolonged actions or those in which specific biological actions of the hormones are enhanced (see Chapter 15).

3. BIOLOGICAL ACTIONS OF NEUROHYPOPHYSIAL HORMONES IN MAMMALS

3.1 Oxytocin

The main effect of oxytocin in mammals is induction of contraction of smooth muscle, especially in the uterus and mammary gland. In inducing uterine contraction, oxytocin may play a role in parturition; its effects on the mammary gland lead to milk ejection.

The precise physiological role of oxytocin in parturition is not clear.[16] It undoubtedly induces contractions of the uterus of (some) pregnant mammals, but the fact that in some species parturition can occur quite normally after total hypophysectomy casts some doubt on the true role of the hormone. Oxytocin is widely used to induce labour in women, though quite high doses are used. It seems likely that the hormone plays a part in regulating parturition in many species, but that it acts as one of a complex of factors rather than independently.

Some experimental evidence suggests that oxytocin may induce contractions of the fallopian tubes and play a part in facilitating transport of spermatozoa in the female tract of some mammals. There is evidence from several mammalian species for increased serum levels of oxytocin during or immediately after mating.

The role of oxytocin in promoting milk ejection in mammals is more clearly established. Here it is important to distinguish between milk secretion and milk ejection. The former relates to the synthesis of milk components and their secretion into the lumen of the mammary alveolus, while the latter is the evacuation of the alveolar contents into the cisternae and larger ducts of the mammary gland, together with the removal of the milk from the gland itself. Milk secretion is under the control of several different hormones, including prolactin (q.v.). Milk ejection is regulated primarily by oxytocin.[17]

Ejection of milk from the alveolus is achieved by contraction of the myoepithelial cells ('basket cells') which surround this structure (Figure 4.4). This forces the contents of the alveolus into the duct system of the mammary gland. Such contraction is stimulated by oxytocin, which is itself released at the onset of suckling as a result of the so-called 'milk ejection reflex'. The act of suckling, or other stimuli associated with it, stimulates receptors within the mammary gland which give rise to a nervous impulse passing back to the brain and eventually to the supraoptic nucleus of the hypothalamus and thence to the neurohypophysis. This stimulates release of oxytocin into the blood stream which causes contraction of the myoepithelial cells. Activation of the complete reflex takes only about one minute, and the consequence is a sudden rise in intramammary duct pressure and a consequent free flow of milk (sometimes referred to as 'let-down' of milk). The importance of the milk ejection reflex is demonstrated experimentally by anaesthetizing a nursing mother—the suckling young are then unable to obtain any appreciable quantity of milk from the mammary glands. The myoepithelial cells round the mammary alveoli are presumably controlled hormonally because separate innervation of the individual cells, distributed as they are throughout the gland, would be extremely complex.

Oxytocin may also play a role in regulation of the oestrous cycle, causing luteolysis in some mammals. In this connection the discovery of substantial

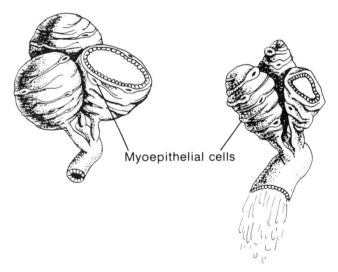

Figure 4.4 Diagram showing how the contraction of myoepithelial cells forces milk out of the mammary alveoli into the ducts. (Reprinted with permission from Schmidt, G. H. (1971) *Biology of Lactation*. Copyright (1971) G. H. Schmidt)

amounts of oxytocin in the corpus luteum is of particular interest;[18] concentrations in ovarian venous blood are greater than those in arterial blood, suggesting that the hormone is in fact formed at this site, not just concentrated there from the circulation. It has been suggested that oxytocin produces its effects on the corpus luteum by stimulating prostaglandin $F_{2\alpha}$ release from the uterine endometrium, but the demonstration that secretion of oxytocin from the ovary is stimulated by a prostaglandin $F_{2\alpha}$ analogue[19] casts some doubt on this.

3.2   Vasopressin

Two main effects of vasopressin can be recognized in mammals: the antidiuretic effect and the vasopressor effect (increase in blood pressure). The second of these is only observed at very high vasopressin concentrations, and it is likely that in many mammals it has little physiological significance. The antidiuretic actions of vasopressin reflect effects mainly on the kidney. When a partially isolated kidney is perfused it produces large quantities of dilute urine. Inclusion of small amounts of vasopressin in the perfusate leads to a marked reduction in the amount of urine, and an increase in its concentration. The hormone appears to stimulate the reabsorption of water by acting on the epithelial cells of the distal part of the renal tubule.[20,21]

Removal of the capacity of an animal to produce vasopressin (for example, by lesions in the hypothalamus severing the fibres to the neurohypophysis) leads to development of the condition *diabetes insipidus*—secretion of large amounts of very dilute urine (usually combined with *polydipsia*—extreme thirst). The condition can be corrected by treatment with vasopressin. Hereditary diabetes insipidus has also been recognized in man and in laboratory rats of the Brattleboro' strain. It can result from either defective production of vasopressin, or inability of the kidney tubules to respond to the hormone.[20]

Conservation of water is, of course, particularly important in animals living in desert conditions, and also in hibernating animals. In many such species the need for strict water conservation is met partly by a particularly efficient use of vasopressin, often associated with an increased size of the neurohypophysis.

Vasopressin has also been reported to have interesting behavioural effects. In particular, it appears to enhance memory and possibly learning in rats subjected to avoidance and other tests.[22,23] The physiological significance of such observations remains unclear, and their molecular basis is completely unknown. The occurrence of both vasopressin and oxytocin within the central nervous system has been established, however.[23]

4.   ACTIONS OF NEUROHYPOPHYSIAL HORMONES IN NON-MAMMALIAN VERTEBRATES[12,24]

Most detailed work on the neurohypophysial hormones has been done with systems derived from a small number of vertebrate species. A considerable amount of comparative work has been done, but we still do not understand the full physiological role of neurohypophysial hormones in several of the vertebrate groups.

Little is known of the effects of neurohypophysial hormones on cyclostomes or elasmobranchs; in the studies that have been carried out, little effect was observed on either water/electrolyte relationships, or on smooth muscle contraction (except at very high concentrations). In teleosts, more definite effects have been found, including pronounced actions on water and salt balance in several species, as well as vasopressor actions in the eel, and possible effects on oviduct contraction in some species which show viviparity. Lungfish also respond to neurohypophysial hormones, particularly with regard to water and salt balance.

Considerable work has been carried out on the actions of neurohypophysial hormones on amphibia, particularly the anura (frogs and toads). This is largely because their very potent actions on water and salt transport through frog and toad bladder and skin (using *in vitro* systems) have provided a useful model system for studying the biochemical actions of these hormones (see below). It is notable that arginine vasotocin, which is found naturally in the frog, is much more active than any other neurohypophysial hormone. Arginine vasotocin also causes contraction of the anuran oviduct but there has been relatively little work on actions on other physiological functions in amphibia.

The neurohypophysial hormones also play an important part in controlling water and salt relationships in reptiles and birds. In these groups, as in mammals, they act mainly at the level of the kidney. The hormones (especially vasotocin) also cause contraction of the oviduct in reptiles and birds. Oxytocin causes a *fall* in blood pressure in the hen.

Neurohypophysial hormones may be involved in regulation of secretion of adenohypophysial hormones. Vasopressin can stimulate secretion of ACTH, and may act synergistically with CRH. Regulation of MSH secretion may be effected mainly by factors produced by fragmentation of oxytocin (see Chapter 3). A role for the neurohypophysial hormones in regulating the secretion of the other adenohypo-physial hormones, especially in some groups of lower vertebrates, cannot be ruled out.

5. ASSAY OF NEUROHYPOPHYSIAL HORMONES[25,26]

A range of bioassay methods for the neurohypophysial hormones has been described. These are based on each of the main biological effects, including both *in vivo* and *in vitro* actions. The most important bioassays for vasopressin are those which employ the antidiuretic and pressor (raising of blood pressure) effects in the intact rat. Assays for oxytocin utilise the effects of this hormone on milk ejection or on contraction of the uterus. In both cases techniques *in vivo* and *in vitro* have been described. A very sensitive assay has been devised for vasotocin which measures the transport of water and sodium across isolated frog bladder.

Some of the bioassays are sufficiently sensitive to measure the hormones in physiological fluids, but they are mostly rather non-specific, and liable to interference by other substances in blood. Radioimmunoassays have now been described for both oxytocin and vasopressin and these are the methods of choice for many studies on physiological levels and on secretion. For such small peptide hormones the best antibodies are often produced by using an immunogen in which the peptide has been conjugated on to a larger protein, such as serum albumin.

The availability of a range of bioassays for the neurohypophysial hormones has enabled a spectrum of activities to be determined for each of the naturally-occurring hormones as well as for the large number of synthetic analogues that have been prepared. Table 15.1 shows the activities of various peptides in some of these different assays. Two main points emerge from such information.

(1) There is considerable overlap between the different hormones—in particular oxytocin shows some activity in the vasopressin bioassays, and vasopressin shows some activity in oxytocin assays.

(2) Some derivatives show quite different potencies in the two vasopressin assays (vasopressor and antidiuretic) or the two oxytocin assays (milk ejection and uterus contraction). As is discussed in more detail in Chapter 15, this suggests strongly that the receptors for vasopressin in the kidney are not identical to those in blood vessels, and that those for oxytocin in the mammary gland are not identical to those in the uterus.

6.   BIOSYNTHESIS AND SECRETION OF NEUROHYPOPHYSIAL HORMONES[7]

6.1   Biosynthesis[27]

Small peptides such as oxytocin and vasopressin can in principle be synthesized by enzymic pathways not involving ribosomes and the normal mechanisms of protein synthesis. It is now clear, however, that the neurohypophysial hormones are synthesized as part of much larger protein precursors, which are subsequently processed by enzymic cleavage to give the active peptides.

An early indication that vasopressin is produced as the product of a single gene was provided by work on an allelic polymorphism of the peptide seen in some members of the pig family, which is most easily explained as the result of a simple base substitution in the gene coding for the peptide or its precursor.[10] Subsequently a great deal of experimental evidence has accumulated indicating that oxytocin and vasopressin are synthesized, separately, as protein precursors, in the cell bodies of specific hypothalamic neurones. They are then incorporated into granules and transported in this form from the hypothalamus to the neurohypophysis, down the axons of the neurones in which they are synthesized. During the process of granule formation, each precursor protein is converted to the peptide hormone and residual peptides/proteins by enzymic action. One enzyme involved may be a transamidase, which would explain the presence of the amidated C-terminus of oxytocin and vasopressin. Each precursor includes not only a neurohypophysial peptide, but also a neurophysin. As has been discussed, neurophysins are proteins associated with the peptide hormones in storage granules, and probably play a role in the storage and axonic transport of the hormones.

The concept that the neurohypophysial hormones are produced as protein precursors, together with neurophysins, was first proposed by Sachs in the 1960s, on the basis of experiments in which radioactive amino acids were infused into the third ventricle of the brain of dogs. Production of labelled oxytocin and vasopressin occurred only after a considerable lag period, and was paralleled by the appearance of labelled neurophysins. Labelled proteins containing both neurophysin and peptide hormones were subsequently detected. Formation of such proteins is puromycin sensitive, but their conversion to peptide hormone and neurophysin is not. Experiments of this type led to partial characterization of the precursors. Confirmation was provided by extraction of mRNA from appropriate regions of the hypothalamus, translation of this in cell-free systems and characterization of the proteins so formed which were precipitated by antibodies to neurophysin and the peptide hormones.[28]

The most complete characterization of such a precursor protein has been provided, however, by cloning the cDNA corresponding to mRNA for the precursor containing bovine vasopressin[29] (see Chapter 18 for an account of the approach used). The complete nucleotide sequence of this cDNA has been determined, and provides the amino acid sequence of the precursor. Figure 4.5 shows the overall organization of the precursor that can be deduced from this. The precursor contains 166 amino acids, including a signal peptide of 19 amino acids followed by the sequence of arginine vasopressin. This is linked to neurophysin II by a Gly–Lys–Arg sequence; the glycine residue probably provides the NH_2 group of the C-terminal amide of vasopressin, and the Lys–Arg provides a site for cleavage. The neurophysin is followed by a single arginine and a 39 amino acid sequence corresponding to a glycopeptide which is known to be associated with vasopressin in secretory granules. cDNA cloning has also now provided an almost complete characterization of the precursor of oxytocin. Again, the signal peptide is followed by the oxytocin sequence, and a Gly–Lys–Arg sequence divides this from the sequence of neurophysin I; however, there is no glycopeptide sequence following the neurophysin (Figure 4.5).[30] Ox neurophysins I and II prove to be very similar. It should be noted that although this work appears to provide an almost complete characterization of the vasopressin precursor, there is some evidence for proteins of much greater molecular weight that include the neurohypophysial hormones and the neurophysins; the role of these is as yet poorly understood.

It is now clear from the work on biosynthesis of oxytocin and vasopressin that the two peptides are produced in separate neurones, the cell bodies of which are located in the supraoptic and paraventricular nuclei of the

(a) VASOPRESSIN– NEUROPHYSIN II

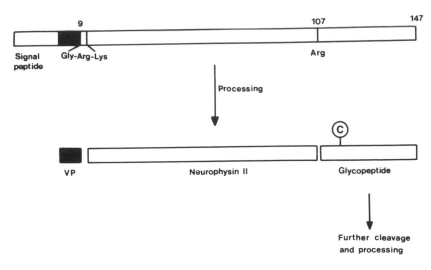

(b) OXYTOCIN – NEUROPHYSIN I

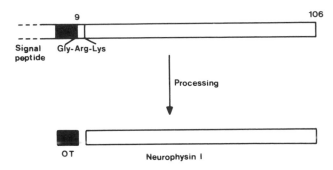

Figure 4.5 Diagram illustrating the nature of the polypeptide precursors of vasopressin (a) and oxytocin (b), and the ways by which they are processed to give mature hormones. C: carbohydrate moiety

hypothalamus, and that they are associated with different neurophysins. The presence of two neurophysins in the neurohypophysis has been recognized for some time, and these have been characterized in detail from several species.[7,8] In some cases additional neurophysins have been detected, but these appear to represent degradation products of one or other of the two main forms.

6.2 Secretion[7,31]

Oxytocin and vasopressin are transported from the hypothalamus to the neuro-hypophysis down the axons of the neurones in which they originate, as granules, by a process of fast axonal transport. They are then stored in the neuro-hypophysis, prior to secretion into the blood stream. Small amounts may also be secreted directly into the cerebrospinal fluid. In the neurohypophysis the hormones are stored in the form of membrane-bound secretory granules, inside the expanded nerve endings of the neurones in which they were originally synthesized. These neurosecretory terminals are closely juxtaposed to a capillary plexus which permeates the neurohypophysis, and are separated from the blood by only a thin basement membrane and the capillary walls, both of which are freely permeable to the secreted hormones.

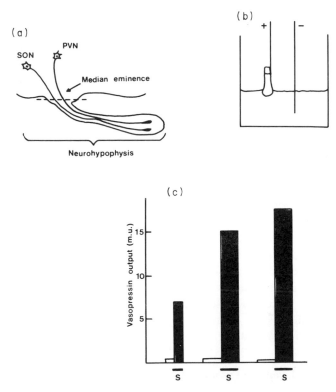

Figure 4.6 Diagram illustrating one way by which the regulation of the secretion of neurohypophysial hormones can be investigated *in vitro* (a) The posterior gland of the pituitary gland is isolated, attached to the pituitary stalk. (b) The gland is partially immersed in an appropriate buffer, and attached to an electrode. (c) Electrical stimulation (S) of the pituitary stalk leads to secretion of vasopressin, which can then be assayed in the buffer/medium

The mechanism whereby oxytocin and vasopressin are secreted from the nerve endings of the neurohypophysis has been a matter for some dispute. Some authors have suggested that release, signalled by calcium moving into the nerve ending, involves dissociation of the neurophysin-hormone complex, and diffusion of the free hormone out of the cell. However, an alternative theory, now almost universally accepted, is that secretion here, as in the case of most other protein and peptide hormones, occurs by exocytosis of the secretory granules. Such a mechanism is supported by the observation that neurophysin is released along with the hormones, and by electron microscope studies in which exocytosis can be shown to be occurring at the cell membrane.

The regulation of secretion of vasopressin and oxytocin appears to be largely nervous, secretion occurring as the result of nerve impulses travelling from the hypothalamus, via the neurones and axons in which the hormones are synthesized and transported. The regulation of the secretion of these hormones can be studied by simple *in vitro* techniques using the type of experimental design illustrated in Figure 4.6. In preparations where the neurohypophysis plus its 'stalk' are incubated *in vitro*, secretion of oxytocin and vasopressin can be stimulated by electrical impulses applied to the cut end of the stalk. The electrical stimulus is accompanied by an influx of calcium into the neurohypophysis (as can be demonstrated using $^{45}Ca^{2+}$) and it is likely that this influx, which is the result of depolarization of the plasma membrane, consequent on the passage of nervous impulses, is the signal for hormone release.

Depolarization of the cell membrane by high potassium concentrations also leads to an influx of Ca^{2+} and to hormone release, as does the presence of calcium ionophores (which facilitate passage of Ca^{2+} across cell membranes). Various other pieces of evidence also indicate that calcium plays a major role in controlling secretion from the neurohypophysis. Precisely how an increased calcium concentration within the nerve endings leads to increased exocytosis of hormone-containing granules is not yet clear, but the mechanism is likely to be a general one, as is discussed in detail in Chapter 1. Some evidence suggests that vasopressin (and possibly oxytocin) are stored as more than one pool, part being readily releasable and part less so.

6.3   Physiological control of vasopressin secretion[20,32]

The main physiological factors regulating vasopressin secretion are blood osmolarity and blood volume. An increase in the former or a decrease in the latter gives rise to increased vasopressin secretion, which leads in turn to reduced secretion of water by the kidney.

An increase of blood osmolarity (induced by infusion of hypertonic saline) of as little as 1–2% initiates an increased vasopressin secretion. The detection of such small changes in blood osmolarity appears to be carried out by osmoreceptors, probably situated in or near the supraoptic nucleus of the hypothalamus. It will be recalled that this is also the site of biosynthesis of vasopressin. The osmoreceptors are particularly sensitive to changes in Na^+

concentration; non-electrolytes such as glucose or glycerol have little effect. Some workers believe that the osmoreceptors are situated in the neurones in which vasopressin is synthesized, but others consider that these receptors are associated with other neurones in the supraoptic nucleus, separated from the hormone-producing neurones by one or two synaptic junctions. Wherever the osmoreceptors are situated, however, it is clear that once activated they induce firing of the neurone carrying them, and that the signal is eventually carried down the axons of the neurosecretory cells to the neurohypophysis, where it gives rise to depolarization of the membrane of the nerve ending, and hormone release. The properties of the receptors, though not their location, are similar to those of the thirst receptors situated in the ventromedial nucleus of the hypothalamus. It should be noted, however, that the term 'receptor' in this context is not necessarily equivalent to the conventional type of hormonal receptor.

The receptors linking change of blood volume to vasopressin secretion are located in the left atrium (detecting changes in the stretching of the atrial wall) and in the carotid sinus (detecting changes in blood pressure). Nerve signals from these receptors are carried to the brain, and thence, presumably, to the supraoptic nucleus of the hypothalamus, but the precise path of the signal through the brain has not been delineated.

Other receptors concerned in regulating vasopressin secretion undoubtedly exist, although they have not been as fully described as those above. Angiotensin, catecholamines, temperature and stress can all affect vasopressin release. The receptors detecting changes in blood pressure, volume and osmolarity appear to play an additive, or sometimes synergistic, role, but it is not clear whether blood osmolarity or volume is the dominant factor. It has been suggested that extreme reduction of blood volume, by severe haemorrhage, has a more dramatic effect in increasing vasopressin levels than does prolonged dehydration. On the other hand, the osmoreceptors may be more sensitive to small changes than the volume receptors.

6.4 Physiological control of oxytocin secretion[33,34]

Oxytocin secretion in the female mammal is induced mainly by the stimulus of suckling or milking, the so-called milk-ejection reflex, although genital stimulation (during mating or parturition, or experimentally) can also induce release of the hormone.

Exteroreceptors in the mammary gland are of various kinds, including probably stretch, touch and heat receptors. In response to suckling these combine to send a nervous signal to the brain. The path of the signal through the brain has been traced in part; the nervous signal eventually reaches the oxytocin-producing neurones of the paraventricular and supraoptic nuclei of the hypothalamus, and thence the neurosecretory termini of the neurohypophysis.

Similar neuroendocrine reflex arcs probably operate in the induction of oxytocin release by genital manipulation.

7.   THE CELLULAR MODE OF ACTION OF NEUROHYPOPHYSIAL HORMONES

7.1   Actions on the frog bladder[35-38]

Transport of water and solutes across the frog or toad bladder has provided a particularly convenient system for studying the biochemistry of the actions of neurohypophysial hormones. Although the most potent natural hormone in this system is vasotocin (the natural hormone of amphibia), much of the experimental work has been carried out using arginine vasopressin. The system has been used as a model for the actions of vasopressin on the mammalian kidney, although there are undoubtedly some important differences.

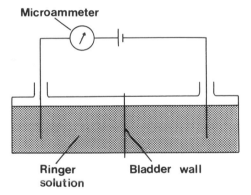

Figure 4.7 Simplified diagram of apparatus used to study the action of vasopressin and related hormones on the frog bladder or skin. A section of bladder wall separates two compartments containing Ringer solution. The effect of vasopressin added to one compartment on the permeability of the bladder wall is studied by measuring the change in electrical conductivity

The type of apparatus frequently used is shown in Figure 4.7. A piece of frog bladder wall separates two chambers containing a suitable medium, usually based on frog Ringer solution. The system is essentially asymmetric in that one side (the mucosal or luminal) is normally bathed by the urine while the other (the serosal) is bathed by body fluid.

7.1.1 Water transport

In the absence of neurohypophysial hormones the bladder acts as a tight bag containing the urine; the *bulk transfer* of water across the wall is very small, although, in the type of experimental arrangement shown in Figure 4.7, tritiated water diffuses quite rapidly through the wall, in either direction. If vasopressin is added to the fluid bathing the serosal side of the membrane, diffusion of tritiated water through the membrane, in either direction, increases considerably, and bulk flow of water through the membrane, in response to a hydrostatic head or an osmotic gradient, is greatly increased. No bulk transfer of water occurs in the absence of an external osmotic or hydrostatic driving force, indicating that water moves passively across the bladder. Water transport occurs in the complete absence of sodium ions, so the sodium pump does not contribute directly to this process. *In vivo* the osmotic gradient is due to active, energy-requiring reabsorption of sodium from the urine, and this is the energy-requiring step which precedes vasopressin-controlled water reabsorption from the urine. Vasopressin thus induces large net transfer of water with only a modest increase in non-directional diffusion permeability.

These observations indicate that vasopressin increases the permeability of the luminal membrane to water. It may do so by increasing the size or number of pores available for water transport, or by enhancing the solubility or diffusion of water in the lipid phase of the membrane. Which of these two mechanisms applies is not known. Early studies suggested that the hormone caused an increase in the size of pores available for water transport from about 6 to 20 Å diameter. However, it now seems likely that much of this early evidence for such enlargement reflects the presence of undisturbed layers next to the membrane. If such layers are dispersed by vigorous stirring, physicochemical studies suggest that the transport of water across the luminal membrane occurs by a solubility–diffusion process through lipid regions,[37] or through very restricted channels.[36] Vasopressin could promote diffusion through the lipid bilayer of the region by altering lipid composition or arrangement, and the observation that the hormone also promotes transport of moderately lipophilic compounds, such as butyramide, supports such a mechanism. However, the effect of vasopressin on such compounds is much smaller than that on water transport, and much recent evidence favours the idea that the hormone increases the number (or possibly size) of narrow aqueous pores;[36] most notably it has been demonstrated that it markedly increases the permeability of the luminal membrane to protons, which could traverse the membrane by jumping from one water molecule to another through a pore, but could not readily cross a lipid bilayer, due to their charge.

Studies using electron microscopy have demonstrated that aggregates of particles, presumably proteins, increase in number in the luminal membrane of frog or toad urinary bladder cells in response to vasopressin treatment.[39] They occur even when the hormonal treatment is not associated with increased water transport—in the absence of a hydrostatic head or osmotic gradient. Vasopressin increases the number of aggregates, without affecting their size distribution, which would accord with their playing a role in the formation of small aqueous pores (Figure 4.8). In unstimulated cells intracellular membranous vesicles containing similar aggregates can be seen. It is possible that vasopressin stimulates fusion of such vesicles with the luminal membrane, and some direct evidence for such fusion has been obtained by electron microscopy. In kidney collecting ducts, vasopressin also stimulates formation of intramembrane particle clusters, and here these have been suggested to be endocytotic coated pits.[39a]

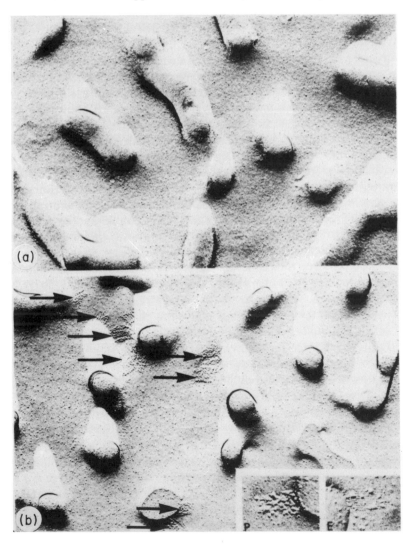

(c)

Figure 4.8 Freeze-fracture electron micrographs showing the effect of vasopressin on
aggregation of membrane particles (probably
proteins) in the luminal membrane of toad
bladder cells. (a) Control bladder ($\times$36,125).
(b) Bladder exposed to vasopressin ($\times$36,125);
note the intramembrane particle aggregates
(arrows). Inserts: higher magnification micrographs ($\times$72,250) of the particle aggregates
viewed from the two complementary fracture
faces. (c) Model illustrating the possible
organization of many small ($\sim$2°A radius)
channels within intramembrane particle aggregates. (Reprinted with permission from
Wade, B. (1980). *Curr. Topics Membrane
Transport* **13**, 123–147. Copyright (1980)
Academic Press Ltd.)

7.1.2 Solute transport

As mentioned, neurohypophysial hormones also alter the permeability of the
frog bladder to certain small solutes, including urea, acetamide and some low
molecular weight alcohols. The effect is very selective; permeability to urea may
be increased ten fold while that to thiourea is unaffected. The passage of such
solutes through the bladder is passive, as for water, can occur in either direction,
and is probably not carrier-mediated. To some extent their transport is affected
by increased passage of water through the membrane, suggesting use of a common channel. However, this does not accord with all the evidence: after treatment of the bladder with vasopressin, passage of urea is retarded by about 80%
relative to water, while passage of thiourea remains extremely low. If the small
solutes have the same channels as water, it is difficult to see how their transport
can be so selective, and it seems likely that it is in fact independent of that of
water.

7.1.3   Transport of sodium

The transport of sodium ions across the frog bladder is active (i.e. energy requiring, and against the concentration gradient) unlike that of water or urea. Active transport of sodium *in vivo* occurs from urine to body fluids, and it is this process that gives rise to the osmotic gradient which drives resorption of water. Sodium transport from luminal to serosal surfaces of the bladder is stimulated by vasopressin, acting on the serosal side. The action of the hormone appears to be directly on the sodium transport system. The hormone causes increased oxygen consumption by the bladder wall, but this seems to be a consequence of increased (active) transport of sodium rather than a primary effect. If sodium transport is prevented by excluding this ion from the medium, the effects of vasopressin on oxygen consumption are no longer observed, although effects on water and urea transport are still apparent.

7.1.4   Site of action of vasopressin

The neurohypophysial hormones act on the frog bladder to alter its permeability to water and various solutes. The location of the permeability barrier is thought to be in the plasma membrane of the cells, adjacent to the bladder lumen. Thus, if a hypotonic solution is applied to the luminal side of the bladder wall, the cells are unaffected; if vasopressin is added, the cells swell considerably, indicating that the barrier to water flow has been removed. On the other hand, hypotonic medium applied to the serosal side of the bladder causes the whole tissue, including mucosal cells, to swell, even in the absence of vasopressin. The location of the barrier to ions has also been located at the luminal plasma membrane of the mucosal cells by measuring the electrical resistance across the bladder wall. When vasopressin is added to the medium on the serosal side of the wall, the resistance across the wall drops markedly; almost all the drop in resistance occurs across the outer plasma membrane of the mucosal cells.

The primary binding site of vasopressin appears to be distant from its ultimate site of action, in that it is only active when applied on the serosal side of the bladder wall. This 'action at a distance' is probably mediated by cyclic AMP as a second messenger within the cell. The presence of a simple permeability barrier at the mucosal surface of the bladder does not explain how the organ can retain selective permeability to small molecules, while allowing bulk transfer of water. A theory to explain this effect proposes two permeability barriers in series, one with selective permeability to solutes, but completely permeable by water, the second presenting a barrier to water but not solutes. The two barriers would have to be independently responsive to vasopressin (or its second messenger). Quite how these two barriers are organized at the luminal membrane of the mucosal cells is not clear, however. The need for their existence is removed if water transport across the membrane is by diffusion or through very narrow pores, as many authors now believe, and if vasopressin-dependent transport of Na^+ and organic solvents is by processes independent of that for water transport.

7.1.5   Receptors for vasopressin/vasotocin[38]

The cellular receptors for these hormones have been shown to occur in the plasma membrane of the mucosal cells, presumably on the serosal side. In plasma-membrane preparations from bladder epithelium binding of vasopressin and related peptides can be demonstrated, and under appropriate conditions such binding can be correlated with activation of adenylate cyclase. The binding affinity of peptides to such receptors has been studied in detail and correlates well with their potency in stimulating water transport and activating adenylate cyclase.

7.1.6   Mechanism of action; the role of cyclic AMP[40]

Various attempts have been made to explain the action of neurohypophysial hormones on the amphibian bladder at the biochemical level. An early suggestion, now no longer tenable, was that the hormone increased permeability by stimulating hyaluronidase, which broke down mucopolysaccharides in the bladder wall. Another hypothesis proposed an interaction between receptors and hormone by means of disulphide bridges, but this too has now been discredited, mainly because hormone analogues in which one of the sulphur atoms is replaced by a methylene retain some activity, although they could not be involved in disulphide formation.

The main modern hypothesis concerning the action of these hormones on the bladder proposes that they stimulate cyclic AMP production, and that this mediates the hormonal action. The second messenger here would provide a vital link between hormone–receptor interaction on the serosal side of the bladder wall, and altered permeability on the luminal side. In support of this hypothesis, cyclic AMP and theophylline (applied at the serosal side of the bladder wall, *in vitro*) can mimic the effects of vasopressin on water, solute and ion transfer across the membrane. Vasopressin causes a rise in the tissue levels of cyclic AMP. As in other cyclic AMP-mediated systems the action of the hormone appears to be primarily to activate adenylate cyclase, although it has been suggested that it also modulates the activity of phosphodiesterase.

As in other systems in which cyclic AMP acts as a second messenger, it is thought that it serves to activate a protein kinase. Precisely how this then leads to the observed changes in transport of water, solutes and sodium ions is not yet established, although phosphorylation of specific membrane proteins has been proposed. Alternatively, or additionally, phosphorylation of key proteins could lead to fusion of vesicles with the luminal membrane and insertion of proteins able to form aqueous pores or other types of transport system. There is evidence that elements of the cytoskeleton, including microtubules and microfilaments, are required for the action of vasopressin, and these too may be among the targets for protein kinase.

7.1.7   Other factors affecting transport across toad bladder

Various factors may modulate the actions of vasopressin on toad or frog bladder. These include α and β adrenergic agents, prostaglandins and adrenal steroid hormones. The actions of these are complex and their physiological significance is unclear. They will not be discussed in detail here.

7.1.8   Actions on amphibian skin

Transport of water and sodium across the skin is important in osmoregulation in amphibians. Such transport is regulated by neurohypophysial hormones, and a similar experimental procedure to that used for the bladder has been used for its study. The hormonal regulation of these processes in skin seems to be similar in general to that seen in the bladder, although differing in some details.

7.2   Actions of vasopressin on the kidney[36,41]

The kidney is a complex organ; the key unit concerned with excretion is the nephron, the structure of which is illustrated in Figure 4.9. For an account of the physiology of the nephron and its excretory function, a textbook of physiology should be consulted. Briefly, the non-macromolecular components of the plasma of blood passing through the capillary plexus of the glomerulus are filtered into Bowman's capule. From there they pass through the nephron to the renal pelvis and thence to the bladder. Sodium, solutes and water are re-absorbed selectively during passage of the filtered blood through the nephron. Vasopressin acts largely on the epithelial cells lining the collecting ducts; in the absence of the hormone the luminal plasma membrane of these cells is almost impervious to water, but in its presence passage of water is facilitated and re-absorption occurs along the osmotic gradient due to a high salt concentration in the interstitial fluid surrounding the collecting tubule.

Despite the complex physiology and anatomy of the system, the biochemistry of the actions of vasopressin in the kidney is similar to that in amphibian bladder or skin. The hormone appears to bind to receptors on the plasma membrane of the epithelial cell that is opposite the luminal membrane, resulting in activation of adenylate cyclase and subsequent increased water transport across the luminal membrane. The mechanism by which increase in water permeability occurs is probably very similar to that in amphibian bladder, the latter therefore providing a satisfactory model for biochemical studies, in which the anatomy and physiology is relatively simple.

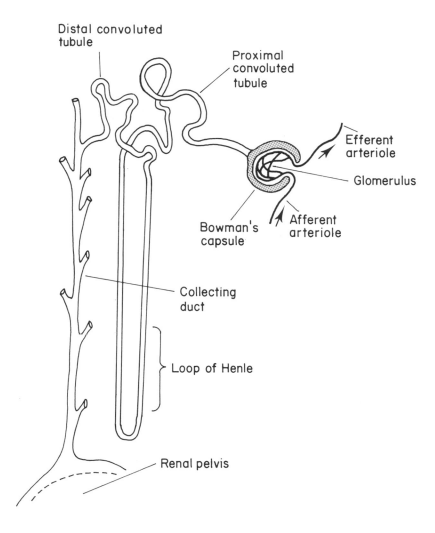

Figure 4.9 The nephron

Although its main actions are on the collecting duct, vasopressin also stimulates adenylate cyclase in various other regions of the nephron, but the significance of this is not clear. It also has effects on sodium transport and some other solutes, but these are probably less marked than in amphibian bladder.

7.3   Vasopressin and liver metabolism

Vasopressin, like several other hormones, acts on the liver to cause glyco-genolysis. The effect can be seen to occur very rapidly in hepatocytes *in vitro*, and involves an activation of the enzyme phosphorylase, probably via a phos-phorylation cascade. Cyclic AMP does not appear to be the messenger involved here, however, and mobilization of internal calcium stores leading to activation of calcium-dependent protein kinase via calmodulin is a possible mechanism of action. Vasopressin also causes rapid changes in liver phosphatidyl inositol metabolism, but some evidence suggests that this is not involved in the actions on glycogenolysis. The physiological significance of the actions of vasopressin on the liver is not clear.

7.4   Actions on the uterus and mammary gland[33,38]

7.4.1   Site of action

The main actions of oxytocin in the mammal are to stimulate contraction of the smooth muscle of the mammary gland and uterus. *In vitro* systems utilizing strips of, or whole, uterus have been used for many years for the study of the actions of oxytocin; such muscle can be induced to contract by application of the hormone. Smooth muscle in the mammary gland is closely associated with other tissues (see Figure 4.4), but that the smooth muscle is the target here of oxytocin is shown by use of dispersed cell preparations. [³H]oxytocin can be shown to bind specifically to the smooth muscle cells and the hormone causes contraction of such cells in dispersed cell preparations.[38]

7.4.2   Receptors

Oxytocin and related neurohypophysial hormones bind specifically to receptors on their target tissues, but (with a few exceptions) other tissues possess few, if any, such receptors. Receptors for oxytocin on the smooth muscle cells of uterus or mammary gland are of high affinity (K_d 1–5 nM). Binding of analogues to the receptors is generally fairly well correlated with the biological activity of these analogues, but there are some analogues which bind but do not have activity. Such analogues are often antagonists of the natural hormones; derivatives modified on Tyr^2 are often of this type.

Binding of oxytocin to its receptors requires the presence of Mg^{2+} or related cations (Mn^{2+}, Co^{2+}, Ni^{2+} or Zn^{2+}, but not Ca^{2+}) and this effect has been studied extensively. The chemical nature of oxytocin receptors has not been determined in detail, although like most peptide hormone receptors they do appear to be proteins associated with the plasma membrane. The numbers of oxytocin receptors in target tissues, particularly the uterus, are increased by the actions of oestrogens, and this effect is counteracted by progesterone. Most noticeable is the marked increase in the numbers of uterine oxytocin receptors during the oestrous cycle in many species. A marked increase occurs around oestrus. Binding of [^{3}H]oxytocin to both uterus and mammary gland increases during pregnancy, largely due to an increase in concentration of receptors rather than an increase in the affinity of existing receptors.

7.4.3 Post-receptor events

The biochemical mechanisms whereby the signal produced by binding of oxytocin to its receptor stimulates contraction of the smooth muscle cell are poorly understood. There is no evidence to suggest involvement of cyclic AMP as a second messenger, and the mechanism of action may therefore be very different from that of vasopressin on water transport. Involvement of Ca^{2+} as a mediator of the actions of oxytocin seems likely, but a detailed mechanism is difficult to formulate.

Oxytocin causes a small depolarization of the myometrial cell membrane, an effect which could lead to (or be caused by) entry of Ca^{2+} into the cell. Such a depolarization could also serve to transmit the signal for contraction to adjacent cells. When Ca^{2+} is removed from the extracellular medium, the ability of oxytocin to stimulate contraction of myometrial or mammary smooth muscle decreases within a few minutes. The effect is gradual rather than abrupt, however, suggesting that the Ca^{2+} dependence may not be for external Ca^{2+} but for internal stores which are depleted gradually when Ca^{2+} is excluded from the external medium. The effects of oxytocin may thus involve elevation of cytosolic Ca^{2+} levels, but whether this involves increased influx (or decreased efflux) of the cation or mobilization of intracellular stores in mitochondria or sarcoplasmic reticulum is not yet clear. If intracellular stores of Ca^{2+} are mobilized in response to oxytocin, then there remains the need for a second messenger to carry the signal from the hormone–receptor complex at the plasma membrane to the appropriate store. The nature of this remains unclear, although it should be noted that oxytocin does stimulate prostaglandin $F_{2\alpha}$ production and secretion by endometrial (but not myometrial) cells of the uterus, which probably accounts for the luteolytic action of oxytocin seen in some species. $PGF_{2\alpha}$ could also serve as an intracellular messenger for oxytocin action, although no direct evidence for this has been provided, and some evidence suggests that oxytocin can stimulate uterine contraction *in vitro* without increasing prostaglandin production, and even if prostaglandin production has been inhibited by indomethacin. Cyclic GMP may also function as such a second messenger, and some, but not all, studies have provided evidence for an increase in cyclic GMP levels in the uterus after oxytocin stimulation.

Oxytocin can induce phosphorylation of myosin light chains in dispersed mammary myoepithelial cells,[42] which could play a major role in promoting cellular contraction. Phosphorylation of myosin appears to play an important role in control of contraction of some other types of smooth muscle. Exclusion of Ca^{2+} from the medium markedly lowered this effect of oxytocin.

8. NEUROHYPOPHYSIAL HORMONES IN BLOOD: TURNOVER AND METABOLISM[20,33]

Most current evidence suggests that vasopressin and oxytocin circulate in the plasma as free peptides rather than in association with binding protein. In man, circulating concentrations of vasopressin vary from <0.1 µU/ml (<0.25 pg/ml) in overhydrated individuals to up to 20 µU/ml (50 pg/ml) in dehydrated individuals. Circulating concentrations of oxytocin in women during suckling may be up to 15 µU/ml (35 pg/ml), but in the absence of such stimulation they are low (less than 1 µU/ml; 2 pg/ml). The half-life ($t_{1/2}$) of vasopressin in the circulation in man is 10–20 minutes; that of oxytocin appears to be rather lower (about 5 minutes). Degradation occurs mainly in the kidney and liver, where degradative enzymes attack at the disulphide bond and the amide groups. In pregnancy circulating vasopressinase–oxytocinase, produced by the placenta, increases the rate of degradation by cleaving the peptide bond between residues 1 and 2.

REFERENCES

1. Oliver, G. and Schäfer, A. E. (1895) *J. Physiol., Lond.,* **18**, 277–279.
2. Heller, H. (1974) In: *Handbook of Physiology: Endocrinology,* Vol. IV, Part 1 (Eds. Knobil, E. and Sawyer, W. H.). American Physiological Society, Washington, pp. 103–117.
3. van Dyke, H. B., Chow, B. F., Greep, R. O. and Rothen, A. (1942) *J. Pharmacol. Exp. Ther.,* **74**, 190–209.
4. Kamm, O., Aldrich, T. B., Grote, I. W., Rowe, L. W. and Bugbee, E. P. (1928) *J. Amer. Chem. Soc.,* **50**, 573–601.
5. du Vigneaud, V. (1954–55) *Harvey Lectures,* **50**, 1–26.
6. du Vigneaud, V. (1969) *Johns Hopkins Med. J.,* **124**, 53–65.
7. Pickering, B. T. (1978) *Essays in Biochemistry,* **14**, 45–81.
8. Breslow, E. (1979) *Ann. Rev. Biochem.,* **48**, 251–274.
9. Acher, R. (1980) *Proc. Roy. Soc. B.,* **210**, 21–43.
10. Ferguson, D. R. (1969) *Gen. Comp. Endocr.,* **12**, 609–613.
11. Wallis, M. (1975) *Biol. Rev.,* **50**, 35–98.
12. Sawyer, W. H. (1977) *Fed. Proc.,* **36**, 1842–1847.
13. Katsoyannis, P. G. and du Vigneaud, V. (1958) *J. Biol. Chem.,* **233**, 1352–1354.
14. Rudinger, J., Pliska, V. and Krejci, I. (1972) *Recent Progr., Horm. Res.,* **28**, 131–172.
15. Hope, D. B., Murti, V. S. and du Vigneaud, V. (1962) *J. Biol. Chem.,* **237**, 1563–1566.
16. Marshall, J. M. (1974) In: *Handbook of Physiology: Endocrinology,* Vol. IV, Part 1 (Eds. Knobil, E. and Sawyer, W. H.). American Physiological Society, Washington, pp. 469–492.
17. Folley, S. J. and Knaggs, G. S. (1970) In: *Pharmacology of the Endocrine System and Related Drugs: The Neurohypophysis,* Vol. 1 (Eds. Heller, H. and Pickering, B. T.). Pergamon Press, Oxford, pp. 295–320.

18. Wathes, D. C. and Swann, R. W. (1982) *Nature*, **297**, 225–227.
19. Flint, A. P. F. and Sheldrick, E. L. (1982) *Nature*, **297**, 587–588.
20. Kleeman, C. R. and Berl, T. (1979) In: *Endocrinology*, Vol. 1. (Eds. DeGroot, L. J. *et al.*). Grune & Stratton, New York, pp. 253–275.
21. Handler, J. S. and Orloff, J. (1981) *Ann. Rev. Physiol.*, **43**, 611–624.
22. Koob, G. F. and Bloom, F. E. (1982) *Ann. Rev. Physiol.*, **44**, 571–582.
23. Bloom, F. E. (1981) *Scientific American*, **245** (no. 4), 114–124.
24. La Pointe, J. (1977) *Amer. Zool.*, **17**, 763–773.
25. Edwards, C. R. W. (1979) In: *Hormones in Blood*, Vol. 2, 3rd edn. (Eds. Gray, C. H. and James, V. H. T.). Academic Press, London, pp. 400–421.
26. Edwards, C. R. W. (1979) In: *Hormones in Blood*, Vol. 2, 3rd edn. (Eds. Gray, C. H. and James, V. H. T.). Academic Press, London, pp. 423–450.
27. Brownstein, M. J., Russell, J. T. and Gainer, H. (1980) *Science*, **207**, 373–378.
28. Sachs, H. (1969) *Adv. Enzymol.*, **32**, 327–372.
29. Land, H., Schutz, G., Schmale, H. and Richter, D. (1982) *Nature*, **295**, 299–303.
30. Land, H., Grez, M., Ruppert, S., Schmale, H., Rehbein, M., Richter, D. and Schütz, G. (1983) *Nature*, **302**, 342–344.
31. Nordmann, J. J. (1983) In: *The Neurohypophysis: Structure, Function and Control* (Eds. Cross, B. A. and Leng, G.). Elsevier, Amsterdam, pp. 281–304.
32. Share, L. and Grosvenor, C. E. (1974) In: *MTP International Review of Science, Physiology Series 1, Vol. 5, Endocrine Physiology* (Ed. McCann, S. M.). Butterworths, London, pp. 1–30.
33. Vorherr, H. (1979) In: *Endocrinology*, Vol. 1 (Eds. DeGroot, L. J. *et al.*). Grune & Stratton, New York, pp. 277–285.
34. Grosvenor, C. E. and Mena, F. (1974) In: *Lactation*, Vol. 1 (Eds. Larson, B. L. and Smith, V. R.) Academic Press, New York, pp. 227–276.
35. Jard, S. and Bockaert, J. (1975) *Physiol. Rev.*, **55**, 489–536.
36. Handler, J. S. and Orloff, J. (1981) *Ann. Rev. Physiol.*, **43**, 611–624.
37. Andreoli, T. E. and Schafer, J. A. (1976) *Ann. Rev. Physiol.*, **38**, 451–500.
38. Soloff, M. S. and Pearlmutter, A. F. (1979) In: *Biochemical Actions of Hormones*, Vol. VI (Ed. Litwack, G.). Academic Press, New York, pp. 266–333.
39. Wade, J. B. (1980) *Current Topics in Membranes and Transport*, **13**, 123–147.
39a. Brown, D. and Orci, L. (1983) *Nature*, **302**, 253–255.
40. Strewler, G. J. and Orloff, J. (1977) *Adv. Cyclic Nucleotide Res.*, **8**, 311–361.
41. Morel, F. (1983) *Recent Progr. Hormone Res.*, **39**, 271–304.
42. Dubin, N. H., Ghodgaonkar, R. B. and King, T. M. (1979) *Endocrinology*, **105**, 47–51.

Chapter 5

Hormones of the adenohypophysis: Corticotropin, melanotropins and opioid peptides

1. STRUCTURAL AND BIOSYNTHETIC RELATIONSHIPS

It has been known for many years that there is structural homology between corticotropin and the melanotropins. This was first thought to be a consequence of evolutionary relationships (divergent evolution from a common ancestor). Although such considerations are still important, it is now realized that the relationships between the members of this family of peptide hormones mainly reflect their origin from a common precursor, pro-opiocortin or pro-opiomelanocortin. The picture of relationships based on recent biosynthetic studies will be summarized briefly at the start of this chapter since it provides a convenient and elegant framework on which to hang some of the older work.

The biologically active peptides now included in this family are listed in Table 5.1. They include corticotropin, several melanotropins and the endorphins and

Table 5.1 The peptide hormones of the corticotropin–melanotropin-endorphin family

Hormone	No. of residues	Main actions
Corticotropin (ACTH)	39	control of corticosteroidogenesis
α-melanotropin (α-MSH)	13	control of skin-darkening
β-melanotropin (β-MSH)	18	control of skin-darkening
β-lipotropin	91	precursor of β-MSH and β-endorphin
γ-lipotropin	58	precursor of β-MSH?
β-endorphin	31	opiate-like actions (neurotransmitter?)
α-endorphin	16	opiate-like actions (may be degradation products of β-endorphin)
γ-endorphin	17	
[Met]-enkephalin	5	opiate-like actions (neurotransmitter?)
[Leu]-enkephalin	5	

other recently discovered opioid peptides. The nature of the large precursor, pro-opiocortin, which incorporates several of these peptides, has been elucidated by various authors[1-3] (see also Section 5). The current view is shown in Figure 5.1. The precursor (31K) has a molecular weight of 31 000 daltons. During processing it is cleaved to a glycosylated N-terminal fragment of 16 000 daltons (16K fragment), a C-terminal fragment (9K) which is equivalent to the peptide β-lipotropin (known for some time) and a central fragment (4.5K) which is equivalent to corticotropin. β-lipotropin can be further processed to give β-MSH, γ-lipotropin, β-endorphin and perhaps α-endorphin and γ-endorphin. The corticotropin-like peptide can be processed further to a α-MSH and CLIP (corticotropin-like intermediate lobe peptide). Not all the active peptides are, or could be, produced in equal quantities in any one pituitary cell; in particular, the MSH-like peptides are produced primarily in the intermediate lobe of the pituitary and corticotropin is produced primarily in the anterior lobe.

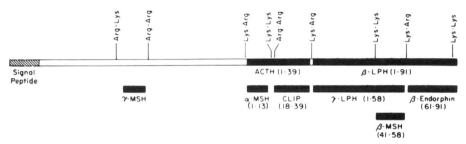

Figure 5.1 Prepro-opiocortin: The precursor of the corticotropin/melanotropin/ endorphin peptide family. The paired basic residues (Arg–Lys, Arg–Arg etc.) are positions where proteolytic cleavage can occur to convert the precursor to active peptides (ACTH, β-lipotropin, γ-lipotropin, α-MSH, β-MSH, γ-MSH, CLIP, β-endorphin). (Reproduced, with minor modifications, with permission from Krieger, D. T. *et al.* (1980) *Recent Prog. Hormone Res.* **36**, 277–344. Copyright (1980) Academic Press)

The complete nucleotide sequence of the gene coding for pro-opiocortin has now been determined,[3] and the amino acid sequence of the 31K precursor is thus known. It is clear that there is considerable internal sequence homology within the precursor, which explains the marked structural homology between α-MSH and β-MSH. It is also likely that other biologically active peptides remain to be discovered in this family—no clear-cut function has yet been ascribed to the 16K fragment, or to β-lipotropin. The 16K fragment is probably processed to give another MSH-like peptide, γ-MSH.

Some of the peptides listed in Table 5.1 cannot be derived from pro-opiocortin. Furthermore, it is unlikely that [Met]enkephalin is derived from this precursor; although such a conversion is structurally possible, no direct biosynthetic evidence has been adduced that it actually occurs and alternative precursors have been discovered. There exist at least two additional precursor molecules which are processed to give the two enkephalins and other active peptides.[4-6]

It must be stressed that many of the peptides in this family are found not only in the pituitary gland, but also in various parts of the brain, including the hypothalamus. The opioid peptides in particular may be primarily brain peptides rather than pituitary peptides, and their function as hormones, circulating in the blood, has not been unequivocally demonstrated as yet.

2.   CHARACTERIZATION AND BIOLOGICAL ACTIONS OF CORTICOTROPIN

2.1   Isolation

Hypophysectomy causes atrophy of the adrenal cortex and injection of adeno-hypophysial extracts reverses this atrophy. The pituitary factor responsible for the control of the growth and secretion of the adrenal cortex is corticotropin. Work on its isolation began in the 1930s, and by the early 1940s a 'pure' protein hormone which had corticotropin activity had been isolated. However, as in the case of the neurohypophysial peptides, it was subsequently shown that this preparation represented a relatively small active polypeptide bound to an inert 'carrier' protein. When the two were separated, using acid extraction, and the corticotropin-peptide was purified, using oxycellulose adsorption and then chromatography, an extremely active, pure peptide was obtained;[7] subsequent work has confirmed this as the major active species *in vivo*. Whether the inert protein to which corticotropin was bound in the earlier work is a true carrier protein, or whether it was an extraction artefact, has not been determined.

2.2   Assay

Early assays were based on the trophic effects of the hormone on the adrenal cortex of hypophysectomized rats. A more satisfactory *in vivo* assay was devised by Sayers and his colleagues in 1944, and depended on the depletion of the ascorbic acid content of the adrenal glands after corticotropin injection. Important bioassays that have been developed more recently mostly utilize the promotion of steroidogenesis by corticotropin, using intact animals (i.e. *in vivo*) and *in vitro* systems (quartered adrenal glands or adrenal cell suspensions).[8] Radioreceptor assays have also been developed for corticotropin, utilising receptor preparations made from adrenal cortex.[9] Several good immunoassays have also been developed; use of antibodies specific for particular regions of the corticotropin molecule enables the investigator to obtain information about the structural integrity of the corticotropin-like molecules being investigated, and to distinguish intact corticotropin from other, structurally related, peptides.[10]

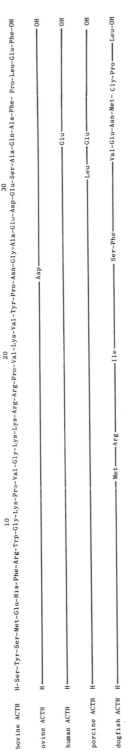

Figure 5.2 Amino acid sequences of corticotropins (ACTHs). The complete sequence of bovine ACTH is shown. For ovine, human, porcine and dogfish ACTHs the solid line indicates identity with the sequence of bovine ACTH, and differences from the bovine ACTH sequence are indicated

2.3   Structure[11,12]

Corticotropin obtained by acid extraction of anterior pituitaries was shown to be a polypeptide of molecular weight about 4500. In fact, a family of such peptides was recognized quite early on, but almost all of the subsequent work has been carried out on the most abundant form (β-corticotropin or corticotropin A). The full significance of the minor corticotropin-like peptides, some of which are glycosylated, has not been fully investigated, and although some of them probably represent extraction artefacts, others may be naturally-occurring components. In the following discussion only the major component, corticotropin A, will be considered.

The amino acid sequence of pig corticotropin was determined by Bell and co-workers in 1954 (Figure 5.2). Sequences for sheep, ox and human corticotropin have now also been determined. More-recent work indicated that the sequences described earlier incorporated several errors, and the structures shown in Figure 5.2 take these revisions into account. The amino acid sequences of corticotropins from these four species are very similar; differences occur only at residues 25, 31 and 33. The sequence has clearly been conserved strongly during the course of evolution. The amino acid sequence of a dogfish corticotropin has now been determined and is also shown in Figure 5.2; not surprisingly, this differs substantially from the mammalian corticotropins.

Complete chemical synthesis of pig and human corticotropins has now been achieved (although the earlier work was based on sequences which have now been shown to be slightly wrong). Much synthetic work has been carried out on peptides representing parts of the corticotropin sequence;[11,12] they provide a basis for extensive studies on the relationships between structure and function (see Chapter 15).

2.4   Biological actions of corticotropin[13,14]

The main actions of corticotropin are on the adrenal cortex. Various extra-adrenal effects have also been described, including melanocyte-stimulating, lipolytic and behavioural actions, but doubts remain as to their physiological significance.

2.4.1   The adrenal cortex and its hormones[14]

The adrenal glands are small paired glands (about one-thirtieth the size of the kidney in man) situated close to the kidney. They consist of an inner medulla, which originates embryologically from the neural crest, and an outer cortex, which is mesodermal in origin. Only the cortex is regulated by corticotropin, the medulla is largely under nervous control.

The adrenal cortex is a complex steroid-secreting organ. It consists of three zones: the zona glomerulosa, zona fasciculata and zona reticularis (Figure 5.3).

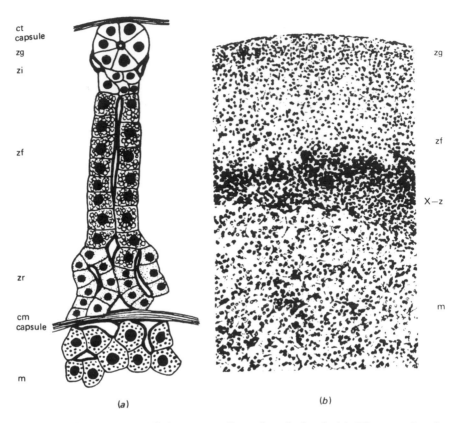

Figure 5.3 The structure of the mammalian adrenal gland. (a) Diagram showing zonation of the adrenal cortex. (b) Section through the adrenal cortex of a castrated male mouse. ct capsule, connective tissue capsule; cm capsule, circummedullary capsule; m, medulla; x-z, x-zone (appears in the absence of androgens); zf, zona fasciculata; zg, zona glomerulosa; zr, zona reticularis. (Reproduced from Gorbman, A. and Bern, H. A. (1962). *A Textbook of Comparative Endocrinology*.

In hypophysectomized rats the last two are much reduced in size, but the zona glomerulosa is relatively little affected. The adrenal cortex produces many different steroids (over 40) and the main ones vary from one species to another. The main adrenal steroids are the mineralocorticoid aldosterone, produced largely in the zona glomerulosa, concerned primarily with regulation of salt and water relationships and little affected by corticotropin, and the various glucocorticoids (cortisol, cortisone, corticosterone, etc.) which are produced mainly in the zona fasciculata and zona reticularis and which are concerned with the regulation of carbohydrate metabolism—they promote gluconeogenesis and deposition of glycogen. The structures of some of the more important adrenal

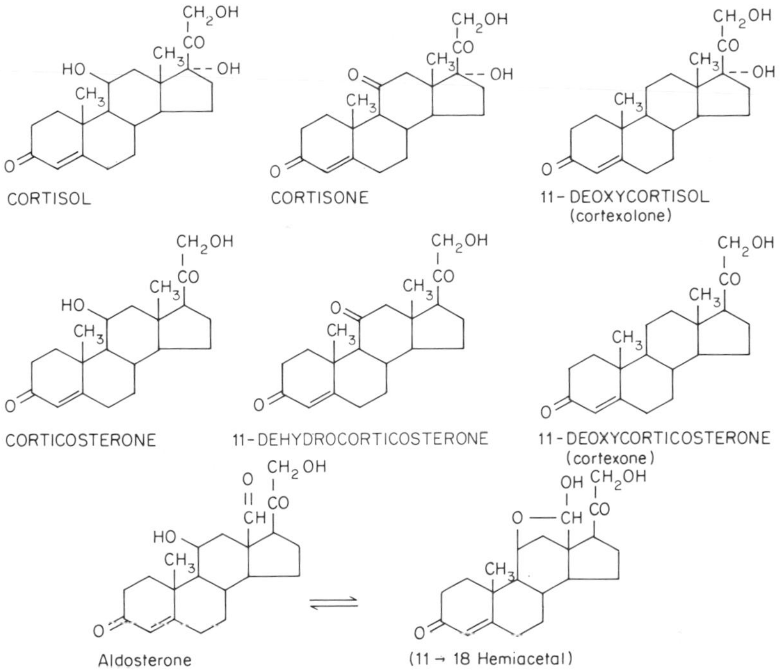

Figure 5.4 Structures of some biologically active corticosteroids. (Reproduced from Schulster, D., Burstein, S. and Cooke, B. A. (1976) *Molecular Biology of the Steroid Hormones.*

steroids are shown in Figure 5.4. In addition to the mineralocorticoids and glucocorticoids, the adrenal cortex also produces substantial quantities of androgens, oestrogens and progestagens.

Like other steroid hormones, the glucocorticoids and mineralocorticoids are produced in the adrenal gland from cholesterol, mainly via pregnenolone and progesterone (Figure 5.5). Regulation of these hormones occurs primarily at the level of biosynthesis rather than secretion; there is very little storage of the steroids in the gland.

2.4.2   Actions of corticotropin on the adrenal cortex[13,14]

2.4.2.1   Corticosteroidogenesis. Stimulation of corticosteroid production by corticotropin can be demonstrated using a range of preparations, including intact and hypophysectomized animals, perfused adrenal glands, adrenal fragments or slices incubated *in vitro*, and preparations of adrenal cells. Increased levels of corticosteroids can be detected in the adrenal vein of hypophysectomized rats or dogs 3–6 minutes after injection of corticotropin, and rise to a maximum after 6–12 minutes. Increased output from rat adrenal quarters

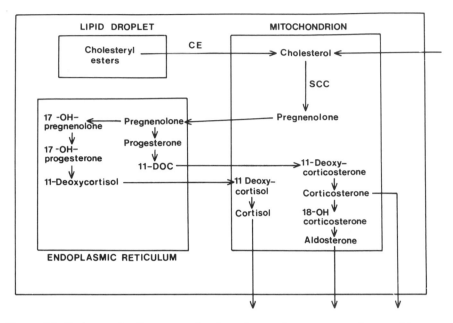

Figure 5.5 Pathways leading to production of some important corticosteroids. Most of the reactions involved in the conversion of cholesterol to cortisol, aldosterone and corticosterone occur in the mitochondria or the endoplasmic reticulum. CE, cholesteryl esterase; SCC, side chain cleavage complex; 11-DOC, 11-Deoxycorticosterone

or cells incubated *in vitro* can be detected within 1 minute of applying corticotropin.

The corticosteroids responding to corticotropin treatment are glucocorticoids, primarily corticosterone and cortisol. Little or no effect is seen on aldosterone production in mammals; corticotropin may play an important part in regulating aldosterone in the frog, but in mammals aldosterone seems to be mainly regulated by the renin–angiotensin system though MSH-like peptides may also play a role. The action of corticotropin on glucocorticoids is mainly on production of newly-formed steroids—the resting glucocorticoid content of 0.2 μg in the adrenal of the hypophysectomized rat is less than the output during the first 2 minutes of corticotropin stimulation.

After hypophysectomy, the ability of the adrenal cortex to respond to corticotropin is maintained for about 4 hours. Glucocorticoid levels in the blood decrease progressively during this time. Subsequently the responsiveness of the adrenal cortex decreases, until about 2 days after hypophysectomy when an acute corticosteroidogenic response to ACTH can no longer be demonstrated.

An interesting feature of the action of corticotropin in animals whose main glucocorticoid is corticosterone is that it increases the proportion of cortisol that is produced, presumably by inducing or activating the 17-hydroxylase. A

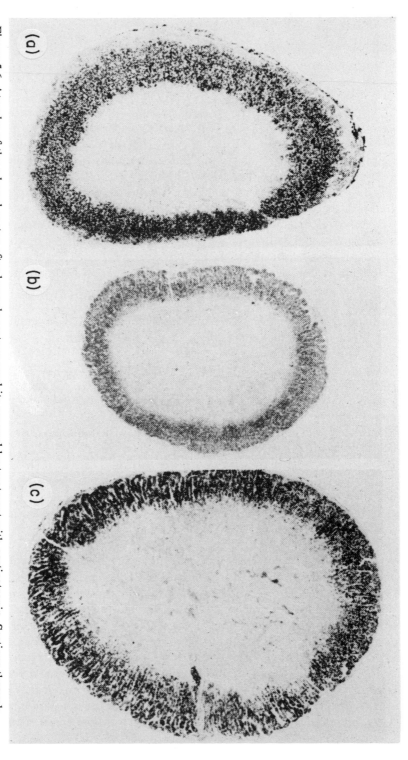

Figure 5.6 Atrophy of the adrenal cortex after hypophysectomy and its reversal by treatment with corticotropin. Sections through (a) the adrenal glands of normal, (b) hypophysectomized and (c) corticotropin-treated, hypophysectomized mouse are shown. (Reproduced with permission from Lostroh, A. J. and Woodward, P. (1958) *Endocrinology* **62**, 498–505. Copyright (1958) The Endocrine Society)

good deal of work has been carried out on the biochemical mode of action of corticotropin on corticosteroidogenesis, and will be discussed in detail in Section 6.

2.4.2.2   Trophic actions. After hypophysectomy, the adrenal cortices atrophy (particularly the zona fasciculata and zona reticularis; Figure 5.6). This can be reversed by pituitary extracts and has generally been thought to be primarily due to loss of corticotropin. Atrophy of the adrenal cortices can also be induced by injecting large doses of glucocorticoids into an intact animal (presumably due to feedback inhibition of corticotropin secretion) and, conversely, injection of large doses of corticotropin can induce hypertrophy of the adrenal cortices.

Doubts have been expressed, however, as to whether corticotropin is the sole, or indeed the main, trophic factor for the adrenal cortex. The hormone *inhibits* division of adrenocortical cells in culture, under conditions where corticosteroid production is enhanced.[15] Furthermore, antiserum to corticotropin did not cause regression of the adrenal cortex in intact rats, although it did inhibit corticosteroidogenesis.[15] Some earlier work suggested that pituitary growth hormone (or a growth hormone-dependent factor such as a somatomedin) plays a major role in maintaining growth of the adrenal cortices, possibly acting synergistically with corticotropin.

One trophic role of corticotropin does seem to be well established— maintenance of the ability of the adrenal cortex to show an acute (corticosteroidogenic) response to the hormone. About 2–4 days after hypophysectomy, the adrenal cortices of the rat can no longer respond to corticotropin treatment by enhanced corticosteroid production. If the animal is maintained on low doses of corticotropin, responsiveness is retained.

2.4.2.3   Other actions on the adrenal cortex. Administration of corticotropin *in vivo* causes depletion of adrenal cholesterol. This is presumably partly a consequence of utilization of cholesterol for corticosteroidogenesis, but the quantity of cholesterol that disappears is considerably greater than the quantity of corticosteroids produced.

Another well-documented effect of corticotropin on the adrenal cortex is ascorbic acid depletion, which has been used as the basis of several bioassays. Ascorbic acid levels in the adrenal venous blood rise rapidly after infusion of corticotropin into hypophysectomized rats, but fall to baseline after 5–8 minutes (when corticosterone production is rising). Ascorbic acid release is always accompanied by corticosteroidogenesis, but once ascorbic acid in the adrenals has been depleted, re-stimulation with corticotropin can still give corticosteroidogenesis, unaccompanied by further ascorbic acid release. The significance of ascorbic acid depletion by corticotropin is not clear.

Corticotropin also has various other effects on the metabolism of the adrenal cortex, presumably associated with its effects on steroidogenesis. Thus, adrenal glycogen depletion and increased glucose oxidation both follow administration of corticotropin to hypophysectomized rats.

Finally, corticotropin has important effects on the blood supply to the adrenal cortex, causing a marked increase in the flow of blood through the gland. This presumably serves to carry extra nutrients to the adrenal cortex (including cholesterol, which is taken up from the blood to a considerable extent in at least some species) and to speed up removal of corticosteroids. This may be important because a build-up of corticosteroids within the gland would tend to inhibit further steroidogenesis, by a short-loop feedback effect.

2.4.3 Extra-adrenal actions of corticotropin

The predominant actions of corticotropin are undoubtedly on the adrenal cortex, but various other direct actions have been described. The hormone has direct MSH-like actions on skin pigmentation in many lower vertebrates and some mammals, including, in long-term treatment, man. These actions clearly reflect the structural similarities between corticotropin and melanotropin. Corticotropin also stimulates release of free fatty acids from adipose tissue *in vivo* and *in vitro,* an action mediated by stimulation of adenylate cyclase on the fat cells. The hormone has effects on carbohydrate metabolism, including hypoglycaemic and diabetogenic effects, depending on the state of the test animal. A variety of behavioural effects have also been reported, including increased mental performance in man and animals.

Many of these extra-adrenal effects are only observed with very big doses of hormone. It is likely that some of them reflect not interaction of corticotropin with its own natural receptors, but binding to receptors for other members of the corticotropin/melanotropin family. None of the extra-adrenal effects of corticotropin has yet been shown unequivocally to be of physiological significance.

3. CHARACTERIZATION AND BIOLOGICAL ACTIONS OF THE MELANOTROPINS

3.1 Isolation and structure[16,17]

In many amphibia and fish, changes in the colour and (particularly) darkness of the skin are important adaptations in response to the surrounding environment. In some cases such colour changes are very rapid, and are under direct nervous control. In others, however, changes are more gradual and hormonal control is the rule. The hormones controlling such changes are the melanotropins (MSHs) which derive from the anterior and intermediate lobes of the pituitary.

The first demonstrations that the pituitary can control skin colour were made in 1916 by Smith and by Allen. They showed that hypophysectomy of tadpoles caused a lightening of the skin, which could be reversed by injection of pituitary extracts. *In vivo* assays using tadpoles and frogs were devised, and later an

```
                                                                                              10     13
1.  α-MSH                                                 Acetyl-Ser-Tyr-Ser-Met-Glu-His-Phe-Arg-Trp-Gly-Lys-Pro-Val-NH2

2.  MSH (dogfish)                                                       H-Ser-Met-Glu-His-Phe-Arg-Trp-Gly-Lys-Pro-Met-(NH2)

3.  MSH (dogfish, tyrosyl)                                          H-Tyr-Ser-Met-Glu-His-Phe-Arg-Trp-Gly-Lys-Pro-Met-(NH2)
                                                                                      10                           18
4.  β-MSH ('bovine')                                  H-Asp-Ser-Gly-Pro-Tyr-Lys-Met-Glu-His-Phe-Arg-Trp-Gly-Ser-Pro-Pro-Lys-Asp-OH

5.  β-MSH ('porcine')                                 H-Asp-Glu-Gly-Pro-Tyr-Lys-Met-Glu-His-Phe-Arg-Trp-Gly-Ser-Pro-Pro-Lys-Asp-OH

6.  β-MSH (Macacus)                                   H-Asp-Glu-Gly-Pro-Tyr-Arg-Met-Glu-His-Phe-Arg-Trp-Gly-Ser-Pro-Pro-Lys-Asp-OH
                                                                      10                                       20      22
7.  β-MSH (human)                     H-Ala-Glu-Lys-Lys-Asp-Glu-Gly-Pro-Tyr-Arg-Met-Glu-His-Phe-Arg-Trp-Gly-Ser-Pro-Pro-Lys-Asp-OH
                                                                                      10                           18
8.  β-MSH (horse)                                     H-Asp-Glu-Gly-Pro-Tyr-Lys-Met-Glu-His-Phe-Arg-Trp-Gly-Ser-Pro-Arg-Lys-Asp-OH
                                                                                              10                           17
9.  ACTH (res. 1-17)                                          H-Ser-Tyr-Ser-Met-Glu-His-Phe-Arg-Trp-Gly-Lys-Pro-Val-Gly-Lys-Lys-Arg...
                                   37          40                                     50                                   60
10. β-lipotrophin (ovine)         ...Ala-Glu-Lys-Lys-Asp-Ser-Gly-Pro-Tyr-Lys-Met-Glu-His-Phe-Arg-Trp-Gly-Ser-Pro-Pro-Lys-Asp-Lys-Arg...
      (res. 37-60)
```

Figure 5.7 Comparison of the sequences of peptides of the MSH/ACTH family. Solid lines enclose identical sequences. Both 'bovine' and 'porcine' β-MSHs are found in bovine, ovine and pig pituitaries. Each of the two types of dogfish MSH is partially amidated at the C-terminus. (Reproduced with permission from Wallis, M. (1975) *Biol. Rev.* **50**, 35–98. Copyright (1975) Cambridge University Press)

extremely sensitive *in vitro* assay was described, which uses isolated frog skin. Development of these assays enabled purification of the melanocyte stimulating factors, particularly by Lerner and his colleagues.

During the 1950s, melanotropins were isolated from several different mammals. There are two main types, which occur together in single individuals. α-MSH is a very basic peptide containing 13 amino acids and a blocked N and C terminus (Figure 5.7), which occurs mainly in the intermediate lobe of the pituitary. β-MSH is a slightly basic peptide containing 18 amino acids in most species, which occurs at least partly in the anterior lobe of the pituitary. Sequences of MSHs from several different mammals have been determined. That of α-MSH is identical for every mammal investigated, but there are some variations between the β-MSHs from different species. There is also evidence that some species (cow, pig, sheep) contain two slightly different forms of β-MSH (Figure 5.7). The relative amounts of α-MSH and β-MSH vary widely between different mammalian species; rat and mouse contain rather little β-MSH, whereas the sheep contains large amounts.

It can be seen from the structures shown in Figure 5.7 that the sequences of α- and β-MSH share substantial structural homology with each other and with ACTH. Indeed, the sequence of α-MSH is identical to the first 13 residues of ACTH. The explanation of these homologies in terms of biosynthetic relationships has already been considered in the first section of this chapter. A consequence of the structural homologies is that ACTH possesses significant intrinsic MSH-like activity (about 1% of that of MSH) in many assays (see Section 2.4.3). Indeed, long-term therapy with ACTH has in some cases resulted in considerable darkening of patients' skin as an unwanted side-effect.

The adult human pituitary gland is unusual in not having a distinct intermediate lobe. The concentration of α-MSH in the human pituitary is very low and it seems unlikely that significant amounts of this peptide circulate in the normal human adult. However, the pituitary of the human foetus does have an intermediate lobe, and this does contain substantial quantities of α-MSH.[18] α-MSH has also been detected, using immunofluorescence, in some regions of the pituitary of pregnant women. A human β-MSH has been described having a sequence 4 residues longer than that of other mammalian β-MSHs. However, it seems likely that this is in fact an artefact of the extraction methods used, due to an artefactual cleavage of β-lipotropin (see below).[19]

Peptides with melanotropic activity have been isolated from two species of dogfish, and from salmon (Figure 5.7). These have amino acid sequences which show clear homology with the corresponding peptides from mammalian pituitaries. They possess about one-hundredth the potency of mammalian α-MSH (in assays using amphibia). Unlike mammalian α-MSH, none of the dogfish melanotropins is N-acetylated.[19]

The sequence predicted for pro-opiocortin indicated a third melanotropin-like peptide, γ-MSH (Figure 5.1). This has now been synthesized; the synthetic peptide has relatively low melanotropic activity but may potentiate the steroidogenic response to ACTH.[17]

3.2 Biological actions of melanotropins[20,21]

3.2.1 Actions on frog skin

The dark pigment of frog skin consists mainly of melanin, which is contained in granules in pigment cells or melanophores. In a light-adapted (or hypophysectomized) frog these granules are associated in the centre of the melanophores; in a dark-adapted or melanotropin-treated animal they are dispersed throughout the melanophores, allowing maximum light absorption (Figure 5.8). The dispersal of melanin granules appears to be the main way in which melanotropin exerts its skin-darkening actions.

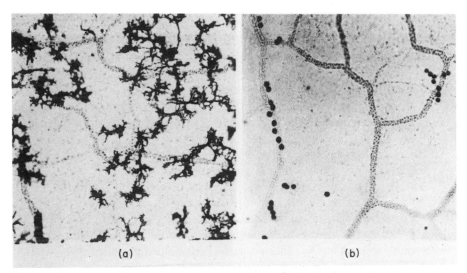

(a)   (b)

Figure 5.8 MSH induces expansion of melanin granules in frog skin. (a) Expanded melanophores from a normal animal; (b) Melanophores from an animal with an inactive pars intermedia. (Reproduced with permission from Driscoll, W. T. and Eakin, R. M. (1955) *J. Exp. Zool.* **129**, 149–175. Copyright (1955) Wistar Institute of Anatomy and Biology)

3.2.2 Actions in other lower vertebrates[21]

Melanotropins appear to be important in controlling skin darkness in many amphibia, fish and reptiles. They also affect cells containing pigments other than melanin: the white iridophores (containing guanine-rich reflecting platelets) and the red or yellow lipophores (with carotenoid or pteridine-containing granules); they can thus cause relatively complex colour changes. The hormone generally causes contraction of iridophores and dispersal of pigment in lipophores. It must be re-emphasized, however, that not all control of skin pigmentation in lower vertebrates is hormonal. In many cases, especially in some teleost fish and reptiles, control is by nervous rather than hormonal signals.

In addition to these melanin- (and other pigment granule) dispersing effects, melanotropins also have longer-term effects, in many species, on the content of pigments in the skin. Prolonged treatment with melanotropins causes increases in numbers of melanophores and melanin content per melanophore in frogs and various teleosts, a decrease in skin guanine (the 'pigment' of iridophores) and an increase in content of pteridine (a pigment of lipophores).

3.2.3   Actions of melanotropins in mammals[20,22–24]

Although most preparations of α- and β-melanotropin that are available have been made from mammalian pituitaries, the role of melanotropins in mammals is not clear. Prolonged injection of melanotropin into man and many other mammals does cause some increased skin pigmentation, apparently as a consequence of both dispersion of melanin granules and increased melanin production.[23] Melanotropins can also increase the amount of pigment laid down in growing hair, which may be an important factor in animals (such as many species of weasel) which have different coloured coats in winter and summer. Presumably increasing day length in spring leads to increased secretion of melanotropin(s) and hence growth of brown hair rather than white.[22]

However, changes in skin or coat colour are clearly not a major adaptive mechanism in most mammals, and alternative roles for the melanotropins have been sought. Many extra-pigmentary effects have been described, on visual adaptation, behaviour, thyroid activity, sebaceous gland activity, lipolysis and calcium and sodium levels in serum. The behavioural effects are of particular interest[25] in view of the occurrence of melanotropins in the brain, and their possible inclusion among the fast-growing list of peptide neurotransmitters. Nevertheless, none of the effects listed here has been shown conclusively to have physiological significance.

As already pointed out, α- and β-melanotropins do not appear to occur in the normal adult human pituitary. However, α-MSH does occur in the foetal human pituitary, and possibly in the pituitary of pregnant women. It has been suggested that secretion of α-melanotropin (or possibly CLIP—see Section 3.3) from the foetal pituitary is important in promoting growth and functioning of the foetal adrenal cortex, and that the rapid disappearance of α-melanotropin from the pituitary and circulation after parturition acts as a trigger for conversion of the adrenal gland to its adult form.

3.2.4   The relationship between structure and function in melanotropins[17]

Structure–function relationships in the corticotropin/MSH peptide family are considered in more detail in Chapter 15. Among naturally-occurring peptides, mammalian α-MSH is the most active melanotropin known (in amphibian skin-darkening assays), β-MSH is about one-third as active and ACTH possesses about one-hundredth the activity of α-MSH.

Structure–function relationships among the MSHs have involved the preparation of large numbers of analogues. Work on the chemistry of such peptides has also provided analogues of great value for the study of hormone–receptor interactions, including derivatives containing radioactive, fluorescent and photolabile labels (see Section 7).

3.3 Corticotropin-like intermediate lobe peptide (CLIP)

The sequence of α-MSH corresponds to the N-terminal 13 residues of corticotropin. Another peptide contained in the intermediate lobe, CLIP, corresponds to residues 18–39 of corticotropin. This occurs in amounts approximately equivalent to those of α-MSH, and it is secreted along with this peptide. The precise role of CLIP remains unclear, but it, and the closely related peptide comprising residues 22–39 of corticotropin, have been shown to stimulate insulin secretion.[26]

4. LIPOTROPIN AND THE OPIOID PEPTIDES: ENKEPHALINS AND ENDORPHINS

In 1975 peptides were discovered with opioid-like actions, and during the past few years such factors have been studied intensively. They are formed from the same or similar precursors as corticotropin and the melanotropins (see Section 1), and are therefore considered in this chapter. However, their hormonal role may be much less important than their role as neurotransmitters or neuromodulators in the central nervous system, and treatment of their role as neuropeptidés must necessarily be very restricted here. Some of the opioid peptides are closely related to the lipotropins, which will also be discussed in this section.

4.1 The lipotropins[24,27]

The lipotropins were discovered by C. H. Li and his colleagues in the 1960s. although various earlier reports had suggested the presence of peptides with lipolytic activity in the pituitary gland. Three lipotropins were first isolated from sheep anterior pituitary glands (α, β and γ) and were subsequently also found and characterized in various other species, including man. β-lipotropin is the largest of these peptides (91 residues in the sheep) and the sequence of γ-lipotropin corresponds to the first 58 residues of this hormone, in the sheep. γ-lipotropin is thus presumably derived from β-lipotropin by proteolytic cleavage (Figure 5.9). The sequence of α-lipotropin has not been determined. Sheep and human β-lipotropins have been synthesized chemically.[24,28]

Ovine: H-Glu-Leu-Thr-Gly-Glu-Arg-Leu-Glu-Gln-Ala-Arg-Gly-Pro-Glu-Ala-Gln-Ala-Glu-Ser-Ala-Ala-Ala-Arg-

Porcine: H ———— Ala ———— Ala-Pro-Pro ———— Asp ———— Pro ———— Gly

10   20

Ovine: -Ala-Glu-Leu-Glu-Tyr-Gly-Leu-Val-Ala-Glu-Ala-Ala-Glu-Ala-Glu-Lys-Asp-Ser-Gly-Pro-Tyr-Lys-

Porcine: ———————————————————— Gln ———— Glu

Human: ———————————————— Gln ———— Glu ———————— Arg-

30   40

Ovine: -Met-Glu-His-Phe-Arg-Trp-Gly-Ser-Pro-Pro-Lys-Asp-Lys-Arg-Tyr-Gly-Gly-Phe-Met-Thr-Ser-Glu-Lys-

Porcine: ——————————————————————————————————————

Human: ——————————————————————————————————————

50   60

Ovine: -Ser-Gln-Thr-Pro-Leu-Val-Thr-Leu-Phe-Lys-Asn-Ala-Ile-Ile-Lys-Asn-Ala-His-Lys-Lys-Gly-Gln-OH

Porcine: ———————————— Val ———————————————————— OH

Human: ———————————— Tyr ———————————————————— OH

70   80   90

Figure 5.9 Amino acid sequences of β-lipotropins. The sequence of the ovine hormone is shown in full. The porcine and human hormones differ at the positions indicated—the solid lines indicate regions of the sequence identical to ovine β-lipotropin. γ-lipotropin corresponds to residues 1–58. (Reproduced with permission from Wallis, M. (1975) *Biol. Rev.* **50**, 35–98. Copyright (1975) Cambridge University Press)

Inspection of the sequence of β-lipotropin (Figure 5.9) reveals that it includes not only the complete sequence of γ-lipotropin (residues 1–58), but also the sequence of β-MSH (residues 41–58). The possibility that β-lipotropin might be a biosynthetic precursor of both β-MSH and γ-lipotropin was recognized at an early stage, and support for this concept was provided by the observation that species-specific variations in the structure of β-MSH have exactly correspond-ing variations in the structure of β- and γ-lipotropin (thus, residue 2 in pig β-MSH, and the equivalent position of β- and γ-lipotropin, is Glu, whereas the corresponding residue in each of the corresponding 3 peptides of the sheep is Ser).

The biosynthetic precursor/product relationship between these three hor-mones has been further confirmed by pulse-chase experiments.[1,27] It is notable that the β-MSH sequence is bounded at both N- and C-terminal ends by a pair of basic residues, a feature which is now recognized as commonly determining the specificity of cleavage/processing of peptide precursors (see Section 5.1). It is now recognized that the sequence of β-lipotropin also includes that of β-endo-rphin (residues 61–91) so its role as a precursor peptide (or rather as an inter-mediate in the processing of the ultimate precursor, pro-opiocortin) is extended. γ-lipotropin is presumably an intermediate formed in the processing of β-lipotropin. It is notable that the N-terminal 30–35 residues of β-lipotropin show considerable species variation, but the C-terminal 56 residues are strongly conserved; this presumably reflects the fact that this C-terminal region functions as a precursor for β-MSH and β-endorphin.

Both β- and γ-lipotropins have lipolytic actions *in vivo* and *in vitro*, and it was recognition of these effects which led to the isolation and naming of these fac-tors.[24] However, their *in vitro* lipolytic activity on rabbit adipose tissue, which has been most extensively studied, is considerably lower than that of corti-cotropin or α-MSH,[24] and the action on rat and mouse adipose tissue is even lower. Growth hormone and TSH also have lipolytic activity, and the role of the lipotropins as physiological regulators of lipid mobilization is thus in some doubt. Their intrinsic MSH-like activity is only about 1% of that of α-MSH. β-lipotropin has also been reported to cause hypoglycaemia in rabbits, lipo-genesis in rat adipose tissue, decrease in blood coagulation time in the rabbit, increased sebum secretion in the rat and increased aldosterone production by rat adrenal capsular cells. However, none of the actions of the 'hormone' has been established as of major physiological significance, and it may well be that its main role is as an intermediate in the processing of pro-opiocortin and formation of β-MSH and β-endorphin. It should be noted, however, that con-siderable quantities of β-lipotropin are secreted into the circulation (see Sections 5 and 8).

Immunocytochemical methods have shown that β-lipotropin occurs in the anterior and intermediate lobes of the pituitary gland and in various parts of the brain (see also Section 5.1).

4.2 The enkephalins and endorphins[29-32]

4.2.1 Discovery and structure

The potent effects of morphine and its agonists and antagonists have been known for centuries. Pharmacologists had, for some time, explained these effects in terms of interactions at one or more specific opioid receptors, but the nature of the natural factor(s) interacting with such receptors was not discovered until 1975. In that year, Hughes and Kosterlitz[33] isolated two penta-peptides from brain tissue with potent opiate-like activity. These were named enkephalins. Their structures were determined (using very small quantities) by mass spectroscopy:

methionine enkephalin: H–Tyr–Gly–Gly–Phe–Met–OH,

leucine enkephalin: H–Tyr–Gly–Gly–Phe–Leu–OH.

Although extraction of these peptides from brain tissue is difficult, their chemical synthesis is relatively easy, and synthesis of substantial quantities has paved the way for a variety of studies on their chemistry and biology. Synthesis of a range of analogues has also been achieved, the biological properties of which are considered below.

Shortly after the structures of the enkephalins were determined, it was realized that the sequence of one (Met-enkephalin) is identical to the sequence of residues 61–65 of β-lipotropin (Figure 5.9), an observation which immediately associated these brain peptides with the ACTH–MSH family of pituitary peptides. β-LPH *could* act as a precursor of [Met]-enkephalin (but not [Leu]-enkephalin). In fact, however, it is now clear that there are two other precursors for [Met]-enkephalin and/or [Leu]-enkephalin (see Section 5). Enkephalins have been shown to occur in the pituitary gland.

The recognition that [Met]-enkephalin could be derived from β-lipotropin suggested further that processing of β-lipotropin might give rise to other peptides, containing the Met-enkephalin sequence, with some opioid activity. In particular a peptide ('C peptide') comprising residues 61–91 of β-lipotropin had already been described. When this was tested it was found to have even greater morphine-like activity than the enkephalins. This 31-residue peptide was renamed β-endorphin ('endogenous morphine') and isolated from the pituitary glands of various species. Smaller peptides encompassing residues 61–76 of β-lipotropin (α-endorphin), residues 61–77 (γ-endorphin) and other regions of the β-endorphin molecule (see Figure 5.10) have also been described; these all possess some morphine-like activity, although this is rather lower than that of β-endorphin and in some cases qualitatively different. Acetylated forms of these peptides have also been described and may represent inactivated forms (perhaps produced when pro-opiocortin is processed to give corticotropin or MSH, when active endorphin production would be disadvantageous).[34]

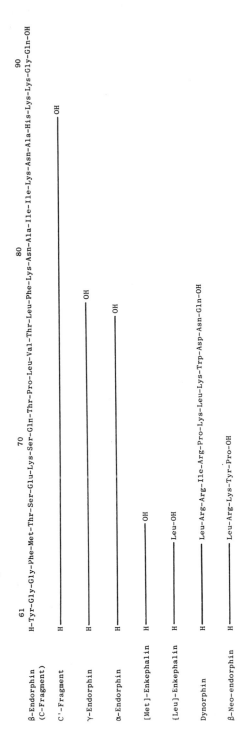

Figure 5.10 Amino acid sequences of various opioid peptides. The complete sequence of sheep β-endorphin is given. Solid lines indicate sequences identical to the β-endorphin sequence, and differences from this sequence are shown in full

β-endorphin and various other endorphins have now been synthesized chemically. Sequences of human, ox, sheep, pig, camel and salmon β-endorphins have been determined; the mammalian peptides are very similar, differing only at three residues. In addition to the enkephalins and endorphins, a considerable number of other opioid peptides has also been extracted from pituitary, adrenal or nervous tissue, including dynorphin and α- and -neoendorphin (Figure 5.10). All of these include the enkephalin sequence and have marked opiate properties. All can be derived from one or other of the three known precursor proteins (Section 5.2).

4.2.2 Biological actions

The biological actions of enkephalins, endorphins and related peptides are similar to those of morphine, though important differences between them have been described. The original assays used to isolate enkephalins utilized the ability of these substances, like morphine, to inhibit the electrically stimulated contraction of guinea pig ileum or mouse vas deferens. The activities of the peptides in these assays is considerably greater than that of morphine; [Leu]-enkephalin is rather less potent than [Met]-enkephalin.

Like morphine, enkephalins and β-endorphin induce analgesia (resistance to pain) in cats and rats. They are far more potent in this respect if injected intraventricularly (i.e. into the ventricles of the brain) than if injected intravenously, presumably because they cannot easily pass the blood-brain barrier.[35] Analgesia induced by β-endorphin is of rather longer duration than that induced by enkephalins. On a molar basis, β-endorphin is 20–50 times more potent than morphine, depending on the particular assay used.

β-endorphin has a range of other biological actions. Thus, it induces hypothermia and salivation in treated animals and behavioural effects in rats and cats, particularly prolonged catatonia (a condition in which the animal is reduced to a rigid state, unresponsive to external stimuli, although completely conscious). β-endorphin and the enkephalins injected intraventricularly also stimulate release of pituitary growth hormone and prolactin, a response presumably mediated by hypothalamic factors, although similar effects on prolactin release may be produced also by direct action at the pituitary.

Like morphine, β-endorphin and [Met]- (although possibly not [Leu]-) enkephalin can induce tolerance and dependence. Rats treated for prolonged periods with β-endorphin become resistant to the analgesic effect and show withdrawal symptoms when administration of the peptide is ceased or when they are treated with the morphine–antagonist naloxone. Rats rendered tolerant (resistant) to morphine are also unresponsive to β-endorphin, and vice versa.

α- and γ-endorphins have been less fully characterized than β-endorphin. Their properties appear to differ in some respects from those of β-endorphin, and in most assays they are less potent. Whether they have physiological functions clearly separate from those of β-endorphin is not known.

Dynorphins and α-neoendorphin (Figure 5.10) have biological activities which resemble those of other opiate peptides. However, they bind preferentially to receptors different from those mainly associated with enkephalins and β-endorphin, suggesting a different biological role.

4.2.3   Receptors[36]

Receptors which bind labelled morphine, morphine agonists or, in some cases, antagonists such as naloxone have been known for some time. They are found in the brain and other morphine target tissues. β-endorphin and the enkephalins will displace labelled ligands from such receptors, and show greater affinity for them than morphine itself. Labelled β-endorphin or enkephalins will also bind specifically to such receptors.

It is now clear that these receptors can be classified into at least three types: μ, κ and δ. The enkephalins possess a preferential affinity for δ receptors; dynorphins and α-neoendorphin bind preferentially to κ receptors; while β-endorphin has a high affinity for all three types of receptor. These observations provide a basis for differential actions of the various opioid peptides, but the precise roles of each of these peptides, and of their receptors, is not yet clear.

The opioid peptides in general appear to have predominantly inhibitory effects on their target cells. A consequence of the binding of enkephalins to receptors on neuronal membranes within the brain is inhibition of action potential discharges in individual nerve cells, possibly as a result of opening of K^+ channels.

4.2.4   The physiological role of enkephalins and endorphins

These peptides have a remarkably wide range of biological activities, but their true physiological role remains unclear. Many of their actions take place in the central nervous system, as shown by the fact that intraventricular injection is necessary for a full response, and it seems likely that they play an important role there as neurotransmitters and neuromodulators.[37] Enkephalins and β-endorphin have been shown to inhibit the release of neurotransmitters, especially dopamine, from brain tissue and the release of acetyl choline from neuromuscular junctions (hence the inhibition of contraction of smooth muscle). A role as inhibitory transmitters accords well with the observed biological effects of the peptides in reducing responses to pain and other stimuli. It has been proposed that release of these (or related) opiate-peptides into the circulation or CNS represents the fundamental basis of acupuncture. β-endorphin shows a distribution in the brain and CNS considerably different from that of the enkephalins, and the relative amounts of [Met]- and [Leu]-enkephalins and the other opioid peptides differ in different neurones. These peptide transmitters (like other such brain peptides) thus show a very specific distribution in the brain, suggesting correspondingly distinct functions, but the precise roles that each plays is not yet clear.

β-endorphin, the larger dynorphins and [Met]-enkephalin also occur in high concentration in the pituitary gland, and are secreted into the circulation. They are thus potentially hormones as well as neurotransmitters (a duality shared with some catecholamines and probably many other peptides). Whether their function as hormones is to inhibit release of transmitters from peripheral neurones or to perform other major roles is unknown, although various alternative functions have been proposed. It is also possible that endorphins or enkephalins may play a role as hypothalamic hormones, regulating pituitary function. Blocking of the inhibitory action of dopamine on prolactin secretion by enkephalins has been demonstrated.[38]

5. DISTRIBUTION, BIOSYNTHESIS AND SECRETION OF THE PEPTIDES OF THE CORTICOTROPIN–MELANOTROPIN FAMILY

5.1   Distribution

The distribution of these peptides has been mainly studied using immunological methods. The results of such studies have to be interpreted with care because peptides with structural similarities may show extensive immunological cross-reaction. As a result, there is considerable controversy about distribution of some of the peptides.

5.1.1   The pituitary gland

Corticotropin occurs mainly in the anterior lobe of the pituitary gland. Various early physiological and pathological observations led to the postulate that the hormone is produced by a group of basophilic cells referred to as corticotrophs, but this assignment was a matter of some controversy until quite recent studies using immunofluorescent antibodies. The use of antibodies specific for corticotropin, showing no cross-reaction with MSH, proved conclusively that corticotropin is produced in stellate basophils containing small granules (200 nm in diameter) located mainly at the cell periphery (Figure 5.11). These corticotrophs represent only a few percent of the total cells in the anterior pituitary. There is also a population of anterior pituitary cells (in the rat) which appears (from immunofluorescence techniques) to contain both corticotropin and the β-subunit of FSH. β-lipotropin and, in some species, β-endorphin and β-MSH also appear to be found in corticotrophs—an observation which is not surprising since these are produced from the same precursor as corticotropin; in some cases, however, the immunological methods used may be detecting peptides inactivated by acetylation or proteolysis rather than active hormones.

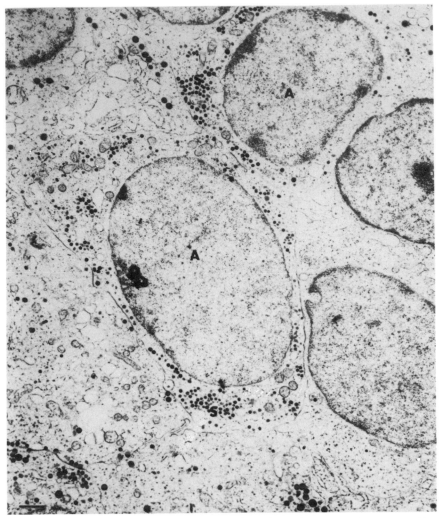

Figure 5.11 Electron micrograph of a section through a metapirone-treated rat anterior pituitary gland showing two corticotrophs (A). Note the large central nucleus and the many small ACTH-containing granules in each cell. (Reproduced with permission from Costoff, A. (1973) *Ultrastructure of Rat Adenohypophysis*. Copyright (1973) Academic Press)

Melanotropins, particularly α-MSH, have always been the main hormones thought to be associated with the intermediate lobe of the pituitary. Several cell types have been described in the intermediate lobe, although these are less distinguishable than those in the anterior lobe. However, immunocytochemical studies suggest that most cells, and indeed most of their granules, contain α-MSH, and also corticotropin-like material (mainly CLIP), β-lipotropin, β-MSH and β- (and other) endorphins. This observation may be related to the fact that all these peptides are formed from the same precursor, but it is also clear that the antisera that have been used for such studies show overlapping specificities. Some studies in which extracts of pig intermediate lobe were fractionated by chromatography and then subjected to immunoassay suggested that there is in fact very little β-lipotropin present in the intermediate lobe and that most of the endorphin present consists of forms which are much less active than β-endorphin itself, such as the C′ fragment, which lacks two amino acids from the C-terminus of β-endorphin, and N-acetylated forms of endorphin.[34,39,40] Thus the bulk of the *active* β-endorphin in the pituitary may be in the anterior lobe.

In general, recent studies using immunological methods of identification suggest that the peptides of the corticotropin–MSH family are more evenly distributed between anterior and intermediate lobes than had been concluded from earlier biological studies. However, since all these peptides are formed from a common precursor, it is clear that individual regions of sequence *must* be equally represented as far as synthesis is concerned, and it is these rather than biologically active peptides that antibodies recognize. It is notable that the distribution of these peptides in the anterior and intermediate lobes varies considerably from one species to another.

[Met]-enkephalin and [Leu]-enkephalin are also found in the intermediate and posterior lobes of the pituitary, but in the former may be mainly confined to the nerve fibres which occur in that lobe.[41]

5.1.2   The brain[43,44]

Enkephalins were first isolated from the brain, and have subsequently been detected, by immunoassay and immunocytochemistry, in many regions of that organ. In the rat, their concentration varies from about 5 n moles/g tissue in the hypothalamus and corpus striatum to 0.01 n moles/g tissue in the cerebellum. The ratio of [Met]-enkephalin : [Leu]-enkephalin varies from about 2 : 1 to 10 : 1,[42] and immunofluorescence studies suggest that the two forms of enkephalin are found in different neurones.[43]

Endorphins are also found in some parts of the brain, but their distribution is different from that of the enkephalins, and they are also found in quite separate neurones. The concentration is greatest in the hypothalamus (approximately 0.15 n moles/g tissue). This is less than 5% of the concentration of endorphin in the intermediate lobe of the pituitary, but a high proportion is in the form of active β-endorphin rather than the inactivated forms which may predominate in the pituitary.

Neurones which contain β-endorphin also appear, by immunocytochemical techniques, to contain corticotropin, β-lipotropin and α-MSH. This is not surprising in view of the biosynthetic relationships of these hormones, but again the possibility of cross-reaction between antisera to the various hormones means that these results must be interpreted with caution. Nevertheless, it does seem clear that at least some active corticotropin and α-MSH is produced in the brain (albeit at much lower concentrations than in the pituitary), and the possibility exists that these hormones can also function as specific neurotransmitters.[43] In this connection it is noteworthy that several behavioural effects have been described when corticotropin and α-MSH are injected intraventricularly into the brain.

All these peptides found in the brain can be still found in this organ in hypophysectomized animals, so they are probably synthesized there rather than imported from the pituitary. In some cases direct biosynthesis of these factors by brain tissue has been demonstrated.[43]

5.1.3   The gastrointestinal tract

Like many other neuroactive peptides, [Met]- and [Leu]-enkephalins are found in the gut as well as the brain.[45] They are localized mainly in terminal nerve fibres occurring in much of the gastrointestinal tract, especially the stomach, duodenum and rectum and (in man and the baboon) in the gastrin-producing endocrine cells. Occurrence of these peptides in the gut presumably relates to the occurrence of opioid receptors in various parts of the gastrointestinal tract and the well-known effects of morphine (and opioid peptides) on gastrointestinal mobility. No evidence has been adduced for the occurrence of β-endorphin, corticotropin, α-MSH or β-lipotropin in the gut.

5.1.4   The adrenal medulla

The adrenal medulla is a rich source of [Met]-enkephalin, [Leu]-enkephalin and related peptides, and was the tissue from which the precursor proenkephalin was isolated.[4,5] The biological role of the peptides in this organ is not clear.

5.2   Biosynthetic precursors

As has already been outlined (Section 1), corticotropin, α- and β-MSH, lipotropin and the various endorphins are all thought to derive from a common precursor, pro-opiocortin. [Leu]-enkephalin cannot be derived from this precursor and although [Met]-enkephalin *could* be derived from pro-opiocortin, it is now clear that a separate precursor is involved in its synthesis.

Three main lines of investigation led to the discovery of pro-opiocortin. Investigation of large forms of corticotropin (detected primarily by use of specific antibodies) suggested that such biosynthetic precursors do exist, although their characterization was delayed because very little material could be isolated.[19,46] Biosynthetic studies, again using specific antibodies combined with pulse-chase type experiments, clearly demonstrated the existence of a series

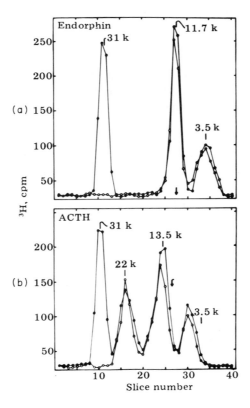

Figure 5.12 Biosynthesis of peptides of the ACTH/MSH/β-endorphin family and their precursor. Cultured pituitary tumour cells were incubated with [³H]phenylalanine and then analysed by immunoprecipitation and gel electrophoresis. (a) A β-endorphin antiserum was used to prepare the immunoprecipitate either before (closed circles) or after (open circles) all the ACTH-containing material had been removed by immunoprecipitation. (b) An ACTH antiserum was used to prepare the immunoprecipitate either before (closed circles) or after (open circles) all the β-endorphin-containing material had been removed by immunoprecipitation. The experiment demonstrates that the same precursor (31K) contains both ACTH and β-endorphin. (Reproduced from Mains, R. E., Eipper, B. A. and Ling, N. (1977). *Proc. Nat. Acad. Sci. USA* **74**, 3014–3018, by permission of the authors)

of large peptides with short half-lives, which contained antigenic determinants for corticotropin, melanotropins, lipotropin and/or β-endorphin, and which could represent a series of molecules in a pathway from the high-molecular weight precursor to the finished hormones.[1,47] Finally, use of the new recombinant DNA technology allowed the cloning and complete sequence determination of the 'gene' corresponding to pro-opiocortin, which enabled complete definition of the location of the various peptide hormone sequences within this precursor.[3,48]

The biosynthetic studies of Mains, Eipper and their colleagues are of particular interest.[1] They studied the biosynthesis of endorphin-like and corticotropin-like peptides in a cultured pituitary tumour cell line. Cells were incubated with a labelled amino acid and extracted, and extracts were then immunoprecipitated with antibodies to β-endorphin or ACTH (which showed no cross-reaction). SDS-gel electrophoresis of the immunoprecipitates revealed three endorphin-like components (of molecular weights 31 000, 11 700 and 3500) and four corticotropin-like components (of molecular weights 31 000, 23 000, 13 000 and <4500) (Figure 5.12). The 31 000 molecular weight component found in each case was the same molecular species (as shown by sequential immunoprecipitation of the same extract with the two antibodies, and also by peptide mapping) and thus represents a common precursor for corticotropin and β-endorphin. The 11 700 molecular weight component is probably equivalent to β-lipotropin, and the 23 000 and 13 000 molecular weight components presumably represent intermediates in processing which include the sequence of corticotropin but not β-endorphin. Several of the higher-molecular weight precursors (31 000, 23 000 and 13 000 molecular weight molecules) are glycosylated, a feature which is likely to mean that the molecular weights determined from mobilities on SDS-gel electrophoresis are not very accurate. Pulse-chase type experiments confirmed that the series of corticotropin and endorphin-like peptides were labelled as would be expected for a precursor-product series.

The order in which the various peptides of the ACTH–endorphin family are arranged in the 31 000 (31K) molecular weight precursor was partly deduced by Eipper and Mains *et al.*,[1] and the definitive organization of this precursor was provided by Nakanishi *et al.*[3] These authors prepared a double stranded cDNA from the messenger RNA for bovine pro-opiocortin, inserted this into the plasmid pBR322 and cloned this in *E. coli* (see Chapter 18). The cloned DNA insert was then subjected to DNA sequence analysis and the complete nucleotide sequence, corresponding to 1091 base pairs was determined. From this DNA sequence the amino acid sequence of pro-opiocortin (or, probably, the pre-pro-opiocortin precursor with an additional N-terminal signal peptide extension) can be deduced. The sequences of β-lipotropin and corticotropin were readily recognized within this precursor; the overall organization was shown in Figure 5.1

Several points stand out. The sequence of β-lipotropin is at the C-terminus. In the nucleotide (mRNA) sequence this is followed by the termination codon UGA. The β-lipotropin sequence is preceded by that of corticotropin, only a pair of basic residues (Lys–Arg) separating the two. Corticotropin and β-lipotropin thus make up the C-terminal half of pro-opiocortin. The N-terminal half of the precursor ('16K fragment') contains no known peptide hormone sequence. It does contain, however, a sequence homologous with that of α- and β-MSHs, including the sequence His–Phe–Arg–Trp which is found in both these peptides; Nakanishi *et al.* therefore proposed that a third MSH-like peptide (γ-MSH) may be produced from this region of the precursor. This has now been synthesized and shown to possess MSH-like activity. The N-terminal half of pro-opiocortin also includes five half-cystine residues, one of which is in the putative signal peptide region.

A striking feature of the amino acid sequence of pro-opiocortin is the distribution of pairs of basic residues (Lys–Lys, Lys–Arg, Arg–Lys, Arg–Arg). The sequences of β-endorphin, β-MSH, β-lipotropin, corticotropin, α-MSH, CLIP (Section 3.3) and the putative γ-MSH are all bounded by such paired basic residues, and it seems indisputable that they provide a recognition signal for the proteolytic enzymes which process the precursor.[49] Different processing schemes must apply in different cell types, presumably due at least partly to the presence of enzymes with different specificities. (Differential inactivation by acetylation and other mechanisms may also contribute to the differences between the products produced by different cells from the same precursor.)[39] Paired basic residues appear to play a similar role in the processing of other prohormones, such as proinsulin and proparathyroid hormone.

The nature of the biosynthetic precursors of the opioid peptides [Leu]-enkephalin, [Met]-enkephalin, α-neoendorphin and dynorphin has also been elucidated using the techniques of recombinant DNA technology, although again the isolation of larger peptides incorporating the sequences of these peptides had indicated the existence of precursors. The precursor of the enkephalins, preproenkephalin A[4,5] (Figure 5.13) is a remarkable molecule incorporating four [Met]-enkephalin sequences, one [Leu]-enkephalin and two further sequences incorporating [Met]-enkephalin but which probably give rise to slightly larger peptides:

Tyr–Gly–Gly–Phe–Met–Arg–Gly–Leu,

Tyr–Gly–Gly–Phe–Met–Arg–Phe.

The precursor of α-neoendorphin and dynorphin, preproenkephalin B, which can also give rise to [Leu]-enkephalin, is also shown in Figure 5.13. This has been identified from studies using mRNA isolated from hypothalamus.[6]

The striking feature about all these precursors is the way in which a single large precursor gives rise to several different biologically active peptides. The way in which such complex molecules have evolved remains to be elucidated.

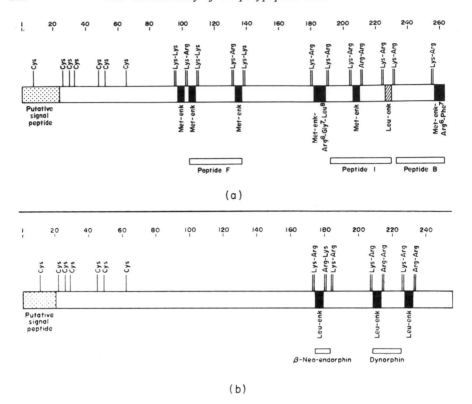

Figure 5.13 Schematic representation of (a) the structure of preproenkephalin A, and (b) the structure of preproenkephalin B. Cys indicates the position of ½-cysteine residues. Lys–Arg, Arg–Lys etc. indicate the positions of paired basic residues, thought to be the sites of cleavage during processing of the precursors to [Met]-enkephalin, [Leu]-enkephalin and other peptides. (Reproduced with permission from Noda, M. *et al.* (1982) *Nature* **295**, 202–206, and Kakidani, H. *et al.* (1982) *Nature* **298**, 245–249. Copyright (©) 1982 Macmillan Journals Limited)

The advantage to the organism may lie in the ability to coordinate the production and secretion of several different peptides, although in the case of pro-opiocortin this appears to be a disadvantage sometimes, since elaborate mechanisms appear to have been evolved to ensure that some of the peptides produced are inactivated before secretion.

5.3   Processing and secretion

Since corticotropin, melanotropins and endorphins are all produced from the same precursor, but different cells contain different hormones, it follows that conversion of precursor to smaller peptides must vary from cell to cell, and presumably that the various cell types contain different proteolytic enzymes.

The specificities of these proteases appear to be similar in that they mainly cleave next to paired basic residues, as previously explained, but further features must influence the specificity in different cells. Particularly interesting is the way that the processing of the region of pro-opiocortin containing the sequence of corticotropin is quite different in corticotrophs (where intact corticotropin is produced) and melanotrophs (where additional cleavages occur to give α-MSH and CLIP, and the former is acetylated and amidated). In addition to differential processing, some peptides may be inactivated in some cells. Thus, as described above, much of the β-endorphin produced in the intermediate lobe of the pituitary may be converted to less active forms by acetylation and/or further cleavage.

The nature of the processing enzymes and their locations within the cells is poorly understood. It does seem, however, that much of the processing may occur within the secretory granules, and the granules in corticotrophs may contain several pro-opiocortin-derived peptides in addition to the main active hormones.[49] This would explain why granules in melanotrophs react with fluorescent antibodies to corticotropin, α-MSH, β-lipotropin and β-endorphin.

Secretion of corticotropin from corticotrophs and α-MSH from melanotrophs is thought to occur by exocytosis of secretory granules, as with most other polypeptide hormones. Since it is likely that such granules contain peptides additional to corticotropin and α-MSH it follows that these should be secreted at the same time. In accordance with this it has been widely reported that β-endorphin-like and β-lipotropin-like peptides (as detected by radioimmunoassay) are released along with corticotropin from the anterior pituitary.[50]

Subcellular localization of enkephalin and endorphin within neurones in the brain and elsewhere has been less fully defined, but [Met]-enkephalin has been reported to occur mainly in large and small vesicles in axons and axon terminals, as one would expect for a neurotransmitter.

5.4   Regulation of secretion

5.4.1   Corticotropin[51,52,56]

As for other anterior pituitary hormones, the secretion of corticotropin is controlled mainly by a factor (or factors) produced in the hypothalamus. The existence of a hypothalamic corticotropin releasing factor (CRF) was recognized in 1955,[53,54] but despite several attempts to isolate this factor, it remained poorly characterized for 25 years. Several CRFs were described from the neurohypophysis, and it was also proposed that vasopressin is a CRF. However, various lines of evidence suggested that none of these factors is the main hypothalamic CRF. Difficulties encountered in isolating CRF included the very small quantities present in the hypothalamus, instability of the factor, difficulties with assays and the presence of corticotropin in hypothalamic extracts.

In 1981 a 41-residue peptide was isolated from sheep hypothalamic tissue that was highly potent in stimulating the release of corticotropin from anterior pituitary cells cultured *in vitro*.[55] Its amino acid sequence was determined (Figure 5.14) and the peptide was synthesized chemically. The sequence is markedly homologous with that of sauvagine, a peptide isolated previously from the skin of a South American frog which has been shown to stimulate secretion of several pituitary hormones. There is also a slight homology with angiotensinogen.

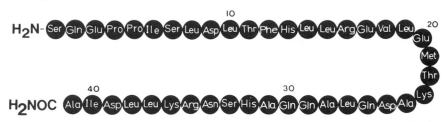

Figure 5.14 The amino acid sequence of sheep corticotropin releasing factor (CRF)

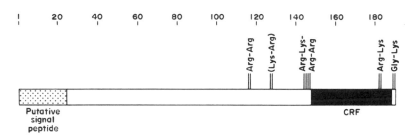

Figure 5.15 Schematic representation of the structure of the precursor of corticotropin releasing factor. The sequence of CRF lies near the C-terminus of the precursor. The glycine residue immediately following the CRF sequence probably provides the amide group on the C-terminus of the mature peptide. (Reproduced with permission from Furutani, Y. *et al.* (1983) *Nature* **301**, 537–540. Copyright © 1983 Macmillan Journals Limited)

This 41-residue peptide is very probably the major hypothalamic CRF. It stimulates secretion of corticotropin and other peptides derived from proopiocortin, including β-endorphin and β-lipotropin, but not unrelated pituitary hormones, and has been shown to act *in vitro* and *in vivo*. It has been shown by radioimmunoassay to occur in hypothalamic portal blood at concentrations sufficient to stimulate corticotropin secretion,[56] and antibody to the peptide blocks release of corticotropin in normal rats.[57] Immunocytochemical studies indicate that it occurs in the nerve fibres of the hypothalamus, particularly in nerve endings abutting the blood capillaries of the portal system, as would be expected for a releasing factor.[58] It appears to be synthesized in the cell bodies of neurones in the paraventricular nucleus of the brain,[58] and a precursor molecule has been identified and characterized using recombinant DNA techniques.[59] As is shown in Figure 5.15, CRF lies near the C-terminus of the precursor; four basic residues precede the peptide sequence, presumably providing the signal for proteolytic processing, and a Gly–Lys sequence follows it which probably signals amidation of the C-terminal alanine by an enzyme that has been identified in the pituitary gland.[60]

The mechanism whereby this 41-residue CRF stimulates corticotropin secretion has not been elucidated in detail, but it has been shown to cause a marked elevation of cyclic AMP concentration in pituitary cells incubated *in vitro*,[61] and this nucleotide may thus mediate at least some of its actions. Although the peptide appears a strong candidate for CRF, some evidence suggests that it is not the only important stimulator of corticotropin secretion. In particular, vasopressin appears to synergize with CRF to cause both corticotropin and β-endorphin secretion and elevation of pituitary cyclic AMP levels.[62,63] There is also evidence that this CRF has effects on behaviour that are not mediated via its actions on corticotropin release. For example, when injected intraventricularly into the brain of rats it caused increased locomotor activity.[64] In accordance with such actions it has been shown to occur in several parts of the brain in addition to the hypothalamus.[65]

It seems clear that this CRF is secreted from neurones and nerve terminals in the median eminence of the hypothalamus, and carried to the anterior lobe of the pituitary by the hypothalamo–hypophyseal portal system. *In vitro* experiments using incubated hypothalamic fragments as well as *in vivo* experiments have been used to study the release of CRF from such neurones, though most such work was carried out before CRF was characterized (and followed biological activity). Various neurotransmitters influence the secretion of CRF, presumably reflecting the nature of the excitatory and inhibitory neurones which make synapses onto the CRF-containing cells (see ref. 66 for a review). Acetyl choline is the main positive stimulus to CRF release; 5-hydroxytryptamine also stimulates CRF release, but probably acts via cholinergic neurones. Noradrenaline (acting via α-adrenergic receptors), γ-aminobutyric acid and, probably, melatonin all inhibit release of CRF (Figure 5.16).

In order to describe the overall physiology of corticotropin release, the hypothalamo–pituitary–adrenocortical axis must be considered as a whole.[66] The main physiological stimulus for corticotropin release is stress. This can be of either neural or systemic origin. 'Neural' stress implies a stimulus which excites corticotropin release by a nervous reflex pathway acting eventually on CRF neurones. 'Systemic' stress results from metabolic changes or tissue damage and may act via a neural reflex leading to CRF neurones, or by direct action (via blood-borne factors) on the pituitary. As mentioned above, in some cases other factors, such as vasopressin, may modulate the actions of CRF on corticotrophs.

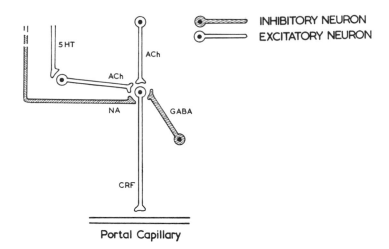

Figure 5.16 A proposed model of the neurotransmitters involved in the release of CRF at the hypothalamus. 5 HT, 5-hydroxytryptamine; ACh, acetylcholine; NA, noradrenaline; GABA, γ-aminobutyric acid. (Reproduced with permission from Jones, M. T. (1978). In *The Endocrine Hypothalamus*, pp. 385–419. Copyright (1978) Academic Press)

   Stress thus causes release (and probably increased synthesis) of CRF and then corticotropin, which will in turn rapidly stimulate glucocorticosteroid production by the adrenal cortex. The response involves a large amplification: release of picogram quantities of the neurotransmitter acetyl choline at synapses on CRF neurones leads rapidly to production of 1–2 μg/minute of glucocorticosteroids in the rat. The glucocorticosteroids then produce changes in carbohydrate, protein and fat metabolism and other physiological responses which represent the physiological adaptation to stress for which the axis was evolved.
   A further important feature of the hypothalmo–pituitary-adrenocortical axis is negative feedback of corticosteroids on corticotropin secretion. This is demonstrated by the enhanced corticotropin secretion which is seen in adrenalectomized animals and which is greatly exaggerated in response to stress and can be reversed by injecting or implanting corticosteroids. The negative feedback by corticosteroids appears to occur at both the pituitary and hypothalamic levels. Thus, direct effects of cortisol on corticotropin secretion from pituitary cells can be demonstrated *in vitro* and *in vivo* (in hypothalamic lesioned animals) and effects on CRF secretion from the hypothalamus can also be shown *in vivo* and *in vitro*. The feedback by corticosteroids shows a biphasic response, which appears to operate at both the hypothalamic and pituitary levels. The rapid response (5–15 minutes) appears to reflect inhibition by corticosteroids of CRF and corticotropin release, while the slow response (more

than 120 minutes) probably represents inhibition of synthesis of these peptides.

Secretion of corticotropin may also be inhibited by a short-loop negative feedback, due to inhibition of CRF release by corticotropin. The various factors regulating the hypothlamo–pituitary–adrenocortical axis are summarized in Figure 5.17.

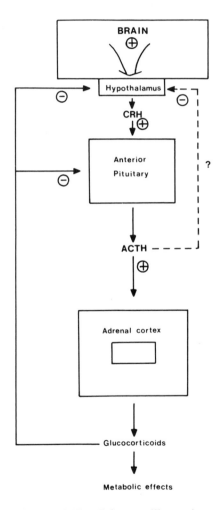

Figure 5.17 Scheme illustrating feedback inhibition of corticosteroids on corticotropin secretion at the level of the hypothalamus and anterior pituitary. CRH, corticotropin releasing factor

5.4.2   α-Melanotropin[50,67]

α-MSH is produced and released from melanotrophs in the pars intermedia. Anatomical investigations have shown that unlike the anterior lobe of the pituitary, the pars intermedia is extensively innervated by nerve axons which have their origin primarily in the paraventricular nucleus of the hypothalamus, and which connect this region to the pars intermedia via the pituitary stalk and the posterior lobe. On the other hand, the pars intermedia is not supplied with blood by the hypophysial portal system. The mechanism for control of α-MSH secretion must therefore be rather different from that for control of the anterior lobe hormones.

If the connection between the hypothalamus and the pituitary is removed by pituitary stalk section or by transplanting the pituitary away from its normal location, secretion, and subsequently biosynthesis, of α-MSH increases, and is sustained. This suggests that α-MSH is under inhibitory control from the brain, probably the hypothalamus. Hypothalamic extracts can themselves inhibit the release of MSH both *in vivo* and *in vitro*. From such extracts several inhibitory peptides (MSH-release inhibiting factors (MIFs)) have been isolated, the most potent and prevalent of which is an amidated tripeptide corresponding to the C-terminus of oxytocin: $H_2N–Pro–Leu–Gly–CONH_2$. An MSH-releasing factor (MRF) also appears to occur in the hypothalamus, and a possible candidate for this is a pentapeptide corresponding to the N-terminus of oxytocin: $H_2N–Cys–Tyr–Ile–Gln–Asn–COOH$. Enzymes capable of cleaving oxytocin to MIF and MRF have been described in the hypothalamus, and levels of these two peptides accord reasonably well with the rate of release of MSH from the intermediate lobe. Nevertheless, considerable reservations remain about the true role of these peptides; oxytocin concentrations are high in the hypothalamus, and fragments of the hormone could be merely degradation products.

Catecholamines, especially dopamine, can also inhibit α-MSH secretion, although at some concentrations catecholamines, especially β-adrenergic agonists, may also stimulate such secretion. Since the pars intermedia is well innervated but not supplied with blood from the hypothalamus, it has been postulated that factors that regulate α-MSH secretion are transported from the hypothalamus to the pars intermedia along nerve axons (in a similar fashion to the transport of oxytocin and vasopressin to the posterior lobe). Most melanotrophs have nerve endings/synapses abutting on them; the majority of such synapses are dopaminergic, and peptidergic synapses, containing granules which are peptide in nature, are rare.[68] Thus, direct neuronal control of α-MSH release via dopamine is probable, but that via oxytocin-related peptides is less likely. It is possible, however, that the oxytocin-derived MIF and MRF are carried to the pars intermedia in the blood, which reaches this lobe mainly via the posterior lobe.

The physiological factors stimulating or inhibiting MSH secretion and production include stress, light (including colour of background and day length), suckling, copulation and dehydration. Since the role of MSH in most mammals is unclear, it is difficult to relate these factors to the biological effects of the hormone. However, in lower vertebrates, where MSH controls the skin-darkening, it is clear that the main stimulus to MSH release comes via the eye, with the pineal gland possibly playing a role in some species. The eye detects changes in light intensity or in the darkness of the background. Light reflected from the background strikes the upper part of the retina; animals on a light background therefore receive more light on this region of the retina which leads to increased production and secretion of MIF and/or dopamine, followed by inhibition of MSH secretion and consequently skin-lightening. Decrease in light intensity produces the reverse effect. In the case of mammals, such as the weasel, which show seasonal changes in coat colour, the effect may be reversed. Here, decreasing photoperiod in autumn (i.e. *decrease* in light detected) leads to increased MIF production, decreased MSH secretion and eventually reduction in the pigment content of the pelage, leading to the white winter coat. Studies with a range of neurotransmitter agonists and antagonists suggest that the neuronal circuits involved in the inhibition of MSH secretion involve synapses operating with 5-hydroxytryptamine, γ-aminobutyric acid and acetyl choline as well as the catecholamines already mentioned.

Finally, it should also be mentioned that there is some evidence for a short-loop feedback by which MSH inhibits its own secretion. Whether this operates at the pituitary or hypothalamic level is not yet clear.

5.4.3 Lipotropin, β-MSH, endorphins and enkephalins

In the pituitary most of the peptides of the corticotropin–MSH–endorphin family (except for enkephalins, dynorphin and α-neoendorphins) are produced mainly in corticotrophs and melanotrophs, from a common precursor, as has been discussed already. Different proportions of the different hormones occur in different cell types, due to variations in processing. A good deal of evidence suggests that granules in corticotrophs contain peptides additional to corticotropin (including β-lipotropin, a large peptide containing γ-MSH and some β-MSH and endorphins) and that granules in melanotrophs contain other peptides (including CLIP, endorphins and in some species β-MSH and glycosylated CLIP) as well as α-MSH. Release of these other peptides appears to occur along with corticotropin (from corticotrophs) and α-MSH (from melanotrophs) and is thus controlled by the same factors as have been discussed for corticotropin and MSH. Thus, Eipper and Mains have found that cells of their mouse tumour line, in response to CRF, secrete not only corticotropin, but also γ-lipotropin, β-endorphin and the 16K fragment from pro-opiocortin.[1,69] Rat intermediate lobe fragments secreted α-MSH, CLIP and endorphins.

Lowry and his colleagues found that perifused cells from rat anterior pituitary lobes secreted corticotropin, lipotropin and some endorphin, and that vasopressin stimulated secretion of all of these; similar results were obtained with cells derived from a human pituitary tumour. They also found that dopamine coordinately inhibited secretion of α-MSH, CLIP, γ-LPH and immunoreactive β- and α-endorphin from rat intermediate lobe cells, while high (depolarizing) potassium concentrations increased secretion of these peptides.[50]

The coordinate release of the various peptides seems to imply a surprising lack of specificity. Quite how the system works at the physiological level is not clear. It remains possible that the only physiologically active peptide being secreted from corticotrophs is corticotropin, and that the other peptides secreted with it are fragments which have to be made at the same time as corticotropin but which either lack significant physiological activity (β-lipotropin?) or have been inactivated by further cleavage or acetylation (endorphin derivatives?). Further work will undoubtedly resolve what is at present a fascinating, but rather perplexing, situation.

Factors controlling the release of enkephalins and endorphins from neurones have not been delineated, although some evidence has been adduced that the dipeptide Tyr–Arg may stimulate release of [Met]-enkephalin in the brain.[70] At this level these opioid peptides must be viewed as neurotransmitters rather than hormones (although the distinction seems less real every day) and as such their release will be at least partly controlled by nervous (electrical) signals rather than by chemical ones. It is clear that release of enkephalin from brain slices can be induced by neuronal depolarization using high potassium concentrations, as would be expected for a neurotransmitter. Stress or pain leads to secretion of enkephalins and endorphins into the cerebrospinal fluid, where they may have an analgesic action, but the mechanisms controlling such secretion are not well understood.[70]

6. THE BIOCHEMICAL MODE OF ACTION OF CORTICOTROPIN ON CORTICOSTEROIDOGENESIS[13,14,71,72]

As described in Section 2, corticotropin has various effects on the adrenal cortex, including stimulation of adrenal corticosteroid formation and content, increased blood flow, trophic effects and depletion of adrenal ascorbic acid content. Whether these (and other) effects are regulated directly and independently (i.e. whether they are all primary effects) is not yet clear. A great deal of effort has been expended on the biochemical basis of the effects of corticotropin on adrenal corticosteroidogenesis and this topic will be considered in some detail here.

6.1 Experimental approaches

The experimental methods used have included both *in vivo* and *in vitro* approaches. Both have advantages and disadvantages. Intact or quartered adrenal glands incubated *in vitro* have proved fruitful systems for biochemical studies, but problems may arise due to restricted diffusion into the gland and to accumulation of products in the incubation medium which may give rise to feedback inhibition on the gland. In order to avoid such problems dispersed-cell systems have been devised in which adrenal cells (mainly from the zona reticularis and zona fasciculata—Figure. 5.3) are dispersed by enzymatic and/or mechanical means. In addition, superfusion (of adrenal cells or glands) has been used to remove products (corticosteroids, etc.) from the vicinity of the cells that have secreted them. Longer-term cell culture of adrenocortical cells has also been valuable in the study of the actions of corticotropin, including both primary culture of normal cells and long-term culture of cell lines derived from adrenal tumours.

Using such *in vitro* systems, corticotropin can be shown to have direct effects on the cells of the adrenal cortex, which clearly reflect the actions of the hormone seen *in vivo*. Stimulation of corticosteroidogenesis is probably the most dramatic of these effects, and in many *in vitro* systems corticotropin can be shown to have rapid effects on this process at concentrations well within the physiological range.

6.2   Corticosteroid biosynthesis

The biochemical pathways of steroid metabolism in the adrenal gland have been elucidated fairly fully, and were summarized in Figure 5.5. The pathways were worked out partly using cell-free systems, in which steroidogenic responses to corticotropin cannot be demonstrated. The main adrenal steroids are formed from cholesterol, which may be obtained from the blood, or synthesized from acetate within the adrenal cortex itself; the relative importance of these two sources of adrenal cholesterol varies considerably according to the species. Considerable amounts of cholesterol are stored in the adrenal gland in esterified form.

Conversion of cholesterol to pregnenolone occurs within the mitochondria and requires NADPH and cytochrome P-450. Further metabolism of pregnenolone occurs extramitochondrially, but the final steps leading to production of corticosterone, cortisol and cortisone (the main mammalian corticosteroids) and aldosterone occur within the mitochondria again.

Examination of the scheme of Figure 5.5 suggests several points at which corticotropin could act; two of these are likely sites of (indirect) action of the hormone. First, availability of steroid precursor (cholesterol) may be increased by hydrolysis of cholesteryl esters (a point of action which appears to be independent of protein synthesis, since it is not blocked by cycloheximide) or by increased uptake from the blood. Secondly, a step in the conversion of cholesterol to pregnenolone, involving side-chain cleavage, is probably accelerated by the hormone—an effect which appears to require protein synthesis, since it can be blocked by cycloheximide. The side-chain cleavage is probably the main rate-limiting step in steroidogenesis; it involves an enzyme complex that includes cytochrome P-450.

6.3 Interaction between corticotropin and the cells of the adrenal cortex; receptors and adenylate cyclase[73,74]

As in the case of many other polypeptide hormones, receptors for corticotropin are thought to be located on the outside of the plasma membrane of cells of the adrenal cortex. The hormone can thus interact with the receptor without entering the cell. The occurrence of specific corticotropin receptors on the membrane of the adrenal cell (or, more precisely, an adrenal tumour cell line) was demonstrated by Lefkowitz and his colleagues in 1970,[9] who have both investigated the properties of such receptors and used them as the basis of a radioreceptor assay for corticotropin, using ^{125}I-labelled corticotropin. The retention of biological activity by corticotropin covalently bound to large, insoluble polyacrylamide or agarose particles has demonstrated that the receptors for the hormone are on the outside of the cell (since the insolubilized hormone could not enter the adrenal cells); such experiments have been subject to some criticism, however, because of the possibility that a small proportion of the hormone, sufficient to stimulate steroidogenesis, may be detached from the particles during the course of an experiment.

Studies on the binding of ^{125}I-labelled corticotropin to subcellular particles have demonstrated that the receptor is associated with the plasma membrane fraction. The corticotropin-responsive adenylate cyclase of the cell is also in this fraction, and the evidence for a direct association between the receptor and this enzyme appears strong. In particular, in some membrane preparations, adenylate cyclase activity can be stimulated directly by corticotropin. Calcium may be involved in the interaction between the receptor and adenylate cyclase. There appear to be at least two types of corticotropin receptor on the adrenal cell, of high affinity (association constant about 10^{12} M^{-1}) and low affinity (association constant about 10^8 M^{-1}).The low-affinity receptors are present in large excess, but their physiological role is in doubt. It remains possible that low- and high-affinity receptors are different forms of the same macromolecule, and not all investigators agree that the two can be distinguished.

The evidence currently available thus suggests that the cellular receptor for corticotropin is similar in its general (but not specific) location and properties to receptors for many other polypeptide hormones. Further light has been shed on corticotropin–receptor interactions by the use of corticotropin analogues (see Chapter 15); correlation between the effects of corticotropin analogues on receptor binding, stimulation (or inhibition) of adenylate cyclase and corticosteroidogenesis provides important evidence of the close functional association between these processes, but some work with analogues does support the possibility that there are at least two distinct receptors on the adrenal cortical cell.

6.4   Cyclic AMP as a second messenger in the action of corticotropin[75-77]

Recognition that cyclic AMP acts as a second messenger in the mode of action of adrenaline and glucagon on glycogenolysis in muscle and liver led to suggestions that this agent mediates the actions of many other hormones. A role for cyclic AMP in the actions of corticotropin was proposed by Haynes and Berthet in 1957, who suggested a mechanism similar to that operating in the case of adrenaline acting on the liver, in which elevated cyclic AMP levels lead to activation of phosphorylase, increased glucose metabolism and increased availability of NADPH, which is required for corticosteroidogenesis. In fact, however, corticotropin has no effect on levels of phosphorylase activity in the adrenal cortex and it now seems unlikely that modulation of NADPH levels plays a major role in the action of the hormone.

The involvement of cyclic AMP as a mediator of corticotropin action has withstood the test of time, however. The following subsections summarize the experimental evidence for this role.

6.4.1   Actions of cyclic AMP on corticosteroidogenesis

In many *in vitro* systems cyclic AMP stimulates steroidogenesis by adrenocortical cells. Very high concentrations have to be used, but this may be due to low permeability of the cell membrane or rapid degradation of cyclic AMP by phosphodiesterase once it has entered the cell; dibutyryl cyclic AMP, which is lipophilic and probably enters the cell more readily, is active at much lower concentrations. The actions of cyclic AMP on corticosteroidogenesis resemble those of corticotropin in many details. Thus, cyclic AMP like corticotropin appears to act on the conversion of cholesterol to pregnenolone, and its action can be blocked by inhibitors of protein biosynthesis such as cycloheximide. Studies with superfused adrenal glands have shown that response to cyclic AMP is more rapid than that to corticotropin, which might be expected if the nucleotide is playing an intermediary role.

Intracellular cyclic AMP levels can also be elevated by inhibitors of phosphodiesterase activity such as theophylline, although this effect is only observed under rather restricted conditions. Theophylline can stimulate steroidogenesis by the adrenal cortex both *in vivo* and *in vitro*, and can markedly potentiate the actions of corticotropin on this process. This provides further evidence for an action of cyclic AMP on adrenocorticosteroidogenesis, although the fact that theophylline can affect many processes within cells, and under some circumstances may inhibit rather than stimulate adrenal steroidogenesis, means that such a conclusion must be a tentative one. Cholera toxin, which stimulates adenylate cyclase, also stimulates corticosteroidogenesis.

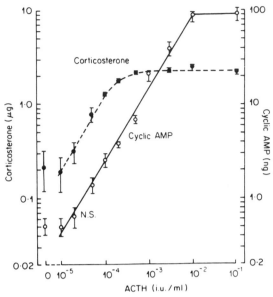

Figure 5.18 Dose–response curves showing the effects of corticotropin on corticosterone secretion by (- - - ● - - -) and cyclic AMP levels in (———○———) isolated rat adrenal cells. Note the logarithmic scales on the ordinates. Corticosterone production is maximal at a concentration of corticotropin that induces only about 5% of the maximal cyclic AMP level. (Reproduced with permission from Mackie, C., Richardson, M. C. and Schulster, D. (1972) *FEBS Lett.* **23**, 345–348. Copyright (1972) Federation of European Biochemical Societies)

6.4.2   *Actions of corticotropin on cyclic AMP levels*

When adrenal slices are incubated with corticotropin, their cyclic AMP content increases, and a similar effect can be shown with various other adrenal systems. The effect is rapid, and elevation of cyclic AMP concentration precedes stimulation of steroidogenesis. The relative potencies of many corticotropin analogues in stimulating cyclic AMP accumulation and steroidogenesis are very similar, although more recently analogues have been found in which the two activities are not closely correlated.[78] Corticotropin also caused elevated levels of cyclic AMP in the adrenal cortex of hypophysectomized rats *in vivo*. It has been shown in many systems that dose–response curves for the stimulation of corticosteroidogenesis or cyclic AMP production by corticotropin are parallel, but it has also been repeatedly demonstrated *in vivo* and *in vitro* that at a corticotropin concentration which induces a maximal corticosteroidogenic response, the cyclic AMP production response is far from maximal (only about 5% of maximum; Figure 5.18). This rather surprising discrepancy may be a consequence of the presence of 'spare receptors'—binding of corticotropin to a small fraction of the total receptor population seems to be sufficient to cause a maximal steroidogenic response. But the biological reason for having such an excess capacity for binding the hormone and producing cyclic AMP is not clear.[74]

6.4.3 Stimulation of adenylate cyclase activity by corticotropin

It has already been mentioned that some adrenocortical membrane preparations that can bind corticotropin (and therefore presumably contain appropriate receptors) also possess adenylate cyclase. Addition of corticotropin to such membranes can stimulate the activity of this enzyme markedly. Corticotropin can thus stimulate the enzyme responsible for producing cyclic AMP, providing further evidence for a role for this nucleotide; this is further supported by the observation that many analogues of corticotropin stimulate the adenylate cyclase to approximately the same extent that they stimulate steroidogenesis. No clear evidence has been proposed for an inhibition of phosphodiesterase in cells of the adrenal cortex; it therefore seems that cyclic AMP levels are modulated mainly by accelerating formation of the nucleotide rather that inhibiting its breakdown.

6.4.4 Cell lines with defective adenylate cyclase

Further evidence supporting the importance of adenylate cyclase and cyclic AMP as mediators of the actions of corticotropin has come from studies on mutants of a cell line derived from a mouse adrenal cortex (Y1 cells).[76] This cell line can be cultured *in vitro*, and responds to corticotropin with increased steroid output and elevated cyclic AMP levels. Mutants of this cell line (Y1(Cyc)) have been obtained which lack the ability to respond to corticotropin. The defect has been shown to lie in the coupling of adenylate cyclase to the corticotropin receptor. Cyclic AMP derivatives can still stimulate steroidogenesis in these mutant cells.

6.4.5 Alternative second messengers

Although the bulk of the evidence suggests that cyclic AMP plays a major role as a secondary messenger for corticotropin, some doubts remain. The dose–response curves for induction of steroidogenesis and cyclic AMP production differ markedly in terms of the sensitivity of the hormone (Figure 5.18), although some authors, at least, claim that even doses of corticotropin causing only a marginal stimulation of steroidogenesis also give some elevation of cyclic AMP levels.[74] Furthermore, some analogues of corticotropin, such as

NPS-corticotropin (corticotropin substituted on Trp[9] with the o-nitrophenyl-sulphenyl group) and corticotropin 6–24, are much more effective as steroidogenic agents than they are in stimulating cyclic AMP production, compared with the natural hormone.[76,78]

These and other observations have led to suggestions that there may be alternative or additional second messengers for corticotropin in the adrenal cell. One such hypothesis proposes that there are two distinct receptors for corticotropin, one linked to adenylate cyclase and the other operating via a different second messenger; stimulation of cyclic AMP production via the first of these receptors would be sufficient to give a maximal steroidogenic response, but a similar response could also be produced without cyclic AMP production, via the second receptor. As an alternative hypothesis, the 'compartment guidance concept'[74] proposes that elevation of cyclic AMP levels is essential for stimulation of steroidogenesis but that a second factor is also required; again, two receptors may be involved, but operation via a single receptor with some distinct binding sub-sites would also be possible.

The nature of the alternative second messenger that may be involved in corticotropin action has not been elucidated, although several candidates have been proposed including cyclic GMP, Ca^{2+}, prostaglandins, and products of phosphatidyl inositol metabolism.

6.5   Are cyclic-AMP-dependent protein kinases involved in the actions of corticotropin?[13,76,80]

It has been clearly established in several systems where cyclic AMP acts as a second messenger of hormone action that it acts primarily by activating a protein kinase. This in turn catalyses the phosphorylation of key enzymes (or other proteins) in the cell, thereby activating (or inhibiting) them and accelerating (or inhibiting) the processes that they control (Chapter 16). There is considerable evidence that such a mechanism operates in the adrenal cortex.

A specific cyclic AMP-binding protein has been identified in the adrenal cortex. This has been purified and characterized and is similar to the corresponding protein in other tissues. It binds tritiated cyclic AMP non-covalently, with a dissociation constant, K_d, of 3×10^{-8}M. It occurs partly associated with a protein kinase (one of several) which can be assayed by its ability to phosphorylate histones and other exogenously supplied proteins, and which is activated by cyclic AMP. The association between the cyclic AMP-binding protein and the protein kinase has been confirmed by isolating the complex. The two appear to dissociate in the presence of cyclic AMP, with activation of the protein kinase. Conversely, addition of binding protein to the kinase subunit (in the absence of cyclic AMP) leads to inhibition of the enzyme.

Further evidence for the importance of cyclic AMP-dependent protein kinase in the action of corticotropin has been provided by work on mutants of the mouse adrenal cell line, Y1, which has already been referred to.[76,80] Mutants of this cell line (Y1(Kin)) have been isolated which have an intact corticotropin-sensitive adenylate cyclase system but in which the cyclic AMP-dependent protein kinase has a very low affinity for cyclic AMP. Such mutants show a good response to corticotropin with regard to cyclic AMP accumulation, but the steroidogenic response is much impaired. Cyclic AMP derivatives show similarly impaired ability to stimulate steroidogenesis.

It thus seems likely that a major role of cyclic AMP within the adrenal cortex is to activate protein kinase by dissociating it from its inhibitory subunit. The picture is complicated by several factors, however, including the observations that several other, cyclic AMP-independent, protein kinases occur in the adrenal cortex and that measured cyclic AMP levels in the *unstimulated* adrenal cortex appear to be high enough to activate much of the dependent protein kinase (although other modulating factors or subcellular compartmentation of cyclic AMP may invalidate this conclusion).

The cyclic AMP-dependent protein kinase presumably activates one or more key proteins in the adrenal cell. The nature of these is as yet unknown, but some candidates will be considered in the next section.

6.6 Sites of action of corticotropin on steroidogenesis

It appears then that a major role of corticotropin is to cause elevation of cyclic AMP levels, which in turn activate protein kinase in the adrenal cortex. Protein kinase then presumably phosphorylates and activates key proteins in the corticosteroidogenic pathway. Evidence has been provided that corticotropin or cyclic AMP derivatives do indeed stimulate phosphorylation of a considerable number (about 1%) of the proteins in the adrenal cortex,[81] and key proteins in the pathway of steroid synthesis may be among these. Two sites in adrenal corticosteroidogenesis have been identified as likely targets for the action of corticotropin, one of which involves enhanced mobilization of cholesterol within the gland while the other involves increased conversion of cholesterol to pregnenolone.

6.6.1 Activation of cholesteryl esterase

Cholesterol is stored as esters in lipid droplets within the cells of the adrenal cortex. Hydrolysis of the esters is the first step in mobilization of cholesterol, and is catalysed by the enzyme cholesteryl esterase. Corticotropin enhances cholesterol mobilization by activating cholesteryl esterase, a step which probably involves phosphorylation of the enzyme by protein kinase.[82] However, although increased cholesteryl ester hydrolysis appears to be an important site of action of the hormone, it is by no means the only one, and such hydrolysis may not be rate-limiting in many circumstances.

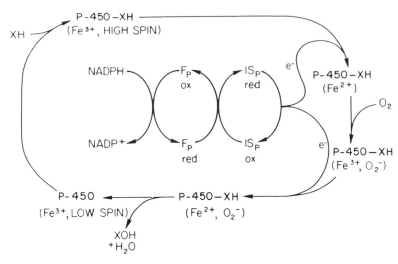

Figure 5.19 Reaction mechanism of hydroxylations catalysed by cytochrome P-450 of mitochondria in the adrenal cortex. XH, substrate; XOH, product; Fp, flavoprotein; ISp, iron-sulphur protein: red, reduced; ox, oxidized. (Reproduced with permission from Simpson, E. R. and Waterman, M. R. (1983) *Can. J. Biochem. Cell. Biol.* **61**, 692–707. Copyright (1983) National Research Council of Canada)

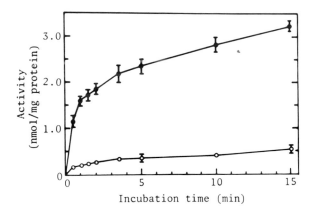

Figure 5.20 Time course of pregnenolone formation by adrenal mitochondria incubated with (●) or without (○) 1 μM activator peptide. (Reproduced with permission from Pederson, R. C. and Browne, A. C. (1983) *Proc. Nat. Acad. Sci. USA* **80**, 1883–1886. Copyright (1983) the authors)

6.6.2 Activation of side-chain cleavage

If adrenal cortex is treated with the protein synthesis inhibitor cycloheximide before corticotropin, the effect of the hormone on corticosteroidogenesis is blocked, despite a clear increase in conversion of cholesteryl esters to cholesterol. Thus, a second site of action of the hormone can be identified. Corticotropin stimulates the conversion of cholesterol to pregnenolone (Figure 5.5), a step which occurs within the mitochondria, so an action at this level can be delineated. Precisely how the hormone acts at this level has not been defined. The main enzyme system involved is the side-chain cleavage enzyme complex, activity of which declines rather slowly after hypophysectomy. However, decline in corticosteroidogenesis is much more rapid than decline in this enzyme activity, so it seems unlikely that direct activation of this enzyme plays a major role in corticotropin action.

Associated with the side-chain cleavage enzyme complex, and involved in conversion of cholesterol to pregnenolone, is cytochrome P-450, which forms a complex with cholesterol and is converted from a 'high spin' to 'low spin' state when cholesterol is converted to pregnenolone.[83] Cytochrome P-450 is associated with an electron transport chain (Figure 5.19) which links oxidation of NADPH to the cholesterol→pregnenolone conversion. There is a good deal of evidence that corticotropin enhances the association of cholesterol with cytochrome P-450, and a cytosolic peptide activator has now been isolated which appears to be involved in this enhanced association.[84] This peptide, of M_r 2200, is labile and stimulates association of cholesterol with cytochrome P-450 in isolated mitochondria. In the presence of NADPH it will stimulate pregnenolone synthesis by isolated mitochondria (Figure 5.20). The concentration (or activity) of the peptide in the adrenal cortex of hypophysectomized rats is enhanced by treatment with corticotropin. It is possible that corticotropin increases the synthesis of this peptide at the transcriptional or translational level, or its formation from a larger precursor. In these cases the hormone could act by stimulating phosphorylation of the key proteins involved in forming the activating peptide. Activation of the peptide directly by phosphorylation does not appear to be completely ruled out yet, however.

Figure 5.21 presents an overall summary of the sites at which corticotropin may act in the cells of the adrenal cortex.

6.7 Biochemical mechanisms of the trophic actions of corticotropin

The information currently available about the mechanism of the action of corticotropin is complex. The complexity is due partly to the multiple, pleiotropic effects of the hormone, which include not only the rapid effects on steroidogenesis considered here but also longer-term trophic effects on the growth of the adrenal cortex and its content of important enzymes and structural components. Some of the effects seen on receptors, second messengers, etc. may be

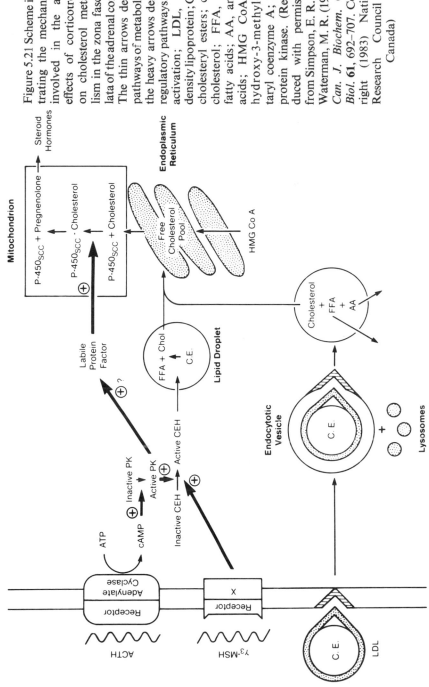

Figure 5.21 Scheme illustrating the mechanisms involved in the acute effects of corticotropin on cholesterol metabolism in the zona fasciculata of the adrenal cortex. The thin arrows denote pathways of metabolism; the heavy arrows denote regulatory pathways. $\oplus$ activation; LDL, low density lipoprotein; C.E., cholesteryl esters; chol, cholesterol; FFA, free fatty acids; AA, amino acids; HMG CoA, 3-hydroxy-3-methyl-glutaryl coenzyme A; PK, protein kinase. (Reproduced with permission from Simpson, E. R. and Waterman, M. R. (1983), *Can. J. Biochem. Cell. Biol.* **61**, 692–707. Copyright (1983) National Research Council of Canada)

involved in mediating effects other than enhanced corticosteroidogenesis. The mechanisms whereby corticotropin stimulates growth of and long-term changes in the adrenal cortex have been less fully studied than the mechanism of its steroidogenic actions. However, long-term actions on the activity of various enzymes, including DNA polymerase and several of the enzymes involved in steroidogenesis, have been demonstrated, as have marked changes in cell shape and cytoskeletal organization.

7.   THE BIOCHEMICAL MODE OF ACTION OF MELANOTROPINS[20,85]

In their actions on skin colour melanotropins act to cause dispersion of pigment granules and synthesis of the pigment melanin. The first effect has been most studied in preparations of amphibian skin. Effects on melanin production are seen also in mammalian systems and have been investigated using cultured mouse melanoma cells. Most work on the mechanism of action of melanotropin has used α-melanotropin.

7.1   Receptors for melanotropin

If α-melanotropin is covalently bound to a large peptide such as tobacco mozaic virus it retains its biological activity, suggesting that receptors for the hormone are located on the outside of the target cell. Indeed, such conjugates sometimes demonstrate enhanced and prolonged potency ('superpotency') probably because internalization and degradation is prevented and because complete dissociation of the whole conjugate, carrying many associated melanotropin molecules, from the target cell is unlikely.[17]

Binding studies using labelled α-melanotropin have also demonstrated the occurrence of receptors for the hormone on melanocytes, although radioiodinated preparations of melanotropin that retain biological activity have proved difficult to prepare. Receptors on melanocytes have also been covalently labelled using photoaffinity labelling, a process which is accompanied by irreversible stimulation of the target cells.[17]

Information about the nature of receptors for melanotropin has also been obtained from studies on structure–function relationships and is discussed further in Chapter 15.

7.2   Actions of melanotropin on pigment granule dispersion in amphibian skin

A good deal of work suggests that cyclic AMP mediates the actions of melanotropins on melanin dispersion in amphibian melanophores. Injection of α-MSH into frogs (*Rana pipiens*) causes both darkening and elevation of cyclic AMP levels of the dorsal skin but not the ventral skin. The relative effectiveness of α- and β-melanotropin and corticotropin on skin darkening correlates well with their effectiveness in raising cyclic AMP levels. In *Xenopus,* the cyclic AMP content of skin from dark-adapted animals was about twice that from light-adapted animals. Furthermore, cyclic AMP and its derivatives will cause dispersion of melanin granules in *in vitro* MSH (frog skin) assays. Melatonin, the pineal hormone which causes aggregation of melanin granules in frog melanophores, and thus antagonizes the effect of melanotropin, inhibits the effects of the hormone on both melanin dispersion and on cyclic AMP levels in isolated frog skin.

Inorganic cations, especially sodium and calcium, also seem to be important in the action of MSH. Sodium may be required in binding of MSH to its membrane receptor, while calcium appears to be necessary for effects subsequent to receptor binding. The calcium ionophore A23187 is a potent stimulator of melanin dispersion. There is also some evidence that microtubules and/or microfilaments are involved in the melanin dispersion response. Presumably cyclic AMP serves to activate a protein kinase which in turn phosphorylates key proteins within the melanophore; these may include components of the cytoskeleton and a calcium pump. It is notable that the processes involved in melanin dispersion are similar to those involved in exocytosis—both processes require movement of subcellular granules within the cell—and similar mechanisms may apply.

7.3   Actions of melanotropin on cultured mouse melanoma cells[35,86]

Melanotropins also have specific biochemical effects on cultured mouse melanoma cells, and these have been investigated in some detail. β-melanotropin induces increased cyclic AMP levels, tyrosinase activity and melanin content in such cells. The order of events has been established as binding of melanotropin to membrane receptors, activation of adenylate cyclase and, after a few hours delay, activation of tyrosinase. Binding of melanotropin to these cells appears to occur in a region of the cell surface overlying the Golgi complex, and is followed by internalization in vesicles—a process which may be an important step in the action of the hormone. The actions of melanotropin on cultured melanoma cells can also be produced by treatment with agents which increase intracellular cyclic AMP concentration (dibutyryl cyclic AMP, theophylline, cholera toxin) and these agents also increase the number of melanotropin receptors on the cells. Clearly, cyclic AMP is implicated in these actions of melanotropin, and presumably protein phosphorylation, but the proteins that are phosphorylated have not yet been identified. Use of inhibitors of protein and nucleic acid synthesis has provided some evidence implicating these processes in the mechanism of action of the hormone in melanoma cells, and the rather slow actions of melanotropin on pigment production in these cells would accord with this.

8. CORTICOTROPIN, MELANOTROPINS AND OPIOID PEPTIDES IN BLOOD[87,88]

Corticotropin occurs in the circulation in several forms, some of which are of greater molecular weight than the 39-residue peptide. These large forms are probably related to the precursor molecule, pro-opiocortin, and may represent intermediate forms in the degradation of the hormone. There is no evidence for the occurrence in blood of carrier proteins for corticotropin or other members of this peptide family. Resting levels of ACTH in human plasma, as measured by various assays, are approximately 10 pM (50 pg/ml), though there is considerable diurnal fluctuation. The half-life ($t_{1/2}$) of the hormone in blood is short, about 3–8 minutes.

α-Melanotropin is barely detectable in the pituitary or blood of the human adult, as previously discussed. Levels of the various opioid peptides in blood have been investigated, although such measurements are made particularly difficult by the immunological cross-reaction seen between the various peptides, and the possibility that some of the circulating forms represent inactivated molecules. Lipotropin, β-endorphin and [Met]-enkephalin have all been measured in normal plasma, resting levels being approximately 20, 20, and 200 pM, respectively (although some authors have found difficulty in detecting β-endorphin in the circulation at all). Circulating levels of corticotropin, lipotropin and β-endorphin are increased about eight-fold by insulin-induced hypoglycaemia, but those of [Met]-enkephalin are unchanged by this stimulus; this accords with the expected occurrence of the first three in the same secretory granules. Similarly, administration of the synthetic glucocorticoid dexamethasone suppresses circulating levels of corticotropin, lipotropin and β-endorphin about eight-fold, but not [Met]-enkephalin. It is notable that most of the peptides of this family can also be detected in quite high concentrations in cerebrospinal fluid.

In some pathological conditions plasma concentration of these peptide hormones may be markedly elevated. Thus, in patients with Addison's disease, levels of corticotropin, β-endorphin and β-lipotropin may be up to 100-fold higher than those in normal individuals, though levels of [Met]-enkephalin are not elevated. Levels may also be elevated in some patients with non-pituitary tumours (ectopic production). Thus, secretion of corticotropin by many lung carcinomas has been observed.

REFERENCES

1. Eipper, B. A. and Mains, R. E. (1980) *Endocrine Rev.,* **1**, 1–27.
2. Numa, S. and Nakanishi, S. (1981) *Trends in Biochem. Sci.,* **6**, 274–277.
3. Nakanishi, S., Inoue, A., Kita, T., Nakamura, M., Chang, A. C. Y., Cohen, S. N. and Numa, S. (1979) *Nature,* **278**, 423–427.

4. Noda, M., Furutani, Y., Takahashi, H., Toyosato, M., Hirose, T., Inayama, S., Nakanishi, S. and Numa, S. (1982) *Nature*, **295**, 202–206.
5. Gubler, U., Seeburg, P., Hoffman, B. J., Gage, L. P. and Udenfriend, S. (1982) *Nature*, **295**, 206–208.
6. Kakidani, H., Furutani, Y., Takahashi, H., Noda, M., Morimoto, Y., Hirose, T., Asai, M., Inayama, S., Nakanishi, S. and Numa, S. (1982) *Nature*, **298**, 245-249.
7. White, W. F. (1953) *J. Amer. Chem. Soc.*, **75**, 503–504.
8. Sayers, G., (1977) *Ann. N.Y. Acad. Sci.*, **297**, 220–241.
9. Lefkowitz, R. J., Roth, J. and Pastan, I. (1970) *Science*, **170**, 633–635.
10. West, C. D. and Dolman, L. I. (1977) *Ann. N.Y. Acad. Sci.*, **297**, 205–219.
11. Hofmann, K. (1974) In: *Handbook of Physiology, Section 7: Endocrinology*, Vol. 4, Part 2 (Eds. Knobil, E, and Sawyer, W. H.). American Physiological Society, Washington D.C., pp. 29–58.
12. Schwyzer, R. (1977) *Ann. N.Y. Acad. Sci.*, **297**, 3–26.
13. Schulster, D., Burstein, S. and Cooke, B. A. (1976) *Molecular Endocrinology of the Steroid Hormones*. Wiley, London.
14. Harding, B. W. (1979) In: *Endocrinology*, Vol. 2 (Ed. DeGroot, L. J. *et al.*). Grune & Stratton, New York, pp. 1131–1137.
15. Ramachandran, J., Rao, A. J. and Liles, S. (1977) *Ann. N.Y. Acad. Sci.*, **297**, 336–348.
16. Li, C. H. (1978) In: *Hormonal Proteins and Peptides*, Vol. 5. (Ed. Li, C. H.) Academic Press, New York, pp. 1–33.
17. Eberle, A. N. (1981) *CIBA Foundation Symposium*, **81**, 13–31.
18. Silman, R. E., Street, C., Holland, D., Chard, T., Falconer, J. and Robinson, J. S. (1981) *CIBA Foundation Symposium*, **81**, 180–190.
19. Lowry, P. J., Silman, R. E., Hope, J. and Scott, A. P. (1977) *Ann. N.Y. Acad. Sci.*, **297**, 49–62.
20. Novales, R. R. (1974) In: *Handbook of Physiology, Section 7: Endocrinology*, Vol. 4, Part 2 (Eds. Knobil, E. and Sawyer, W. H.). American Physiological Society, Washington D.C., pp. 347–366.
21. Baker, B. I. (1981) *CIBA Foundation Symposium*, **81**, 166–175.
22. Geschwind, I. I., Huseby, R. A. and Nishioka, R. (1972) *Recent Progr. Hormone Res.*, **28**, 91–130.
23. Hadley, M. E., Heward, C. B., Hruby, V. J., Sawyer, T. K. and Yang, Y. C. S. (1981) *CIBA Foundation Symposium*, **81**, 244–258.
24. Li, C. H. (1982) In: *Biochemical Actions of Hormones*, Vol. 9. (Ed. Litwack, G.). Academic Press, New York, pp. 1–41.
25. Van Wimersma Greidanus, T. B., de Rotte, G. A., Thody, A. J. and Eberle, A. N. (1981) *CIBA Foundation Symposium*, **81**, 277–289.
26. Beloff-Chain, A., Morton, J., Dunmore, S., Taylor, G. W. and Morris, H. R. (1983) *Nature*, **301**, 255–258.
27. Chrétien, M. and Lis, M. (1978) In: *Hormonal Proteins and Peptides*, Vol. 5 (Ed. Li, C. H.). Academic Press, New York, pp. 75–102.
28. Blake, J. and Li, C. H. (1983) *Proc. Nat. Acad. Sci., U.S.A.*, **80**, 1556–1559.
29. Li, C. H. (Ed.) (1981) *Hormonal Proteins and Peptides*, Vol. 10. Academic Press, New York.
30. Imura, H. and Nakai, Y. (1981) *Ann. Rev. Physiol.*, **43**, 265–278.
31. Beaumont, A. and Hughes, J. (1979) *Ann. Rev. Pharmacol. Toxicol.*, **19**, 245–267.
32. Hughes, J. (Ed.). (1983) *Opioid Peptides. Brit. Med. Bull.*, **39**, 1–106.
33. Hughes, J., Smith, T. W., Kosterlitz, H. W., Fothergill, L. A., Morgan, B. A. and Morris, H. R. (1975) *Nature*, **258**, 577–579.

34. Smyth, D. G., Zakarian, S., Deakin, J. F. W. and Massey, D. E. (1981) *CIBA Foundation Symposium*, **81**, 79–96.
35. Pardridge, W. M. (1983) *Ann. Rev. Physiol.*, **45**, 73–82.
36. Chang, K.-J., Hazum, E. and Cuatrecasas, P. (1980) *Trends in Neurosciences*, **3**, 160–162.
37. Snyder, S. H. (1980) *Science*, **209**, 976–983.
38. Enjalbert, A., Ruberg, M., Arancibia, S., Priam, M. and Kordon, C. (1979) *Nature*, **280**, 595–597.
39. Smyth, D. G., Massey, D. E., Zakarian, S. and Finnie, M. D. A. (1979) *Nature*, **279**, 252–254.
40. Mains, R. E. and Eipper, B. A. (1981) *CIBA Foundation Symposium*, **81**, 32–54.
41. Grossman, A. and Rees, L. H. (1983) *Brit. Med. Bull.*, **39**, 83–88.
42. Cuello, A. C. (1983) *Brit. Med. Bull.*, **39**, 11–16.
43. Krieger, D. T. and Liotta, A. S. (1979) *Science*, **205**, 366–372.
44. Krieger, D. T. (1983) *Science*, **222**, 975–985.
45. North, R. A. and Egan, T. M. (1983) *Brit. Med. Bull.*, **39**, 71–75.
46. Orth, D. N. and Nicholson, W. E. (1977) *Ann. N.Y. Acad. Sci.*, **297**, 27–48.
47. Roberts, J. L., Phillips, M., Rosa, P. A. and Herbert, E. (1978) *Biochemistry*, **17**, 3609–3618.
48. Roberts, J. L., Seeburg, P. H., Shine, J., Herbert, E., Baxter, J. D. and Goodman, H. M. (1979) *Proc. Nat. Acad. Sci., USA*, **76**, 2153–2157.
49. Gráf, L. and Kenessey, A. (1981) In: *Hormonal Proteins and Peptides*, Vol. 10 (Ed. Li, C. H.) Academic Press, New York, pp. 35–63.
50. Jackson, S., Hope, J., Estivariz, F. and Lowry, P. J. (1981) *CIBA Foundation Symposium*, **81**, 141—162.
51. Fink, G. (1981) *Nature*, **294**, 511–512.
52. Yasuda, N., Grear, M. A. and Aizawa, T. (1982) *Endocr. Rev.*, **3**, 123–140.
53. Saffran, M. and Schally, A. V. (1955) *Can. J. Biochem. Physiol.*, **33**, 408–415.
54. Guillemin, R. and Rosenberg, B. (1955) *Endocrinology*, **57**, 599–607.
55. Vale, W., Spiess, J., Rivier, C. and Rivier, J. (1981) *Science*, **213**, 1394–1397.
56. Vale, W., Rivier, C., Brown, M. R., Spiess, J., Koob, G., Swanson, L., Bilezikjian, L., Bloom, F. and Rivier, J. (1983) *Recent Progr. Hormone Res.*, **39**, 245–270.
57. Rivier, C., Rivier, J. and Vale, W. (1982) *Science*, **218**, 377–379.
58. Bugnon, C., Fellmann, D., Gouget, A. and Cardot, J. (1982) *Nature*, **298**, 159–161.
59. Furutani, Y., Morimoto, Y., Shibahara, S., Noda, M., Takahashi, H., Hirose, T., Asai, M., Inayama, S., Hayashida, H., Miyata, T. and Numa, S. (1983) *Nature*, **301**, 537–540.
60. Bradbury, A. F., Finnie, M. D. A. and Smyth, D. G. (1982) *Nature*, **298**, 686–688.
61. Labrie, F., Veilleux, R., Lefevre, G., Coy, D. H., Sueiras-Diaz, J. and Schally, A. V. (1982) *Science*, **216**, 1007–1008.
62. Gillies, G. E., Linton, E. A. and Lowry, P. J. (1982) *Nature*, **299**, 355–357.
63. Giguere, V. and Labrie, F. (1982) *Endocrinology*, **111**, 1752–1754.
64. Sutton, R. E., Koob, G. F., Le Moal, M., Rivier, J. and Vale, W. (1982) *Nature*, **297**, 331–333.
65. Hollander, C. S., Audhya, T., Russo, M., Passarelli, J., Nakane, T. and Schlesinger, D. (1983) *Endocrinology*, **112**, 2206–2208.
66. Jones, M. T. (1978) In: *The Endocrine Hypothalamus* (Eds. Jeffcoate, S. L. and Hutchison, J. S. M.). Academic Press, London, pp. 385–419.
67. Taleisnik, S. (1978) In: *The Endocrine Hypothalamus* (Eds. Jeffcoate, S. L. and Hutchison, J. S. M.). Academic Press, London, pp. 421–439.
68. Stoeckel, M. E., Schmitt, G. and Porte, A. (1981) *CIBA Foundation Symposium*, **81**, 101–122.

69. Mains, R. E. and Eipper, B. A. (1979) *Symp. Soc. Exptl. Biol.*, **33**, 37–55.
70. Hughes, J. (1983) *Brit. Med. Bull.*, **39**, 17–24.
71. Simpson, E. R. and Waterman, M. R. (1983) *Canad. J. Biochem. Cell Biol.*, **61**, 692–707.
72. Ascoli, M. (1982) In: *Cellular Regulation of Secretion and Release* (Ed. Conn, P. M.). Academic Press, New York, pp. 409–444.
73. Glynn, P., Cooper, D. M. F. and Schulster, D. (1979) *Mol. Cell. Endocrinol.*, **13**, 99–121.
74. Schulster, D. and Schwyzer, R. (1980) In: *Cellular Receptors for Hormones and Neurotransmitters* (Eds. Schulster, D. and Levitzki, A). Wiley, Chichester, pp. 197–217.
75. Halkerston, I. D. K. (1975) *Adv. Cyclic Nucleotide Res.*, **6**, 99–136.
76. Schimmer, B. P. (1980) *Adv. Cyclic Nucleotide Res.*, **13**, 181–214.
77. Saez, J. M., Morera, A.-M. and Dazord, A. (1981) *Adv. Cyclic Nucleotide Res.*, **14**, 563–579.
78. Bristow, A. F., Gleed, C., Fauchère, J.-L., Schwyzer, R. and Schulster, D. (1980) *Biochem. J.*, **186**, 599–603.
79. Ontjes, D. A., Ways, D. K., Mahaffee, D. D., Zimmerman, C. F. and Gwynne, J. T. (1977) *Ann. N.Y. Acad. Sci.*, **297**, 295–313.
80. Doherty, P. J., Tsao, J., Schimmer, B. P., Mumby, M. C. and Beavo, J. A. (1981) *Cold Spring Harbor Conferences on Cell Proliferation*, Vol. 8 (Protein Phosphorylation), pp. 211–225.
81. Steinberg, R. A. (1981) *Cold Spring Harbor Conferences on Cell Proliferation*, Vol. 8 (Protein Phosphorylation), pp. 179–193.
82. Boyd, G. S. and Gorban, A. M. S. (1980) In: *Molecular Aspects of Cellular Regulation*, Vol. 1 (Ed. Cohen, P.). Elsevier, Amsterdam, pp. 95–134.
83. Simpson, E. R. (1979) *Mol. Cell. Endocrinol.*, **13**, 213–227.
84. Pedersen, R. C. and Brownie, A. C. (1983) *Proc. Nat. Acad. Sci. USA*, **80**, 1882–1886.
85. Hadley, M. E., Heward, C. B., Hruby, V. J., Sawyer, T. K. and Yang, Y. C. S. (1981) *CIBA Foundation Symposium*, **81**, 244–258.
86. Di Pasquale, A., McGuire, J. and Varga, J. M. (1977) *Proc. Nat. Acad. Sci. USA*, **74**, 601–605.
87. Rees, L. H. and Smith, R. (1982) In: *Recent Advances in Endocrinology and Metabolism*, Vol. 2 (Ed. O'Riordan, J. L. H.). Churchill Livingstone, Edinburgh, pp. 1–15.
88. Rees, L. H. and Lowry, P. J. (1979) In: *Hormones in Blood*, Vol. 3, 3rd edn. (Eds. Gray, C. H. and James, V. H. T.). Academic Press, London, pp. 129–178.

Chapter 6

Hormones of the adenohypophysis: The gonadotropins and thyrotropin (and related placental hormones)

1.   INTRODUCTION

The hormones to be considered in this chapter (LH, FSH, TSH and hCG; see Chapter 3 for nomenclature and abbreviations) are chemically the most complex of the pituitary hormones, and indeed of all well-characterized polypeptide hormones. All contain subunits and are glycoproteins—they have several carbohydrate moieties covalently bound to the polypeptide chains, and these structural similarities justify their inclusion in a single protein family. Biologically the two pituitary gonadotropins have similar actions, primarily on the gonads, but the actions of thyrotropin on the thyroid gland are not obviously related to those of the gonadotropins on the gonads.

In addition to the three pituitary glycoprotein hormones, the family also includes placental protein hormones. A chorionic gonadotropin has been identified in many mammals, the human hormone (hCG) being the most studied. Recent work suggests that this may occur in the pituitary and other tissues as well as the placenta. The human placenta may also produce a chorionic thyrotropin, although this has been less fully characterized.

2.   ASSAY[1-3]

The existence of factors in the pituitary which stimulate the gonads and the thyroid gland was recognized early in the study of the pituitary gland. Bioassays based on these effects were developed and used in the isolation of the hormones, and are still in use. Thus, FSH can be assayed by its effects on ovarian weight *in vivo* in suitable animals. *In vivo* LH bioassays exploit effects on prostate weight (via testosterone production), or depletion of the ovarian content of ascorbic acid or cholesterol. Human chorionic gonadotropin can be assayed by methods similar to those available for LH; the now outmoded pregnancy tests using induction of ovulation or sperm expulsion in frogs or toads were in effect 'assays' for chorionic gonadotropin in human urine. Several good *in vivo* and *in vitro* assays for thyrotropin have been described, using effects on thyroid growth or thyroid hormone production.

More recently good radioimmunoassays and radioreceptor assays have been developed for the gonadotropins and TSH, and these are now the methods of choice for many types of study. Detection of hCG in the urine or blood by immunological methods forms the basis of most modern pregnancy tests. It should be noted, however, that the possession of a common subunit by all of these hormones (see below) can lead to some lack of specificity for the radioimmunoassays. The radioreceptor assays[3] have been developed using receptor preparations from the appropriate target organs, but here too there may be overlapping specificity (especially for gonadotropins) because of the existence of two types of receptor in the same preparation, and because of the structural similarity between the hormones.

3. STRUCTURE AND CHEMISTRY

3.1 Isolation[4,5]

The glycoprotein hormones proved difficult to isolate in pure form, partly because the amounts in the pituitary gland are rather small, partly because they are rather unstable and partly perhaps because they tend to cofractionate with other glycoproteins. Good purified preparations were only reliably obtained in the middle 1960s, and although some structural and chemical work was reported earlier, some of this proved rather inaccurate and misleading. Reliable methods have now been developed for the purification of the glycoprotein hormones from a number of species using precipitation methods, ion-exchange chromatography and gel filtration. Many of the methods described isolate LH, FSH and TSH (and sometimes other pituitary hormones) from the same batch of glands—enabling efficient usage of scarce starting material. Most 'purified' preparations of these hormones still show some microheterogeneity, but this probably reflects heterogeneity as it exists within the gland.

3.2 Subunit structure[5 − 8]

As good preparations of TSH, LH and FSH became available, it was recognized that they are all glycoproteins, of molecular weight about 30 000. It was also soon recognized that each of these hormones contains two different subunits, which can be dissociated by acid pH or 8M urea and then separated by ion-exchange chromatography, counter current distribution or gel filtration. The separated subunits are almost completely without biological activity, but can reassociate in appropriate conditions to give active hormone. Each subunit contains a single, glycosylated peptide chain, and the links between the subunits are non-covalent.

The subunits from the different glycoprotein hormones were isolated and characterized in different laboratories, and as a consequence were given different names. It soon became apparent, however, that one of the two subunits was very similar for the various different hormones, while the other differed considerably. The nomenclature has now been standardized, so that the subunit which is common to the various hormones in a given species is referred to as the α-subunit, whereas that which differs from one hormone to another is called the β-subunit. It is not yet clear whether the α-subunit for LH, FSH, TSH and hCG is identical for a given species; it seems likely that it represents the product of a single gene (and therefore has an identical amino acid sequence) but some variation in post-translational modification (glycosylation and/or proteolytic modification) may occur.

3.3 Amino acid sequence[5,9 – 11]

Amino acid sequences have now been determined for the subunits of glycoprotein hormones from several different species. These confirm that within a species the α-subunit for each hormone is almost identical, but that the β-subunits differ considerably from one hormone to another. This suggests that the β-subunits carry the hormonal specificity, and this is confirmed by hybridization studies (see below). In all cases the polypeptide chains contain several (five or six) internal disulphide bridges, which must contribute considerably to the rigidity of the three-dimensional structure, and several carbohydate moieties, which are attached via the side chains of the asparagine or serine residues. These carbohydrate moieties include sialic acid, L-fucose, D-galactose, D-mannose, D-glucosamine and D-galactosamine, linked in various ways, examples of which are shown in Figure 6.1. Bovine LH and TSH contain about 16% carbohydrate while hCG contains about 30% (by weight).

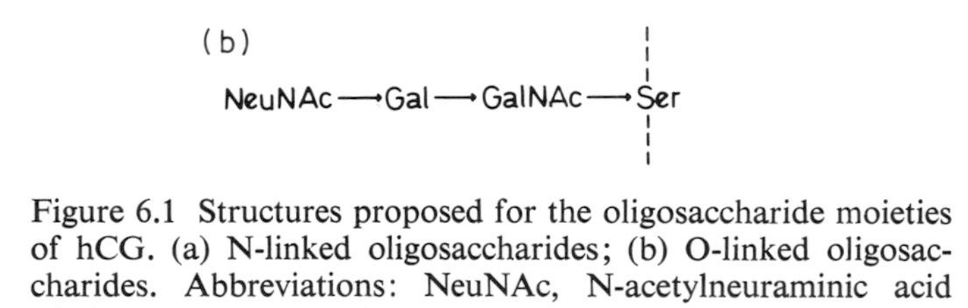

(a)

NeuNAc ⟶ Gal ⟶ GlcNAc ⟶ Man
$$\searrow$$
Man ⟶ GlcNAc ⟶ GlcNAc ⟶ Asn

NeuNAc ⟶ Gal ⟶ GlcNAc ⟶ Man
$$\nearrow$$
$(Fuc)_{0,1}$

(b)

NeuNAc ⟶ Gal ⟶ GalNAc ⟶ Ser

Figure 6.1 Structures proposed for the oligosaccharide moieties of hCG. (a) N-linked oligosaccharides; (b) O-linked oligosaccharides. Abbreviations: NeuNAc, N-acetylneuraminic acid (sialic acid); Gal, galactose; GlcNAc, N-acetylglucosamine; Man, mannose; Fuc, fucose; GalNAc, N-acetylgalactosamine. Based on ref. 5

The amino acid sequence of the α-subunit of the human glycoprotein hormones contains 92 amino acids, 5 disulphide bridges and 2 carbohydrate moieties (Figure 6.2). The β-subunits are rather longer (110–120 residues except

Figure 6.2 Amino acid sequences of ovine, porcine and human glycoprotein hormone α-subunits. The complete sequence of the ovine α-subunit is shown; the sequences of the porcine and human polypeptides are identical except where indicated. *Indicates the position of attachment of an oligosaccharide moiety. Based on refs. 5, 9 and 52

Human LH: H-Ser-Arg-Glu-Pro-Leu-Arg-Pro-Trp-|Cys-His-Pro-Ile-Asn-Ala-Ile-Leu-Ala-Val|-Glu-Lys-Glu-Gly-|Cys|-Pro-Val-|Cys|-Ile-Thr-Val-|Asn-Thr|-
Human CG: H-Ser-Lys-Glu-Pro-Leu-Arg-Pro-Arg-|Cys-Arg-Pro-Ile-Asn-Ala-Thr-Leu-Ala-Val|-Glu-Lys-Glu-Gly-|Cys|-Pro-Val-|Cys|-Ile-Thr-Val-|Asn-Thr|-
Human TSH: H-Phe-Cys-Ile-Pro-Thr-Glx-Tyr-Met-His-Val-Glu-Arg-Arg-Glx-Cys-Ala-Tyr-Cys-Leu-Thr-Ile-|Asn-Thr|-
Human FSH: H-Asn-Ser-Cys-Glu-Leu-Thr-Asn-Ile-Thr-Ile-Ala-Ile-Glu-Lys-Glu-Glu-|Cys-Arg-Phe-Cys-Ile-Thr-Ile-Asn-Thr|-

10   20   30

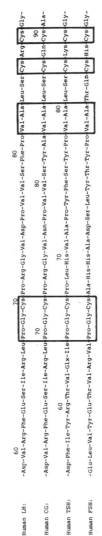

Human LH: -Thr-Ile-Cys-Ala-Gly-Tyr-Cys-Pro-Thr-Met------Arg-Val-Leu-Gln-Ala-Val-Leu-Pro-Pro---------Leu-Pro-Gln-Val-Cys-Thr-Tyr-Arg-
Human CG: -Thr-Ile-Cys-Ala-Gly-Tyr-Cys-Pro-Thr-Met-Thr-Arg-Val-Leu-Gln-Gly-Val-Leu-Pro-Ala----------Leu-Pro-Gln-Val-Val-Cys-Asn-Tyr-Arg-
Human TSH: -Thr-Ile-Cys-Ala-Gly-Tyr-Cys-Met-Thr-Arg-Asp-Ile-Asx-Gly-Lys-Leu-Phe-Leu-Pro-Lys-Tyr-Ala-Leu-Ser-Gln-Asx-Val-Cys-Thr-Tyr-Arg-
Human FSH: -Thr-Trp-Cys-Ala-Gly-Tyr-Cys-Tyr-Thr-Arg-Asp-Leu-Val-Tyr-Lys-Asp-Pro-Ala-Arg-Pro---------Lys-Ile-Gln-Lys-Thr-Cys-Thr-Phe-Lys-

40   50   60

Human LH: -Asp-Val-Arg-Phe-Glu-Ser-Ile-Arg-Leu-Pro-Gly-Cys-Pro-Arg-Gly-Val-Asp-Pro-Val-Ser-Phe-Pro-Val-Ala-Leu-Ser-Cys-Arg-Cys-Gly-
Human CG: -Asp-Val-Arg-Phe-Glu-Ser-Ile-Arg-Leu-Pro-Gly-Cys-Pro-Arg-Gly-Val-Asn-Pro-Val-Val-Ser-Tyr-Ala-Val-Ala-Leu-Ser-Cys-Gln-Cys-Ala-
Human TSH: -Asp-Phe-Ile-Tyr-Arg-Thr-Val-Glx-Ile-Pro-Gly-Cys-Pro-Leu-His-Val-Ala-Pro-Tyr-Phe-Ser-Tyr-Pro-Val-Ala-Leu-Ser-Cys-Lys-Cys-Gly-
Human FSH: -Glu-Leu-Val-Tyr-Glu-Thr-Val-Arg-Val-Pro-Gly-Cys-Ala-His-His-Ala-Asp-Ser-Leu-Tyr-Thr-Tyr-Pro-Val-Ala-Thr-Glx-Cys-His-Cys-Gly-

70   80   90

Human LH: -Pro-Cys-Arg-Arg-Ser-Thr-Ser-Asp-Cys-Gly-Gly-Pro-Lys-Asp-His-Pro-Leu-Thr-Cys-Asp-His-Pro-Gln-NH₂
Human CG: -Leu-Cys-Arg-Arg-Ser-Thr-Thr-Asp-Cys-Gly-Gly-Pro-Lys-Asp-His-Pro-Leu-Thr-Cys-Asp-Asp-Pro-Arg-Phe-Gln-Asp-Ser-Ser-Ser-Lys-Ala-
Human TSH: -Lys-Cys-Asx-Thr-Asx-Tyr-Ser-Asp-Cys-Ile-His-Glu-Ala-Ile-Lys-Thr-Asx-Tyr-Cys-Thr-Lys-Pro-Glx-Lys-Ser-Tyr-OH
Human FSH: -Lys-Cys-Asp-Ser-Asp-Ser-Thr-Asp-Cys-Thr-Val-Arg-Gly-Leu-Gly-Pro-Ser-Tyr-Cys-Ser-Phe-Gly-Glu-Met-Lys-Glu-Tyr-Pro-Thr-Ala-Leu-

100   110   120

Human CG: -Pro-Pro-Ser-Leu-Pro-Ser-Pro-Ser-Arg-Leu-Pro-Gly-Pro-Ser-Asp-Thr-Pro-Ile-Leu-Pro-Gln-NH₂

130   140

Human FSH: -Ser-Tyr-OH

Figure 6.3 Amino acid sequences of the β-subunits of human LH, FSH, TSH and hCG. Regions of sequence that are identical in all four polypeptides are enclosed by 'boxes'; homology is in fact more extensive than this indicates, since in many other places sequences of two or three of the four polypeptides are identical. *Indicates the position of attachment of an oligosaccharide moiety. Based on refs. 5 and 9

for hCG which is considerably longer), contain 6 disulphide bridges and 3 carbohydrate moieties (Figure 6.3). There is considerable homology between the sequences of the β-subunits of TSH, LH and FSH, the sequences being identical at about 40% of all residues (Figure 6.3). Most notably, the positions of the disulphide bridges are completely homologous, and proline residues are strongly conserved, suggesting that the tertiary structures are very similar. The sequences of the β-subunits of human LH and hCG are very similar, differing at only about 20% of all residues (except at the C-terminus), suggesting that these proteins arose as a consequence of a relatively recent gene duplication and divergence. At the C-terminus of the hCG β-subunit, however, there is an extension of the polypeptide chain (compared with hLH) of 30 residues (Figure 6.3). This 'tail' is very rich in proline and serine and carries 4 carbohydrate moieties; these last are attached to side chains of serine residues, whereas those on the rest of the molecule are linked to asparagine. The role of this additional piece of amino acid sequence has not been firmly established, although it is possible that it prolongs the half-life of the hormone in the circulation. Peptides equivalent to this 'tail' of hCG have been synthesized chemically, and have been used to raise antibodies completely specific for this hormone. Their possible use as contraceptive agents has been investigated.

Glycoprotein hormones have been isolated from several different species. Amino acid sequence determination on these shows a moderate amount of species variation (although the data is still fairly limited). Within the artiodactyls (sheep, ox and pig) sequence variation is limited, but comparison of the sequence of the α-subunit of any artiodactyl hormone with the corresponding human α-chain sequence shows about 20–25% of all residues differing. Comparison of human and artiodactyl β-subunit sequences for any one of these hormones shows rather more differences (about 33%). The greater evolutionary conservation of the α-subunit sequence is probably to be expected, since the α-subunit has to play a part in several different hormones.

The application of recombinant DNA technology (Chapter 18) to the hormones of the LH/FSH family has led to the determination of nucleotide sequences corresponding to several mRNAs.[11] These confirm the amino acid sequences that have been reported, and allow extensions of the evolutionary considerations.

3.4 Structure–function relationships[5 – 8,12]

The most revealing experiments carried out on structure–function relationships have involved the formation of hybrid molecules incorporating an α-subunit from one hormone and a β-subunit from another. As has been mentioned, the α- and β-subunits from any one hormone can be dissociated and separated. The separate subunits retain little or no biological activity. Under appropriate conditions the subunits can be recombined to reform α–β dimers, with substantial recovery of biological activity. If hybridization experiments are carried out with the α-subunit prepared from LH and the β-subunit from TSH, the resultant hybrid possesses thyroid stimulating activity. Similar experiments show that such hybrids always possess the biological activity associated with that hormone which donates its β-subunit (Figure 6.4). This accords well with the observation that for any one species the α-subunits have a more or less identical amino acid sequence, although differences in carbohydrate moieties may still provide a basis for significant structural differences.

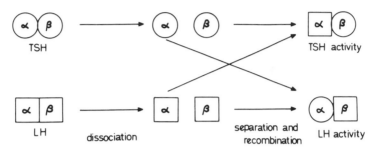

Figure 6.4 Dissociation and recombination of the subunits of the glycoprotein hormones shows that the hormonal specificity is associated with the β-subunit

Relatively little work has been carried out on chemical modification of the glycoprotein hormones and its effects on biological activity. Studies that have been reported suggest that chemical modification tends to destroy biological activity more readily than immunological activity. Thus, modification of three methionine residues in bovine LH led to reduction of biological activity by at least 95%, but full immunological activity was retained. The modified α- and β-subunits still associated normally. Studies on chemical cross-linking of the α- and β-subunits are of particular interest—when the subunits of LH were cross-linked covalently by a carbodiimide there was no loss of biological activity, suggesting that after binding to the receptor no dissociation of subunits is required for expression of biological activity.

Modification of the sugar residues of the gonadotropins has been studied in some detail. Removal of the terminal sialic acid residues from FSH and hCG (but not LH) by neuraminidase leads to drastic loss of *in vivo* biological (but not immunological) activity. This is probably because the half-life of the hormone *in vivo* is very much reduced.

3.5 Comparative studies on the glycoprotein hormones[13]

Comparative structural studies on the hormones of various mammals have already been discussed in Section 3.3. A considerable amount of work has been reported on the gonadotropins and TSHs of lower vertebrates, although characterization is still far from complete. The two-subunit quaternary structure is present and sequence studies on carp gonadotropin α-subunit indicate that this is homologous with mammalian α-subunits. Some interesting studies by Fontaine and his coworkers suggest that hormones in fish may be rather different from those in mammals. A good deal of work suggests that there is only one gonadotropin in teleost fish (and possibly some other lower vertebrate groups), and that this is structurally more different from the mammalian gonadotropins than is fish TSH. Not all workers accept this view, however. Fish TSH and gonadotropin(s) have little activity in mammals. The fish thyroid is stimulated by mammalian TSH, LH or FSH, but gonads of fish respond very poorly to mammalian LH or FSH.

Proteins similar to the glycoprotein hormones have been reported to occur in some micro-organisms, although their significance is not clear. Homology has also been observed between short regions of the α and β chains of the glycoprotein hormones and the peptide chains of cholera toxin, and proposals have been made that there are similarities between the mechanisms of action of these hormones and some such toxins.

4.   BIOLOGICAL ACTIONS OF GONADOTROPINS IN THE MALE

The actions of the gonadotropins are concerned mainly with regulation of the gonads. Consequently, these actions are very different in the male and the female. However, in both sexes the hormones affect both the germ-cell-producing and steroidogenic functions of the gonads, and at the biochemical level many similarities can be seen between the actions of LH and FSH on the testes and ovaries. It must be stressed that there is enormous variation in reproductive cycles, even within mammals, which cannot be considered in any detail here. Fortunately, the differences between seasonal and continuous breeders in the male, and spontaneous and induced ovulators and oestrous and menstrual cycles in the female, are less marked at the biochemical level than at the physiological level.

4.1   The testis[14,15]

The structure of the testis and organization of the rat urogenital system is summarized in Figures 6.5 and 6.6. The bulk of the testis is occupied by seminiferous tubules, which are highly convoluted, and form both the sites for formation and maturation of spermatozoa, and the collecting ducts down which the mature spermatozoa are passed to the rete testis and epididymis, where they are stored. Between the seminiferous tubules are the interstitial cells (Leydig cells), which are the main sites for synthesis of testicular steroids. The accessory ducts and glands of the male, as seen in Figure 6.5, are concerned with the storage and conveyance of spermatozoa to the penis, and the manufacture of components of semen other than spermatozoa. Like the testes, these accessory features regress on hypophysectomy, but this is a consequence mainly of the drastically reduced production of androgens (as a result of removal of LH and FSH) rather than the absence of direct effects of LH and FSH.

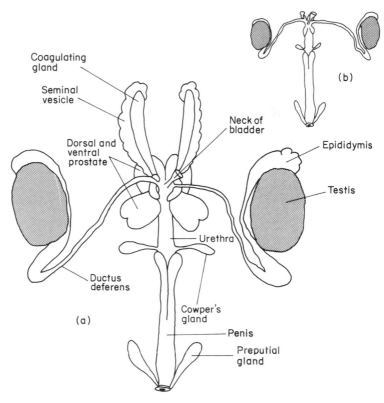

Figure 6.5 The genital system of the male rat. (a) Intact mature rat; (b) littermate of the same age, but hypophysectomized at 30 days. Based partly on Turner and Bagnara

The production of spermatozoa in the seminiferous tubules is illustrated in Figure 6.7. It should be noted that the tubules themselves are avascular, although surrounded by a rich blood supply, so that nutritional and hormonal factors are supplied by diffusion from the outside. The wall of the tubule (the tunica propria) comprises a layer of endothelial cells, a layer of smooth muscle cells and a basement membrane. Within the basement membrane is a layer of very large cells termed Sertoli cells, in which are embedded the germ cells, in

Interstitial
cell

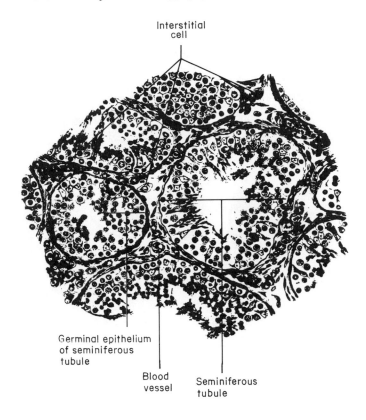

Germinal epithelium
of seminiferous
tubule

Blood
vessel

Seminiferous
tubule

Figure 6.6 Section through the testis, showing distribution of
seminal vesicles and Leydig (interstitial) tissue. (Reprinted from
McClintic, J. R. (1975) *Basic Anatomy and Physiology of the
Human Body*)

various stages of development. The centre of each tubule comprises a lumen in
which are collected the mature spermatozoa, suspended in a fluid which is
secreted by the seminiferous tubules. Spermatozoa are passed from this lumen
to the rete testis and epididymis, where they are stored.

The youngest germ cells, the spermatogonia, lie close to the basement membrane of the seminiferous tubule, largely embedded in Sertoli cells. Some of
these can be considered as stem cells which initiate spermatogenesis in a cyclical
fashion. Once the process has been initiated, cell division continues at a constant rate until mature spermatozoa are formed. At any given time the products
of several waves of initiation of spermatogenesis will be present in a particular
region of the seminiferous tubules, resulting in characteristic association of
maturing cell types.

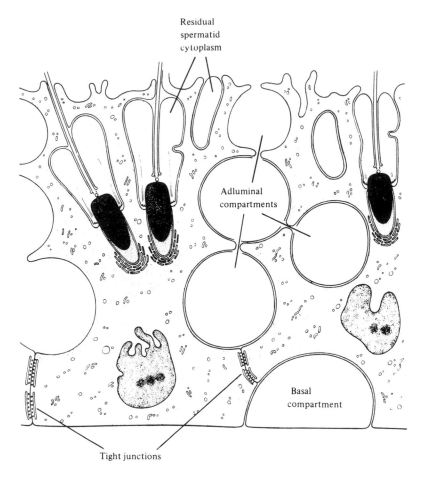

Residual
spermatid
cytoplasm

Adluminal
compartments

Basal
compartment

Tight junctions

Figure 6.7 Diagram showing the fine structure of a Sertoli cell, and part of another. Each Sertoli cell extends from the boundary tissue (bottom) to the lumen (top). The developing germ cells are enclosed in the basal and adluminal compartments between adjacent Sertoli cells and in crypts in the luminal surfaces of the Sertoli cells. Tight junctions restrict passage of solutes between Sertoli cells, and maintain the blood–testis barrier. (Reprinted with permission from Setchell, B. P. (1982) In *Reproduction in Mammals*, Vol. 1, 2nd Edn. (eds. Austin, C. R. and Short, R. V.), pp. 63–101. Copyright (1982) Cambridge University Press)

The first cell division of the stem cells is mitotic, and forms two spermatogonia, one of which remains as a stem cell. The second daughter spermatogonium undergoes four or five further mitotic divisions, giving rise to a population of primary spermatocytes. These then enter meiosis, and each gives rise to four secondary spermatocytes (with halving of the chromosome number) which differentiate to spermatids and then to spermatozoa without undergoing further cell division. The various cells of this spermatogenic process are embedded in the lateral margins of the Sertoli cells (Figure 6.7), and move towards the lumen of the tubules as they mature. The rate at which the spermatogonia are produced and mature appears to be very constant and is probably not altered by hormonal factors. The rate of production of spermatozoa may vary, however, because many developing germ cells degenerate rather than completing their maturation, and the extent of this degeneration may vary.

The developing germ cells lie within a blood–testis barrier formed by tight junctions between (a) the smooth muscle cells (myoid layer) of the tunica propria and (b) the walls of adjacent Sertoli cells. This excludes relatively large molecules, including most peptide and protein hormones, from the interior of the seminiferous tubules. The junctions of the myoid cells are opened in some circumstances, and such hormones may then reach the spermatogonia and Sertoli cells. However, the junctions between the Sertoli cells themselves probably prevent access of large molecules to the germ cells in later stages of maturation.

The Leydig cells of the interstitial spaces between seminiferous tubules are undoubtedly major sources of testicular steroids, particularly androgens. They are characterized by a large Golgi complex, an extensive smooth endoplasmic reticulum and the presence of numerous lipid droplets. The seminiferous tubules themselves appear to be able to generate androgens, and the Sertoli cells also contain extensive smooth endoplasmic reticulum and lipid droplets. In some animals, such as the rat, the seminiferous tubules and interstitial tissue can be fairly readily separated. When the separated components are incubated with labelled androgen precursors, both can synthesize labelled testosterone and other precursors to some extent, although the interstitial cells are by far the most efficient in this respect. If the germinal cells of the testis are eliminated *in vivo*, and seminiferous tubules are subsequently isolated, their ability to synthesize androgens is unimpaired, suggesting strongly that it is the Sertoli cells and not the germ cells that are the sites of steroidogenesis in the tubules. Isolated, cultured, Sertoli cells show very little ability to synthesize androgens, although they can convert testosterone into oestrogens, and probably provide the main source of testicular oestrogens.

4.2 The actions of LH and FSH on the testis[16–18]

After hypophysectomy in the male rat, production of testicular steroids rapidly decreases, spermatogenesis ceases, and the testes and accessory glands decrease in size dramatically (Figure 6.5). Injection of FSH and LH reverses these effects, due to primary actions on two target cells. LH affects mainly the Leydig cells, while FSH has direct actions on the Sertoli cells of the seminiferous tubules.

Stimulation of the Leydig cells by LH results in production of androgens (mainly testosterone) which are carried to the seminiferous tubules, by direct diffusion or via the circulation, where they have important effects, with FSH, on the growth of the Sertoli cells and spermatogenesis. However, it is now generally considered that, in many species anyway, LH has little *direct* effect on the Sertoli cells. Testosterone also maintains virtually all the male accessory organs and glands, and stimulates growth of muscle, bone and hair, so that here too LH has indirect effects.

The primary action of LH on the testis is thus a rather straightforward one, involving actions on the growth of and steroidogenesis by a single cell type, the Leydig cell. It can be studied *in vitro* using dispersed Leydig cells, which in the rat can be separated fairly readily from the other cell types in the testis. There is some evidence from the studies using the rabbit testis perfused *in situ* that FSH may synergize with LH to promote testosterone production (although FSH had no effect by itself). Synergistic effects of FSH on testosterone production *in vivo* have also been shown in the rat, although *in vitro* they are not readily demonstrated.

Through its action on Leydig cells LH takes part in a fairly simple feedback loop. LH secretion is regulated in the male largely via the level of androgens (and to some extent oestrogens) in the circulation. Production of testosterone by the Leydig cells in response to LH stimulation leads to elevated blood levels of this steroid. These have direct effects on the brain which lead to a reduced secretion of LHRH from the hypothalamus, lowered LH secretion from the pituitary gland and hence decreased stimulation of steroidogenesis in the Leydig cells. In this way androgen levels in the circulation are maintained at a rather constant level. Oestrogens, produced in relatively small amounts by the Sertoli cells, are also involved in the feedback mechanism, although their action may be primarily at the pituitary level where they decrease the sensitivity of the gonadotrophs.

The effects of FSH on the Sertoli cell are much more complex than those of LH on the Leydig cell. In part this is because it shares control of the Sertoli cell with testosterone, and at times this steroid hormone appears to be the predominant regulator (testosterone, or LH, alone will maintain spermatogenesis, at a somewhat lowered level, if administered immediately after hypophysectomy in the mature male rat). But also, of course, the Sertoli cell itself has far more complex functions than the Leydig cell, and almost all these functions are regulated, at some stage, by FSH.

That the target cell for FSH acting on the seminiferous tubules, and enhancing spermatogenesis, is the Sertoli cell and not the developing germ cell itself has been demonstrated by studies *in vitro*. FSH will bind specifically to isolated Sertoli cells, and will stimulate cyclic AMP production in such cells. However, it neither binds to, nor stimulates isolated spermatocytes or spermatids. Action on the Sertoli cell rather than on the developing germ cells is to be expected since all the germ cells except the primary spermatogonia are behind the blood–testis barrier and should be inaccessible to polypeptide hormones. Testosterone may also act via the Sertoli cells rather than the germ cells, although direct effects cannot be ruled out since androgen receptors have been demonstrated in spermatids.[19]

Thus, both FSH and testosterone regulate spermatogenesis by actions on the 'nurse cells' rather than on the germ cells themselves. How the Sertoli cells then regulate spermatogenesis is not clear, but, as mentioned in the previous section, the rate of cell division of the germ cells is rather constant, and regulation of the proportion of cells which reach maturity rather than degenerating is probably the determining factor.

The actions of FSH on Sertoli cells include stimulation of protein synthesis, cell division, formation of tight junctions and synthesis and secretion of proteins (including androgen-binding protein and inhibin) and steroids (primarily oestrogens, which are formed from testosterone deriving from the Leydig cells). Precisely how these various actions result in increased spermatogenesis is not clear, but the secretion of both proteins and steroids results in high concentrations of these compounds in the fluid bathing the developing germ cells (now *within* the blood–testis barrier) and could well be of major importance. The protein inhibin produced by the Sertoli cells may be a major factor regulating FSH secretion—it appears to inhibit secretion of the hormone directly at the pituitary level. As yet inhibin has not been fully characterized.

Although LH and FSH are the main pituitary hormones affecting the testis, prolactin too probably has an action in some species. In hypopituitary dwarf mice prolactin increases fertility, probably by synergizing with LH to promote testosterone synthesis. Prolactin may also synergize with testosterone to promote growth of the seminiferous tubules and prostate, possibly by increasing the number of testosterone receptors present. In man, hyperprolactinaemia is frequently associated with infertility, and lowering of prolactin levels by use of specific drugs usually restores fertility. The mechanism of this effect is not known, but the possibility of direct effects of prolactin on the brain has not been excluded.

4.3   The biochemical mode of action of LH on Leydig cell steroidogenesis[20–22]

The actions of LH on Leydig cells have been studied in considerable detail at the biochemical level, using a variety of test systems. Many recent studies have used preparations of isolated, dispersed Leydig cells, which show considerable cellular homogeneity. LH promotes both steroidogenesis by and growth of the Leydig cell, but it is the actions on production of steroids, particularly testosterone, that have been most extensively investigated.

The actions of LH on steroidogenesis in Leydig cells, at the biochemical level, resemble those of ACTH on adrenal steroidogenesis in many respects (see Chapter 5). Binding of the hormone is initially to a membrane receptor, and this binding appears to be linked to activation of a membrane-associated adenylate cyclase. Increased cellular cyclic AMP levels rapidly follow binding of LH to Leydig cells and are thought to activate cyclic AMP-dependent protein kinase by a mechanism similar (and possibly identical) to that already discussed for ACTH. Protein kinase presumably then phosphorylates, and thereby alters the activity of one or more key proteins involved in the steroidogenic pathway. Several proteins that are phosphorylated under the influence of LH have been identified, but the cellular functions of these are as yet unknown. Cholesterol esterase may be a site for activation of the steroidogenic pathway, but activation of later steps, involved in side-chain cleavage, is also probable. LH stimulates accumulation of a labile protein that binds cholesterol and facilitates either its entry into the mitochondria, or its interaction with the cytochrome P-450 moiety of the cleavage enzyme. The main product of the steroidogenic pathway (testosterone) is of course different from that in the adrenal cortex, but much of the pathway is the same, and the site(s) of action may also be the same.

The evidence for this mechanism, which is summarized in Figure 6.8, is very similar in kind to that described earlier for the action of ACTH, and will not be considered in detail. However, as with ACTH, several features do not accord completely with the simple mechanism given here. In particular, LH induces steroidogenesis at concentrations much lower than those necessary to induce a measurable increase in the cyclic AMP level, and use of inhibitors shows that continuing protein synthesis is required for the actions of LH to be manifest. Although it seems likely that cyclic AMP is at least a factor in the regulation of Leydig cell steroidogenesis, it is also likely that other factors, more-or-less independently, also play a role in this regulation and may act as alternative second messengers. The nature of such alternative second messengers has not been established, although there is evidence that prostaglandins and possibly calcium and altered phosphatidyl inositol metabolism may be involved.

LH is also responsible for maintenance of the long-term ability of Leydig cells to produce testosterone, possibly by stimulating the biogenesis of smooth endoplasmic reticulum and enzymes associated with it.[18]

A good deal of attention has been paid to the nature of LH receptors in the testis. These have been solubilized and partially purified, and appear to be similar to LH and FSH receptors from other tissues with regard to size (molecular weight 175 000–225 000) and hormone affinity (K_d 1.5×10^{-10}M).

4.4 The biochemical mode of action of FSH on the Sertoli cell[23-25]

As mentioned, the effects of FSH on the Sertoli cell are more complex than those of LH on the Leydig cell, and it is only recently that substantial progress

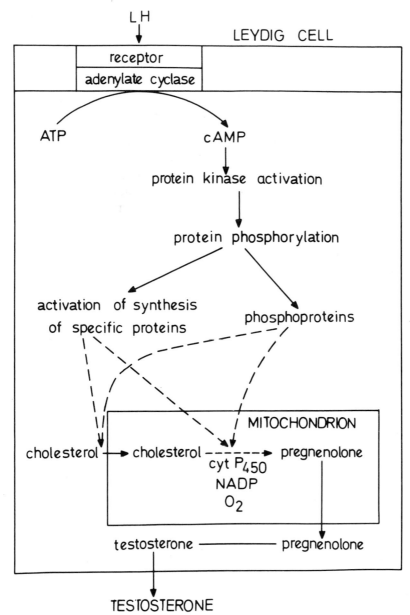

Figure 6.8 A schematic diagram illustrating some of the mechanisms whereby LH may stimulate steroidogenesis in the Leydig cell

has been made in understanding the biochemistry of these effects. The main technical advances that have facilitated this progress include the ability to maintain Sertoli cells in primary culture, and recognition that, in the rat at least,

Sertoli cells become refractory to some of the actions of FSH with age. FSH has its main effects on Sertoli cells of the rat between 10 and 30 days, suggesting that it is concerned primarily with development of the seminiferous tubules and initiation of spermatogenesis, although at least some of its actions can be seen in older animals.

In Sertoli cells, FSH stimulates protein synthesis (including stimulation of synthesis of the specific proteins, androgen-binding protein and inhibin), steroidogenesis, protein secretion, increased cell division, increased cell motility and formation of tight junctions. Many of these effects may involve actions on the cytoskeleton (microtubules and microfilaments). The overall effect of these actions is of course to alter the rate of spermatogenesis, but quite how the observable biochemical effects are linked to the overall biological action is not understood.

Binding of FSH to its cellular receptor on the Sertoli cell causes increased cyclic AMP levels. This appears to be due both to increased activity of adenylate cyclase and decreased activity of phosphodiesterase. Cyclic AMP activates protein kinase, which causes phosphorylation of a large number of proteins within the Sertoli cell. Which of these is important for the subsequent effects is not known, however. Presumably several different phosphorylated proteins may mediate the various actions of the hormone on protein synthesis, exocytosis, steroidogenesis, etc. Among the proteins whose synthesis is stimulated by FSH is a protein kinase inhibitor; this may be responsible for the refractoriness to FSH which develops after periods of prolonged, continuous stimulation.

For the actions of FSH on exocytosis, formation of tight junctions and (*in vitro*) change of cell shape, calcium is required, and may play a part in mediating the actions of the hormone. These effects appear to involve actions on the cytoskeleton (microtubules and microfilaments) and a mechanism involving phosphorylation of myosin light chain kinase, resulting in interaction with calmodulin and activation of myosin (a component of the cytoskeleton) has been proposed (Figure 6.9). Although this represents an interesting hypothesis, specific evidence for such a sequence of events remains fragmentary.

Steroidogenesis by the Sertoli cells involves mainly conversion of testosterone (originating in the Leydig cells) into oestrogens. The biological role of the oestrogens is not clear, but it is well established that their production is regulated by FSH. Again, cyclic AMP is probably involved as a second messenger.

5. ACTIONS OF LH AND FSH IN THE FEMALE

In the female, LH and FSH act primarily on the ovary. As in the case of the testis, the ovary serves both as an organ producing germ cells and as an endocrine organ, and the gonadotropins control both these functions. The situation is complicated, however, by the fact that after ovulation the Graafian follicle

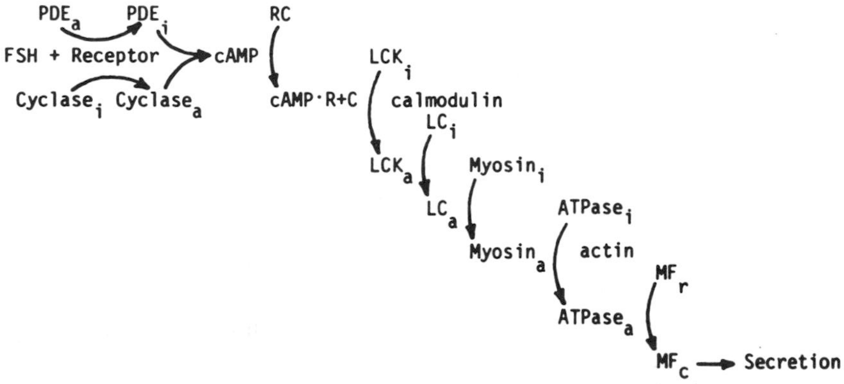

Figure 6.9 A proposed mechanism whereby FSH may regulate events involving the cytoskeleton, including exocytosis, in the Sertoli cell. (Reprinted with permission from Means, A. R., Dedman, J. R., Tash, J. S., Tindall, D. J., van Sickle, M. and Welsh, M. J. (1980) *Ann. Rev. Physiol.*, **42**, 59–70. Copyright (1980) Annual Reviews Inc.)

develops into a further endocrine organ, the corpus luteum, which is also under the control of the gonadotropins; the subsequent development of this organ depends on whether pregnancy ensues. Further complications arise from the very variable nature of ovarian cycles in different species; gonadotropins play an important part in regulating these cycles.

5.1   The ovary[26,27]

The structure of the ovary is summarized diagrammatically in Figure 6.10. The organ consists of a cortex, containing most of the maturing germ cells, and a medulla, containing blood vessels and structural elements. The epithelial layer surrounding the ovary is the germinal epithelium, but production of germ cells (oogonia) from this (or the primordial germ cells giving rise to it) is complete long before birth in most mammals. The oogonia divide to give large numbers of germ cells, and these eventually enter meiosis. However, meiosis is arrested during prophase of the first division, at which stage the cells are known as primary oocytes. Between half a million and two million oocytes are found in the ovary of a new-born human, and it is likely, although not certain, that no further oocytes are produced after birth. Development of oocytes through to the Graafian follicle stage occurs without further progression through meiosis. At ovulation, meiosis resumes, the first polar body is formed and the primary oocyte becomes a secondary oocyte. This enters the fallopian tube, undergoes the second meiotic division (with formation of a second polar body) and forms the mature ovum, now having a haploid chromosome number.

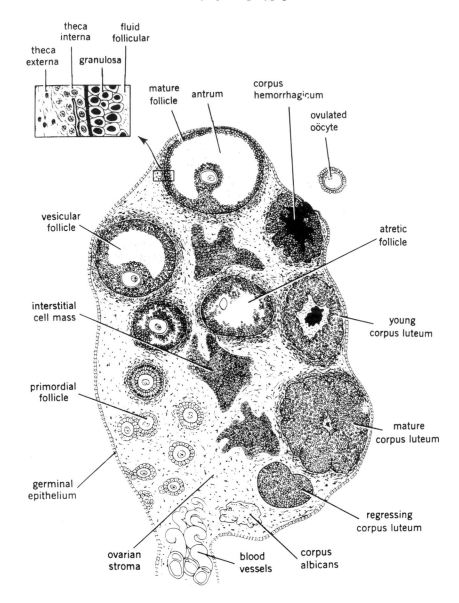

Figure 6.10 Diagrammatic section through a mammalian ovary. Progressive stages in the differentiation of a Graafian follicle are indicated on the left. The mature follicle may become atretic or ovulate and undergo luteinization (right). (Reprinted from Gorbman, A. and Bern, H. A. (1962) *A Textbook of Comparative Endocrinology*)

Each primary oocyte becomes surrounded by a single layer of spindle-shaped cells, which are precursors of granulosa cells. The granulosa cells are surrounded by a membrane (basal lamina) and the whole complex is known as a primordial follicle. In the human foetus incorporation of oocytes into primordial follicles begins in the fourth to fifth month of gestation and is completed about 6 months postpartum. Primordial follicles may survive unchanged for many years, but from the fifth to the sixth month of gestation (in the human foetus) the process of follicular maturation begins; at any time, a very small proportion of primordial follicles is starting on the sequence of developmental changes which results either in the formation of a Graafian follicle followed by ovulation, or in oocyte death and degeneration of the follicular complex (atresia). In practice the vast majority of developing follicles undergo atresia, and this is their sole fate prior to the first ovulation at the time of puberty.

The first step in maturation of the follicle consists of differentiation of the pre-granulosa cells to a single layer of cuboidal cells, which then proliferate to give multiple layers of granulosa cells, with enlargement of the follicle (now a primary follicle). These cells secrete mucopolysaccharide and give rise to a translucent zone around the oocyte—the *zona pellucida*. As this process proceeds, cells from the stroma of the ovary become organized around the outside of the follicle; these 'thecal' cells are highly vascularized, and the innermost layers (*theca interna* cells) differentiate to a typical steroid-secreting cell type.

As the granulosa and thecal cells proliferate, the size of the follicle increases. When it reaches 100–200 μm in diameter (in the human ovary), fluid begins to appear between the granulosa cells of some primary follicles, and eventually forms a fluid-filled cavity, the antrum (Figure 6.10). The oocyte remains attached to one wall of this cavity, which divides the granulosa cells into two types—those still associated with the oocyte (the *cumulus oophorus*) and those attached to the wall of the antrum (the *membranum granulosum*). The follicle is now known as a Graafian follicle.

Although mature Graafian follicles begin to appear in the ovary of the late foetus, they all undergo atresia, rather than ovulation, until sexual maturation. The processes that determine when ovulation begins are not well understood. Although ovulation is generally considered to be mainly under the control of gonadotropins, gonadotropin levels appear in the circulation of very young animals and girls and begin to show cyclic variations long before the first ovulation occurs.

Ovulation involves rupture of the mature Graafian follicle and expulsion of the oocyte with its associated granulosa cells. The remainder of the follicle is converted to a corpus luteum (luteinization), which involves morphological changes in the surviving granulosa cells and an intermingling of these with theca cells and structural and vascular elements. The corpus luteum survives for about 14 days (in the human), and then regresses, unless pregnancy has occurred, in which case it is maintained for a prolonged period by the action of pituitary and/or placental gonadotropins.

Ovarian steroids are produced by several cell types, and in each case their production is influenced by gonadotropins. *In vitro* studies using various ovarian tissues have shown that stromal and thecal cells produce mainly androgens, which serve as substrates for the steroid-producing cells of the follicle and corpus luteum. Follicular cells produce oestrogens (mainly from androgens) and the corpus luteum produces both progesterone and oestrogens.

5.2 The actions of LH and FSH on the ovary. Ovarian cycles[28-30]

As already mentioned, the role of gonadotropins in the development of sexual maturity is not clear. It seems likely that they do play a major role in this process, but the facts that substantial levels of gonadotropins are present in the circulation from birth, and that cyclic secretion of these hormones is established long before ovulations begin, make interpretation of the role of gonadotropins difficult.

In the mature female mammal, reproductive cycles are established (oestrus cycles in most groups, menstrual cycles in primates) in which gonadotropins play a major part. The detailed nature of these cycles varies considerably from one group to another, but these variations cannot be considered in detail here. The purpose of such cycles is to allow ovulation at regular intervals, accompanied, in the oestrus cycle, by a period of sexual receptivity (oestrus). Ovulation is preceded by a period of growth and maturation of one or more Graafian follicles (the follicular phase) and is followed by a period in which the corpus luteum develops and becomes a major steroid-producing organ (the luteal phase). If pregnancy has not resulted from the ovulation, the luteal phase halts rather abruptly with regression of the corpus luteum, accompanied in primates by menstruation, and the follicular phase of the next cycle is initiated. When ovulation does result in pregnancy, the life of the corpus luteum (and the luteal phase) is greatly extended, and production of progesterone by this and subsequently by the placenta, prevents further ovulation until gestation. The variation between different animals in the detailed nature of the cycle is particularly marked with respect to the relative importance of the luteal phase. Thus, in some animals (e.g. the sheep) this constitutes over three-quarters of the cycle, in others (e.g. most primates) follicular and luteal phases are of about equal duration while in some species (e.g. the rat) the luteal phase is very short, the corpus luteum barely becoming established if pregnancy does not result from ovulation. Yet another type of reproductive pattern is seen in induced ovulators, such as the rabbit or cat, which spend a prolonged period in oestrus before ovulation, ovulation being induced by coitus.

LH plays a key role in induction of ovulation. In most species there is a massive peak of LH secretion shortly before ovulation (Figure 6.11), accompanied

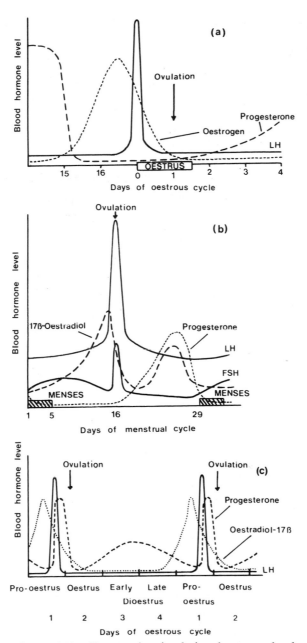

Figure 6.11 Changes in circulating hormone levels during the oestrous and menstrual cycles of (a) the sheep, (b) human and (c) rat. (Based partly on Short, R. V. (1972). In *Reproduction in Mammals*, Vol. 3, 1st Edn. (eds. Austin, C. R. and Short, R. V.), pp. 42–72)

by a smaller peak of FSH and, in some species, prolactin. Experimental injection of LH can induce ovulation in some circumstances, but it is likely that the presence of FSH is also required. LH levels fall rapidly after this peak, but significant levels are retained in the circulation and serve to maintain the growth and steroid-secreting capacity of the corpus luteum during the luteal phase. FSH and prolactin also have luteotrophic actions in some species, but it is notable that the ratio of concentration of LH : FSH is much higher during the luteal phase than during the follicular phase.

As has been mentioned, the life of the corpus luteum varies according to the species, but in all cases it degenerates rather abruptly, and production of progesterone and the other steroids it produces ceases rapidly. This degeneration is triggered in many species (including the sheep) by prostaglandin $F_{2\alpha}$, which is produced by the uterus and carried to the ovary by a local circulation. In the human, however, a different mechanism operates, but the nature of the luteolytic signal is not well understood. Falling levels of LH and FSH during the luteal phase may play a part, although the fall is not marked, and it is possible that intrinsic (preprogrammed) factors are largely responsible.

The rapidly falling progesterone levels occurring after luteolysis give rise to sloughing of the hypertrophied uterine wall and menstruation in women. They also release the hypothalamus and pituitary from feedback inhibition and allow increased secretion of FSH and LH, with initiation of the follicular phase of the next cycle. Increased FSH levels in the circulation stimulate growth of the next Graafian follicles, and secretion of oestrogens by the granulosa cells of these follicles (FSH is the dominant gonadotropin in this phase, although presence of LH is probably also necessary). The resulting surge of oestrogen in the circulation gives rise eventually to the peak of LH secretion that initiates the next ovulation.

It should be noted that the oestrus cycle of the rat is rather exceptional in that the corpus luteum is very short-lived. In the rat, a second peak of progesterone is produced during the cycle, after the LH peak and before ovulation (Figure 6.11). This is produced by the cells of the ovarian stroma rather than the corpus luteum.

If pregnancy occurs, the normal pattern of cycles ceases immediately. The mechanism whereby the maternal system recognizes that pregnancy has occurred again varies widely from species to species. In the sheep, presence of blastocysts in the uterus somehow neutralizes the luteolytic factors (prostaglandins) produced by the uterus, while in women the production by the conceptus of chorionic gonadotropin, which has luteotrophic actions similar to those of LH, prolongs the life of the corpus luteum. In the rat the corpus luteum is maintained largely by prolactin—twice daily surges of this hormone are induced by mating and continue for several days even if pregnancy has not ensued (leading to a condition known as pseudopregnancy). In most species, prolonged

survival of the corpus luteum, under the influence of LH and in some cases prolactin, leads to maintenance of high circulating levels of progesterone, which in
turn inhibits initiation of new ovarian cycles by the pituitary as well as maintaining the hypertrophied uterine wall.

5.3 The mechanism of action of FSH and LH[23,31 – 33,33a]

Most work on the biochemical mechanism of action of gonadotropins has been
carried out in the rat, despite the somewhat atypical reproductive cycles of this
species. Development of methods for short-term primary culture of various
ovarian cell types has helped determine which particular cell types respond to
which hormones, and in what way.

Early experiments by Greep and his colleagues showed that in hypophysectomized rats FSH alone will stimulate growth of the follicles, while LH alone
stimulates growth of stromal and thecal cells. Only if both hormones are
administered is there substantial production of oestrogens by the ovary. The
basis of these observations has now been elucidated, using *in vitro* cultures of
the various ovarian cell types. Thecal and stromal cells respond to LH giving
hypertrophy and conversion of cholesterol to testosterone and other androgens.
In the rat, however, they lack the aromatizing enzyme necessary for conversion
of testosterone to oestrogens, and hence they produce little or no oestrogens
(thecal cells may possibly be an important source of oestrogens in some other
species, however).

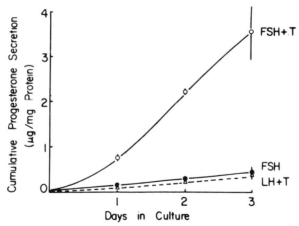

Figure 6.12 Effects of FSH, LH and testosterone (T),
alone or in combination, on progesterone synthesis by
isolated rat granulosa cells. (Reprinted with permission
from Armstrong, D. T. and Dorrington, J. H. (1976)
Endocrinology, **99**, 1411. Copyright (1976) The Endocrine Society)

In the rat, granulosa cells are the main cells producing oestrogens, and they can also produce progesterone. They are also the main cell type responding to FSH. In culture, granulosa cells from pre-antral follicles synthesize little oestrogen unless treated with both FSH and testosterone. The testosterone is required as a substrate for oestrogen synthesis and in its presence FSH dramatically stimulates oestrogen synthesis. LH has no effect. The FSH appears to act at the aromatization step. Thus, the requirement for LH as well as FSH for oestrogen synthesis in the rat reflects the need for testosterone as a substrate, rather than a direct action on the oestrogen-producing granulosa cell.

In addition to its actions on oestrogen production, FSH also has many other effects on granulosa cells and the follicle, including stimulation of growth and formation of the antrum. Many of these effects are effects of the oestrogens produced under the influence of FSH, rather than direct actions of FSH itself. Another important action of FSH (probably synergizing with oestrogens) is induction of LH receptors. These appear on granulosa cells of the Graafian follicle, and then mediate the effects of LH on ovulation and conversion of the follicular remnant to the corpus luteum.

In the rat, granulosa cells can also produce progesterone, and synthesis of this steroid is also controlled by FSH. Again, testosterone is required (Figure 6.12), but in this case not as a substrate, since the progesterone is made largely from cholesterol. The testosterone and FSH act at quite different points in the pathway leading to progesterone production—testosterone acts in the conversion of cholesterol to pregnenolone, whereas FSH stimulates primarily conversion of pregnenolone to progesterone (Figure 6.13).

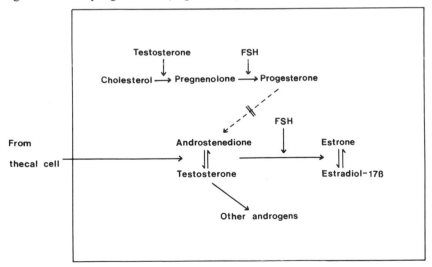

Figure 6.13 The steroidogenic conversions carried out by granulosa cells isolated from immature rats. (Based partly on Dorrington, J. H. and Armstrong, D. T. (1979) *Recent Progr. Hormone Res.*, **35**, 301–342)

As mentioned, in the rat the main luteotrophic hormone, at least in the early stages of pregnancy, is prolactin. FSH induces the appearance of prolactin receptors on granulosa cells. The mechanism whereby prolactin exerts its luteotrophic effects is not clear. In many other species LH plays the major role in maintaining the corpus luteum (as well as initiating its formation), and it probably does so in the rat also after about day 9 of pregnancy.

The steroidogenic activities of LH and FSH on their various target cells in the ovary are probably mediated by cyclic AMP. Cyclic AMP has similar steroidogenic effects to LH and FSH on appropriate cells, and LH and FSH have been shown to stimulate adenylate kinase in many preparations. To some extent the role of cyclic AMP in mediating the actions of LH and FSH on steroidogenesis is similar to the role that it plays in the testis or adrenal gland. It is notable, however, that in many ovarian cellular preparations there is a lag of 12–18 hours between the application of FSH and the initiation of effects, including activation of adenylate cyclase.[33a] It has been proposed that this reflects actions occurring at the level of gene activation.

As in many other tissues, some doubts remain as to whether cyclic AMP can be the only mediator of the action of FSH and LH on steroidogenesis, and whether it plays a role in the other actions of these gonadotropins is not clear.

6.   THE SYNTHESIS AND SECRETION OF GONADOTROPINS

6.1   Synthesis[34,35]

FSH and LH may be produced by a single cell—the gonadotroph—which would provide an exception to the general rule of one hormone, one cell (Chapter 3). Studies using immunofluorescence have suggested that there may be some gonadotrophs which produce predominantly LH and some which produce mainly FSH, but that most gonadotrophs produce both hormones. However, the question as to whether LH and FSH are produced by identical or different cell types remains controversial.

The α and β chains of FSH and LH are produced as precursors, from which a signal peptide is cleaved immediately after secretion into the lumen of the endoplasmic reticulum. Subsequent processing, including glycosylation and possibly proteolytic modification at the N-terminus, occurs while the hormone passes through the secretory pathway, particularly in the Golgi body. The hormones are stored as secretory granules which are mostly 250–300 nm in diameter.

There is no evidence to suggest that the α and β chains of LH or FSH (or the β chains of the two hormones) are synthesized as parts of a single precursor. Indeed, studies on the mRNAs and corresponding cloned cDNAs and genomic DNA for the hormones indicate that each chain has a separate mRNA, indicating that a common protein precursor cannot occur.[11,35]

Similar principles apply to the synthesis of hCG in the placenta. A polypeptide similar (or perhaps identical) to the β chain of hCG may occur in the human pituitary.[36] There may be several different genes giving rise to hCG-like proteins.[11]

6.2 Control of secretion. LHRH[37–38]

Gonadotropins are believed to be secreted by exocytosis, like other protein hormones. Control of secretion is exercised predominantly via the hypothalamus, which produces LH releasing hormone (LHRH) which stimulates secretion of both LH and FSH, although its effects on LH are normally greater than those on FSH. Whether this is the only gonadotropin-release-stimulating hormone remains controversial—there have been several claims for a separate FSH–RH, but the evidence remains inconclusive. Similarly, the exsistence of LH (and FSH)-release-inhibiting factors has been proposed but remains unproven.

LHRH was first isolated from very large quantities of pig and sheep hypothalami by the groups of Schally and Guillemin, respectively (Chapter 3). It is a decapeptide with blocked amino and carboxyl termini (Figure 6.14). The peptide has been synthesized in several laboratories. The synthetic material has identical biological and physicochemical properties to the natural hormone; in particular, it stimulates release of both LH and FSH. A favoured conformation for LHRH has been proposed, based on theoretical calculations of minimal energy requirements. A biosynthetic precursor of LHRH has now been described. LHRH lies close to the N-terminus of this.[53]

Pyro Glu His Trp Ser Tyr Gly Leu Arg Pro Gly **-CONH$_2$**

Figure 6.14. The primary structure of LHRH

A considerable effort has been devoted to the synthesis of analogues of LHRH. Derivatives with enhanced activity or inhibitory activity are potentially of great importance as drugs which might stimulate or inhibit fertility. An analogue with a 20–30-fold enhanced potency is (D-Ala^6, des-Gly^{10})-LHRH ethylamide. An analogue that inhibits the actions of LHRH is (D-Phe^2, Pro^3, D-Trp^6)-LHRH; this completely inhibited ovulation in rats, and is therefore a potential contraceptive agent. Interestingly, infusion of LHRH itself can block implantation in mated rats, presumably because the normal cyclic secretion of endogenous LH and other hormones is disrupted. Use of LHRH and its agonists as human contraceptives has been explored and the potential seems considerable.[39]

Although LHRH is generally considered to act via the pituitary gland, it is becoming clear that the hormone may have other actions. Thus, direct actions of LHRH on the ovary have been demonstrated, although their physiological significance remains uncertain. LHRH also occurs in various parts of the brain and CNS apart from the hypothalamus, and a function as a neurotransmitter or neuromodulator is possible,[39a] and may be associated with effects of the hormone on behaviour.

The actions of LHRH on the gonadotroph are thought to be mediated by cyclic AMP (see also Chapter 3). The specific actions may be modulated by other factors acting on the gonadotrophs, particularly gonadal steroids. Since it is probable that LH and FSH are secreted by the same cell, such modulation may explain how differential secretion of the two hormones is produced (in the oestrus cycle for example).

Receptors for LHRH have been located on gonadotrophs, and it has been proposed that dimerization of such receptors following binding of LHRH is essential for activity.[39b]

6.3   The hypothalamic–pituitary–gonadal axis[40,41]

In both the male and the female the overall physiological control of LH and FSH secretion is maintained by complex feedback loops involving hormones produced by the gonads and hypothalamus. Progesterone, oestrogens and androgens may all be involved in the 'long' feedback system from the gonads. The picture is too complex for a full account here, and just a few major features will be described.

In the female, positive feedback by oestrogens is of major importance in producing the peak of LH which produces ovulation. This peak can be blocked by antisera which neutralize oestrogens. Receptors for oestrogens are found in various parts of the brain, including the hypothalamus, and their effect on LHRH secretion may be either direct or mediated by various neural pathways through the brain. The pituitary also contains oestrogen receptors, and these may enhance the actions of LHRH.

Whether the pre-ovulatory peak of FSH is also regulated by positive feedback from oestrogens is not clear. It should be remembered that the pre-ovulatory LH–FSH peak is preceded in most species by a surge of circulating oestrogens (Figure 6.11). It must also be noted that in some circumstances oestrogens, acting either via the CNS or at the pituitary level, can *inhibit* secretion of LH and FSH. Thus, ovariectomy results in increased levels of gonadotropins in many species and this effect can be reversed by administration of oestrogens. The actions of oestrogens probably vary according to the general hormonal milieu, the state of the oestrus cycle, etc.

Progesterone, too, may stimulate or inhibit gonadotropin release in female animals and women, depending on the general physiological status. The inhibiting effects, which are only seen in the presence of some oestrogen, are probably responsible for suppressing the pre-ovulatory LH peak (and hence ovulation) during the early stages of pregnancy in at least some species, and provide the basis for inclusion of progesterone in most oral contraceptives. Progesterone probably acts by suppressing the actions of oestrogens at the hypothalamic level, and the actions of LHRH (and possibly oestrogens) at the pituitary level.

Androgens, particularly testosterone, inhibit the release of LH (but probably not FSH) in male animals and man. They may act at both the pituitary and hypothalamic levels. FSH secretion in the male appears to be inhibited by oestrogens (as androgen metabolites) rather than androgens themselves. The absence of complex reproductive cycles in males makes for a much more straightforward feed-back inhibition loop in the male than in the female.

In addition to these complex 'long-loop' feedback systems, release of gonadotropins may also be controlled partly by 'short-loop' feedback (e.g. levels of LHRH being modulated directly by circulating levels of LH) and even 'ultra-short-loop' feedback (e.g. circulating or local levels of LHRH directly inhibiting further secretion of LHRH).

7.   THE BIOLOGICAL ACTIONS OF TSH

TSH is the main regulator of thyroid gland growth and function, although its actions in this respect may be modulated by various other factors, including iodide, thyroid hormones themselves, catecholamines and growth hormone. TSH also has some extra-thyroidal effects, including stimulation of lipolysis, but the physiological significance of these is doubtful.

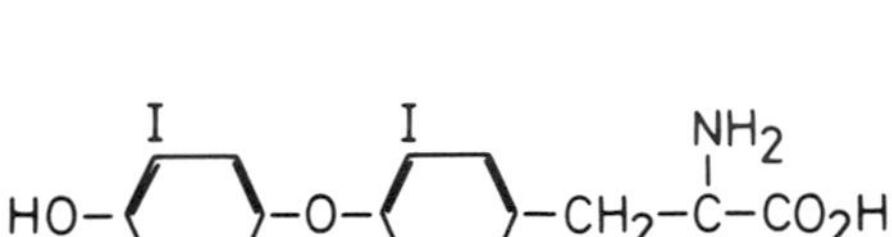

$$HO-\underset{I}{\overset{I}{\bigcirc}}-O-\underset{I}{\overset{I}{\bigcirc}}-CH_2-\underset{H}{\overset{NH_2}{\underset{|}{C}}}-CO_2H$$

L-thyroxine (T$_4$)

$$HO-\overset{I}{\bigcirc}-O-\underset{I}{\overset{I}{\bigcirc}}-CH_2-\underset{H}{\overset{NH_2}{\underset{|}{C}}}-CO_2H$$

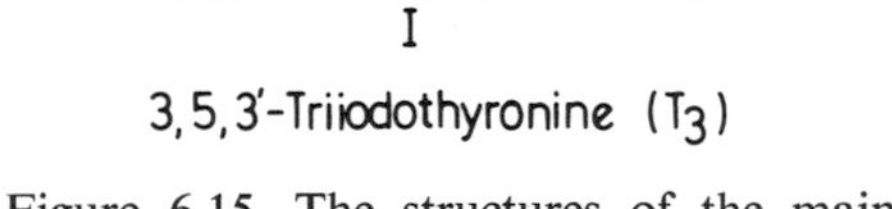

3,5,3'-Triiodothyronine (T$_3$)

Figure 6.15 The structures of the main thyroid hormones

7.1   Thyroid hormone production[42-44]

The production of thyroxine (T4) and tri-iodothyronine (T3) (Figure 6.15) in the thyroid gland is a complex process which involves production of a protein precursor, thyroglobulin, iodination of tyrosine residues in this, coupling of pairs of iodinated tyrosine residues and proteolytic degradation of the iodinated protein (see also Chapter 1).

Thyroglobulin is a large glycoprotein (molecular weight approx. 680 000) containing two polypeptide chains, a large number of cystine residues, but not an exceptional proportion of tyrosine. It is secreted from the thyroid gland cell into an extracellular space, the 'colloid space'. Iodination appears to be largely localized to the cell–lumen interface, but it is not clear whether this is intracellular or extracellular. Iodination of tyrosine residues (which are presumably situated in the tertiary structure to favour accessibility for iodination) is catalysed by a peroxidase, which in the presence of hydrogen peroxide promotes conversion of iodide to iodine. It may be that an enzyme-bound iodinium ion (I^+) or iodine radical $(I^.)$ is actually the active iodinating species. This then iodinates tyrosine residues to mono- or di-iodotyrosine.

It is not yet clear whether coupling of two di-iodotyrosine molecules occurs intermolecularly or intramolecularly. The coupling can occur spontaneously, but it is likely that catalysis by thyroid peroxidase can speed the process.

Whatever the site of iodination of thyroglobulin, it is clear that the iodinated protein is largely stored as extracellular colloid. Prior to conversion of this to T3 and T4, it is engulfed by endocytosis (facilitated by the formation of pseudopods), giving rise to colloid droplets within the cell. These subsequently fuse with lysozomes, which contain a variety of proteolytic enzymes and digest the thyroglobulin, forming free T3 and T4. Much of the products of digestion, including uncoupled iodotyrosines, is reutilized by the cell, but a large proportion of the T3 and T4, together with some free iodide is secreted. The mechanism of secretion is not clear, but it seems likely that exocytosis is not involved.

Although T3 is much more potent than T4 in most target cells, T4 is the predominant form secreted by the thyroid gland. However, there is considerable conversion of T4 to T3 by the peripheral tissues.

7.2   The actions of TSH[45,46]

TSH stimulates many aspects of thyroid metabolism. A partial distinction can be made between fairly rapid actions which mainly relate to increased secretion of thyroid hormones (e.g. activation of secretion, stimulation of iodide uptake, stimulation of energy metabolism) and relatively delayed effects which mainly relate to increased production of the hormones (e.g. increased protein and RNA synthesis, increased cell size, increased cell division).

The most striking rapid effect of TSH is increased uptake of colloid by endocytosis. Within 2 minutes of treating a thyroid slice *in vitro* with TSH, pseudopod formation occurs, and intracellular accumulation of colloid droplets rapidly follows. Stimulation of iodide uptake by the thyroid is a slower process, but is seen within 1 hour of treatment with TSH. Iodination of thyroglobulin is not normally limiting, but may become so at high iodide concentrations; TSH can then be shown to stimulate iodination too, possibly by increasing the rate of production of H_2O_2.

The rapid effects of TSH are not blocked by inhibitors of protein or RNA synthesis, but the long-term effects are, and it seems likely that these effects on protein synthesis and growth of the thyroid gland reflect actions at the transcriptional level. The growth-promoting effects of TSH on the thyroid are dramatically demonstrated by the clinical condition of goitre. Underproduction of T3 and T4, usually due to inadequate dietary iodine, results in oversecretion of TSH (due to lack of negative feedback) and gross overproliferation of thyroidal tissue.

7.3   The biochemical mode of action of TSH[47,48]

Cyclic AMP is thought to be the main mediator of at least the rapid actions of TSH on the follicular cells of the thyroid gland. The effects of cyclic AMP are modulated by iodide and cyclic GMP, and intracellular calcium levels may also play a part in regulating thyroid metabolism.

TSH binds to a specific membrane receptor and stimulates adenylate cyclase in much the same way as many other hormones activate this enzyme. Increased adenylate cyclase activity leads to increased cyclic AMP levels, which in turn lead to activation of cyclic AMP-dependent protein kinase. As with other comparable hormone systems, the identification of the key substrates for this protein kinase, whose phosphorylation leads to stimulation of thyroid metabolism and secretion of T3 and T4, has proved difficult. Most of the protein phosphorylation occurring in the thyroid gland appears to be catalysed by protein kinases which are not cyclic-AMP-dependent, and is not affected by TSH. Two proteins whose phosphorylation is stimulated by TSH are histone H_1 and a putative 'contractile protein'. Phosphorylation of histone H_1 could be involved in the actions of TSH on growth and transcription. The contractile protein that is subject to phosphorylation could be a part of the cytoskeletal system involved in endocytosis of colloid.

The production of cyclic AMP appears to be inhibited by iodide. This is probably a physiological mechanism designed to limit thyroid hormone production if excess iodide is present in the diet. It has been proposed that oxidation of iodide by thyroid peroxidase and its trapping by thyroglobulin is accompanied by formation of a factor ('compound XI') which inhibits adenylate cyclase or stimulates phosphodiesterase. However, the nature and mechanism of action of compound XI is not known.

The TSH receptor of the thyroid gland has been characterized in some detail, and shown to be a protein complex of molecular weight about 200 000. Various studies indicate that it is a glycoprotein and contains a disulphide bridge, the integrity of which is essential for TSH binding. Serum from patients with

Graves' disease, a condition characterized by excessive secretion of thyroid hormones, usually contains antibodies directed against the TSH receptor; it appears that these autoantibodies interact as agonists with the receptor and stimulate the gland in the same way as the hormone itself. In some cases autoantibodies that block the actions of the hormone are also found.[48]

8. SYNTHESIS AND SECRETION OF TSH[49–51]

8.1 Synthesis and secretion

TSH is produced in thyrotrophs in the anterior pituitary. Little is known about the biosynthesis of the β-subunit but it is most unlikely that the α- and β-subunits are synthesized on a common precursor.

TSH is stored in rather small granules, which are thought to be secreted by exocytosis, as in the case of other pituitary hormones.

8.2 Control of secretion. Thyrotropin releasing hormone, TRH

The main hypothalamic factor regulating TSH secretion is the tripeptide TRH. This was the first hypothalamic peptide hormone to be isolated, by the groups of Schally and Guillemin (see Chapter 3). It has the structure,

$$\text{pyroGlu–His–Pro–NH}_2.$$

Since the preparation of synthetic TRH in 1969/1970, its biological and other actions have been the subjects of intensive investigation. Various analogues have been prepared, a few of which have activity greater than that of natural hormone, while a few inhibit the actions of TRH itself.

TRH stimulates secretion of TSH both *in vitro* and *in vivo*, and is being used clinically as a test of pituitary function. It also stimulates secretion of prolactin in many species, although the physiological significance of this effect remains obscure. Different TRH analogues show parallel abilities to stimulate release of TSH and prolactin, suggesting that the TRH receptor is similar on thyrotrophs and lactotrophs. TRH also occurs extensively throughout the brain and CNS, and has been shown to have various behavioural effects. It thus may act as a neurotransmitter or neuromodulator as well as a regulator of pituitary functions.

Despite its small size, TRH appears to act like other peptide hormones via binding to a membrane receptor. This may then cause activation of adenylate cyclase and increase in intracellular cyclic AMP levels, although there is some doubt about this, mainly because thyrotrophs constitute only a small proportion of the cells in most *in vitro* pituitary preparations. Enhanced cyclic AMP production does not seem to be obligatory for the effects of TRH on prolactin secretion (see Chapter 8).

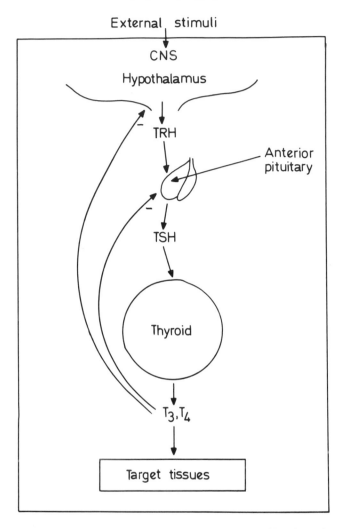

Figure 6.16 The hypothalamic–pituitary–thyroid axis and
its control by feedback mechanisms

The overall physiological regulation of TSH secretion is maintained by a
hypothalamic–pituitary–thyroid axis, which constitutes a typical feedback
loop.[41] Secretion of TRH from the hypothalamus is inhibited by thyroid hormones, which cause lowered secretion of TSH and hence less production of T3
and T4. However, various features complicate the picture (Figure 6.16). In
particular, T3 and T4 also act at the level of the anterior pituitary gland to
inhibit TSH production and secretion. Inhibition is probably effected at the
transcriptional level by binding of the hormones to receptors in chromatin, and

appears to result in reduced production of TSH and of TRH receptors (thus lowering the sensitivity of the thyrotrophs to TRH). Secretion of TRH may also be inhibited by TSH (a 'short-loop' feedback) and possibly by TRH itself (an 'ultra-short-loop' feedback).

REFERENCES

1. Wide, L. (1976) In: *Hormone Assays and their Clinical Applications*, 4th edn (Eds. Loraine, J. A. and Bell, E. T.). Churchill Livingstone, Edinburgh, pp. 87–140.
2. Butt, W. R. (1979). In: *Hormones in Blood*, 3rd. edn., Vol. 1. (Eds. Gray, C. H. and James, V. H. T.). Academic Press, London, pp. 411–471.
3. Reichert, L. E., Leidenberger, F. and Trowbridge, C. G. (1973) *Recent Progr. Hormone Res.*, **29**, 497–532.
4. Butt, W. R. (1975) *Hormone Chemistry*, 2nd edn., Vol. 1. Ellis Horwood, Chichester, pp. 140–194.
5. Pierce, J. G. and Parsons, T. F. (1981) *Ann. Rev. Biochem.*, **50**, 465–495.
6. Sairam, M. R. and Li, C.H. (1978) In: *Hormonal Proteins and Peptides*, Vol. 6 (Ed. Li, C. H.). Academic Press, New York, pp. 1–56.
7. Pierce, J. G., Faith, M. R., Guidice, L. C. and Reeve, J. R. (1976) In: *Polypeptide Hormones: Molecular and Cellular Aspects*. Elsevier, Amsterdam, pp 225–250.
8. Sairam, M. R. (1983) In: *Hormonal Proteins and Peptides*, Vol. 11 (Ed. Li, C. H.). Academic Press, New York, pp. 1–79.
9. Wallis, M. (1975) *Biol. Rev*, **50**, 35–98.
10. Mise, T. and Bahl, O. P. (1980) *J. Biol. Chem.*, **255**, 8516–8522.
11. Talmadge, K., Vamvakopoulos, N. C. and Fiddes, J. C. (1984) *Nature*, **307**, 37–40.
12. Bahl, O. P. and Moyle, W. R. (1978) In: *Receptors and Hormone Action*, Vol. 3 (Eds. Birnbaumer, L. and O'Malley, B. W.). Academic Press, New York, pp. 261–289.
13. Farmer, S. W. and Papkoff, H. (1979) In: *Hormones and Evolution*, Vol. 2 (Ed. Barrington, E. J. W.). Academic Press, New York, pp. 525–559.
14. Setchell, B. P. (1982) In: *Reproduction in Mammals*, 2nd. edn., Vol. 1 (Eds. Austin, C. R. and Short, R. V.). Cambridge University Press, pp. 63–101.
15. Steinberger, E. (1979) In: *Endocrinology*, Vol. 3 (Eds. DeGroot, L. J., Cahill, G. F., O'Dell, W. D., Martini, L., Potts, J. T., Nelson, D. H., Steinberger, E. and Winegrad, A. I.). Grune & Stratton, New York, pp. 1501–1509.
16. Bardin, C. W. (1978) In: *Reproductive Endocrinology* (Eds. Yen, S. S. C. and Jaffe, R. B.). Saunders, Philadelphia, pp. 110–125.
17. Hansson, V., Calandra, R., Purvis, K., Ritzen, M. and French, F. S. (1976) *Vitamins and Hormones*, **34**, 187–214.
18. Ewing, L. L. and Zirkin, B. (1983) *Recent Progr. Hormone Res.*, **39**, 599–635.
19. Wright, W. W. and Frankel, A. I. (1980) *Endocrinology*, **107**, 314–318.
20. Catt, K. J. and Dufau, M. L. (1978) In: *Receptors and Hormone Action*, Vol. 3 (Eds. Birnbaumer, L. and O'Malley, B. W.). Academic Press, New York, pp. 291–339.
21. Payne, A. H., Chase, D. J. and O'Shaughnessy, P. J. (1982) In: *Cellular Regulation of Secretion and Release* (Ed. Conn, P. M.). Academic Press, New York, pp. 355–408.
22. Cooke, B. A., Dix, C. J., Magee-Brown, R., Janszen, F. H. A. and Van der Molen, H. J. (1981) *Adv. Cyclic Nucleotide Res.*, **14**, 593–609.
23. Dorrington, J. H. and Armstrong, D. T. (1979) *Recent Progr. Hormone Res.*, **35**, 301–342.
24. Means, A. R., Dedman, J. R., Tash, J. S., Tindall, D. J ., van Sickle, M. and Welsh, M. J. (1980) *Ann. Rev. Physiol.*, **42**, 59–70.

25. Reichert, L. E. and Abou-Issa, H. (1978) In: *Receptors and Hormone Action*, Vol. 3 (Eds. Birnbaumer, L. and O'Malley, B. W.). Academic Press, New York, pp. 341–361.
26. Ross, G. T. and Schreiber, J. R. (1978) In: *Reproductive Endocrinology* (Eds. Yen, S. S. C. and Jaffe, R. B.). Saunders, Philadelphia, pp. 63–79.
27. Baker, T. G. (1982) In: *Reproduction in Mammals*, 2nd. edn. Vol. 1 (Eds. Austin, C. R. and Short, R. V.). Cambridge University Press, pp 17–45.
28. Short, R. V. (1984) In: *Reproduction in Mammals*, 2nd edn. Vol. 3 (Eds. Austin, C. R. and Short, R. V.). Cambridge University Press, pp. 115–152.
29. Yen, S. S. C. (1978) In: *Reproductive Endocrinology* (Eds. Yen, S. S. C. and Jaffe, R. B.). Saunders, Philadelphia, pp. 126–151.
30. Goodman, A. L. and Hodgen, G. D. (1983) *Recent Progr. Hormone Res.*, **39**, 1–73.
31. Richards, J. S. (1979) *Recent Progr. Hormone Res.*, **35**, 343–373.
32. Leung, P. C. K. and Armstrong, D. T. (1980) *Ann. Rev. Physiol.*, **42**, 71–82.
33. Salomon, Y. (1980) In: *Cellular Receptors for Hormones and Neurotransmitters* (Eds. Schulster, D. and Levitzki, A.). Wiley, Chichester, pp. 149–161.
33a. Erickson, G. F. (1983) *Mol. Cell. Endocrinol.*, **29**, 21–49.
34. Chatterjee, M. and Munro, H. N. (1977) *Vitamins and Hormones*, **35**, 149–208.
35. Fiddes, J. C. and Goodman, H. M. (1980) *Nature*, **286**, 684–687.
36. Matsuura, S., Ohashi, M., Chen, H. C., Shownkeen, R. C., Stockell Hartree, A., Reichert, L. E., Stevens, V. C. and Powell, J. E. (1980) *Nature*, **286**, 740–741.
37. Jutisz, M., Berault, A., Debeljuk, L., Kerdelhue, B. and Theolayre, M. (1979) In: *Hormonal Proteins and Peptides*, Vol. 7 (Ed. Li, C. H.). Academic Press, New York, pp. 55–122.
38. Sharp, P. J. and Fraser, H. M. (1978) In: *The Endocrine Hypothalamus* (Eds. Jeffcoate, S. L. and Hutchinson, J. S. M.). Academic Press, London, pp. 271–332.
39. Fraser, H. M. (1982) *Nature*, **296**, 391–392.
39a. Jan, Y. N. and Jan, L. Y. (1982) *Frontiers in Neuroendocrinology*, **7**, 211–230.
39b. Gregory, H., Taylor, C. L. and Hopkins, C. R. (1982) *Nature*, **300**, 269–271.
40. Knobil, E. (1980) *Recent Progr. Hormone Res.*, **36**, 53–88.
41. Piva, F., Motta, M. and Martini, L. (1979) In: *Endocrinology*, Vol. 1 (Eds. DeGroot, L. J., Cahill, G. F., O'Dell, W. D., Martini, L., Potts, J. T., Nelson, D. H., Steinberger, E. and Winegrad, A. I.). Grune & Stratton, New York, pp. 21–33.
42. Taurog, A. (1979) In: *Endocrinology*, Vol. 1 (Eds. DeGroot, L. J., Cahill, G. F., O'Dell, W. D., Martini, L., Potts, J. T., Nelson, D. H., Steinberger, E. and Winegrad, A. I.). Grune & Stratton, New York, pp. 331–342.
43. Nunez, J. and Pommier, J. (1982) *Vitamins and Hormones*, **39**, 175–229.
44. Taurog, A. (1974). In: *Handbook of Physiology*. Section 7, Vol. III (Eds. Greer, M. A. and Solomon, D. H.). American Physiological Society Washington, D.C., pp. 101–133).
45. Dumont, J. E. and Vassart, G. (1979) In: *Endocrinology*, Vol. 1 (Eds. DeGroot, L. J., Cahill, G. F., O'Dell, W. D., Martini, L., Potts, J. T., Nelson, D. H., Steinberger, E. and Winegrad, A. I.). Grune & Stratton, New York, pp. 311–329.
46. Van Herle, A. J., Vassart, G. and Dumont, J. E. (1979) *New England J. Med.*, **301**, 307–314.
47. Dumont, J. E., Takeuchi, A., Lamy, F., Gervy-Decoster, C., Cochaux, P., Roger, P., Van Sande, J., Lecocq, R. and Mockel, J. (1981) *Adv. Cyclic Nucleotide Res.*, **14**, 479–489.
48. Rees Smith, B. and Buckland, P. R. (1982) *Ciba Foundation Symposium*, **90**, 114–132.

49. Reichlin, S., Martin, J. B. and Jackson, I. M. D. (1978) In: *The Endocrine Hypothalamus* (Eds. Jeffcoate, S. L. and Hutchinson, J. S. M.). Academic Press, London, pp. 229–269.
50. Morley, J. E. (1981) *Endocrine Rev.* **2**, 396–436.
51. Jackson, I. M. D. (1982) *New England J. Med.,* **306**, 145–155.
52. Fiddes, J. C. and Goodman, H. M. (1979). *Nature,* **281**, 351–356.
53. Seeburg, P. H. and Adelman, J. P. (1984) *Nature,* **311**, 666–668.

Chapter 7

Hormones of the adenohypophysis: Growth hormone

1. INTRODUCTION

Growth hormone, prolactin and placental lactogen are members of a family of protein hormones showing clear structural resemblances, despite the fact that the first two originate in the pituitary gland while the third is placental in origin (see Chapter 3). The structural resemblances are to some extent paralleled by similarities in biological actions, although the rather general actions of growth hormone on the body contrast somewhat with the specialized actions of prolactin on the mammary gland. It seems likely that the biochemical modes of action of growth hormone and prolactin will also be found, eventually, to be similar, but as yet such similarities are not very evident. Growth hormone will be considered in this chapter, but an account of the structural and biological relationships between growth hormone, prolactin and placental lactogen will be deferred to the next chapter, and discussed there in the light of what is known of the molecular evolution of these hormones.

2. THE CHEMISTRY OF GROWTH HORMONE[1]

2.1 Isolation

The ability of pituitary extracts to stimulate growth was recognized in the 1920s, and attempts to purify the factor responsible soon followed. For some time in the 1930s many workers considered that a unique growth hormone might not exist, since synergism between (impure) prolactin and thyrotropin promoted substantial somatic growth in some animals. However, by the middle 1940s a protein hormone was isolated from beef anterior pituitary glands by Li and his coworkers, and independently by Wilhelmi and his colleagues, which had potent growth-promoting activity and which was almost free of both prolactin and thyrotropin. This was designated growth hormone or somatotropin.

Since that time pituitary growth hormone has been isolated from a wide variety of mammals, and from some non-mammalian vertebrates. Human growth hormone was isolated in 1956, and has proved important for clinical purposes. The earlier methods for purification of growth hormone utilized mainly selective precipitation techniques, but more recently a range of alternative purification methods has been described which use the more powerful methods of chromatography, electrophoresis, etc. These give the hormone in greater yields and with considerably increased purity. It is also worth noting that many of the more recent methods described for isolation of growth hormone also allow purification of other pituitary hormones in the course of the purification scheme—permitting more economical use of what is usually rather scarce starting material.

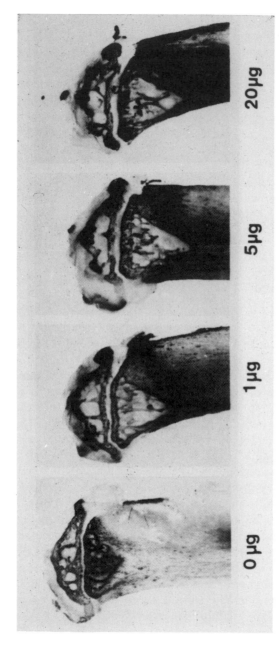

Figure 7.1 The 'tibia-test' assay for growth hormone. Injection of growth hormone (doses shown) into hypophys-ectomized animals increases the width of the epiphyseal cartilage of the tibia. (Reprinted with permission from Lostroh, A. J. and Li, C. H. (1957) *Endocrinology* **60**, 308–317. Copyright (1957) The Endocrine Society)

2.2   Assay[3,4]

Biological assays for growth hormone mostly use as their endpoint increase in body weight or some other aspect of growth. The earlier assays, which are still widely used, measured the increase of body weight of adult female rats or hypophysectomized rats after injection of growth hormone; a good relation can be established between the dose of hormone administered and the increase in body weight. Hypopituitary dwarf mice (a genetic strain of mice in which a defect in the pituitary leads to underproduction of growth hormone and several other pituitary hormones) can also be used as the test animals, increase in body weight again being followed. An alternative bioassay for growth hormone is to follow the increase in the width of the proximal tibial epiphysial cartilage (the tibia test) in hypophysectomized rats (Figure 7.1). Growth hormone increases the width of this cartilage (a corollary of increased cartilage, and hence skeletal, growth in many parts of the body), and the increase in width is proportional to the amount of growth hormone administered. This assay is rather quicker and more sensitive than the assays utilizing body weight, but involves more experimental manipulation and is probably rather less specific.

These *in vivo* bioassays have been available, and used, for 30–40 years, during which time various improvements have been described. They have the usual disadvantages associated with *in vivo* assays, including high cost, low sensitivity and variability, and there have been many attempts to devise *in vitro* bioassays which might replace them. Unfortunately, none of these alternatives has proved generally satisfactory, and the *in vivo* assays remain the ones used most commonly by those wishing to measure the biological activity of growth hormone preparations.

Immunoassays for growth hormone have been successfully developed, and are the methods of choice for measuring growth hormone levels in serum and in secretion studies.[4] The first immunoassay for growth hormone used the red blood cell haemagglutination inhibition principle, but this has now been completely superseded by radioimmunoassay methods (Chapter 2), of which many variants are available. The now widely used chloramine-T method for iodination of protein for use in radioimmunoassay was first used in connection with the radioimmunoassay of growth hormone. Nevertheless, despite the large amount of work that has been devoted to the technique, radioimmunoassay of growth hormone remains a rather more difficult procedure than radioimmunoassay of some other polypeptide hormones.

Radioreceptor assays have also been described for growth hormone.[5] One, using binding to a human lymphocyte cell line, is quite specific for human growth hormone, whereas others, which utilize receptors made from rabbit or rat liver, show a much wider responsiveness (Figure 7.2). The receptor assays combine to some extent the sensitivity and precision of immunoassays with the

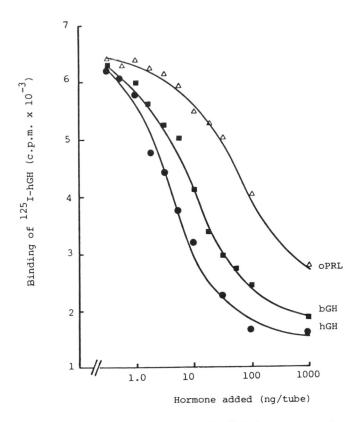

Figure 7.2 Displacement of ^{125}I-labelled human growth hormone from solubilized receptors from pregnant-rabbit liver by various hormones. hGH, human growth hormone; bGH, bovine growth hormone; oPRL, ovine prolactin. (Data from Cadman, H. F. (1981) D.Phil. thesis, University of Sussex)

biological relevance of bioassays, although as yet there remains some doubt as to whether the receptors really are the mediators of biological actions.

An important reservation which applies to all immunoassays and, to some extent, receptor assays, is whether the material that is being measured (in plasma, etc.) really does correspond to the biologically active hormone. In the case of growth hormone this is a considerable problem. As will be discussed below, several workers have found a serious discrepancy between the amount of growth hormone detected in blood by radioimmunoassay and the amount detected by bioassay. The significance of this remains to be fully understood, but the need for caution in interpreting results obtained by radioimmunoassay must be emphasized (see also Chapter 2).

2.3   Structure and chemistry[1,6]

2.3.1   *Primary structure*

Early work on the chemistry of growth hormone led to proposals for unusual two-chain molecules. More recent results have shown that the hormone in all species studied is a protein of molecular weight about 20 000 containing a single polypeptide chain of about 200 amino acid residues. There is a tendency for (non-covalent) aggregation of molecules to occur, and in some species, such as the ox and sheep, dimers predominate. The protein is not conjugated with carbohydrate or lipid moieties.

The complete amino acid sequence of human growth hormone was first described by Li and his coworkers in 1966.[7] Several errors were subsequently discovered in this sequence, and the definitive sequence must be considered to be that determined from the corresponding nucleic acid sequence.[8] Complete sequences of growth hormones from several other species have now been described including the ox (Figure 7.3), sheep, rat and horse, and partial sequences for growth hormones from additional species have also been determined (summarized in ref. 9). In all cases, the primary structure contains two disulphide bridges, one of which links distant parts of the molecule while the other forms a small loop near the C-terminus.

The detailed three-dimensional structure for growth hormone is not known, mainly because crystals suitable for X-ray diffraction studies have not been prepared, despite considerable effort. Use of the physical techniques of circular dichroism and optical rotatory dispersion has provided some information about the secondary structure of growth hormones from various species. All contain a good deal of α-helix, 50–60% of the polypeptide chain having this conformation.

2.3.2   *Species variation*

Comparison of the amino acid sequences of growth hormones from different species reveals a considerable amount of variation,[9] and such variation is also expressed in many other physicochemical properties. Thus, bovine growth hormone is a fairly basic protein (isoelectric point about 8), whereas human growth hormone is acidic (isoelectric point about 5); bovine growth hormone exists in concentrated solution mainly as a dimer, whereas the human hormone exists as a monomer. These differences mean that one has to be very careful in extrapolating from one species to another—for example, isolation procedures highly successful for human growth hormone may be quite inappropriate for bovine growth hormone.

Species variation is also apparent with regard to the immunological and biological properties of growth hormone. Thus, there is virtually no immunological cross-reaction between the bovine and human hormones. Non-primate growth

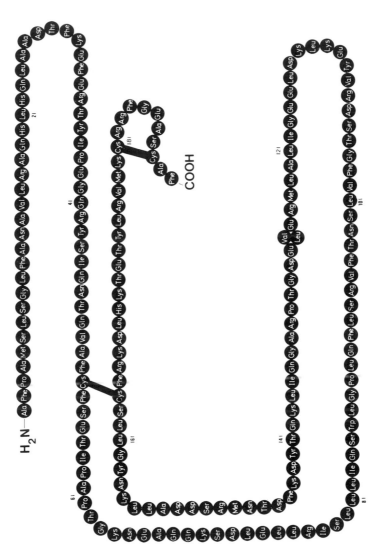

Figure 7.3 The amino acid sequence of bovine growth hormone

hormones have little biological activity in man or primates—an important aspect of the biological specificity of the hormone which will be discussed further below. Primate growth hormones possess lactogenic activity, unlike the non-primate proteins.

The species variation seen in growth hormone can be explained in terms of the molecular evolution of the hormone, as will be discussed further in Chapter 8.

2.4 The relationship between structure and function[1]

Lack of a known three-dimensional structure for the growth hormone molecule and of a simple *in vitro* bioassay precludes a full understanding of the features of the hormone which are responsible for binding to the receptor and producing a biological response. Nevertheless, there has been a good deal of work on the hormone aimed at determining which features of the primary structure are (or are not) required for activity. Such studies also provide evidence that the multiple actions of growth hormone may reside in different parts of the molecule. Only a few aspects of this work can be described here.

2.4.1 *Chemical modification*

Oxidation of the cystine bridges of growth hormone, or reduction (with mercaptoethanol) followed by alkylation with iodoacetic acid, giving carboxymethylcysteine:

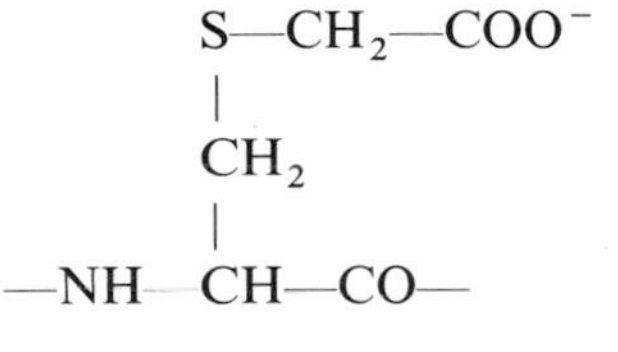

$$S—CH_2—COO^-$$
$$|$$
$$CH_2$$
$$|$$
$$—NH—CH—CO—$$

completely inactivates the molecule. However, reduction followed by alkylation with iodoacetamide, giving carbamidomethylcysteine:

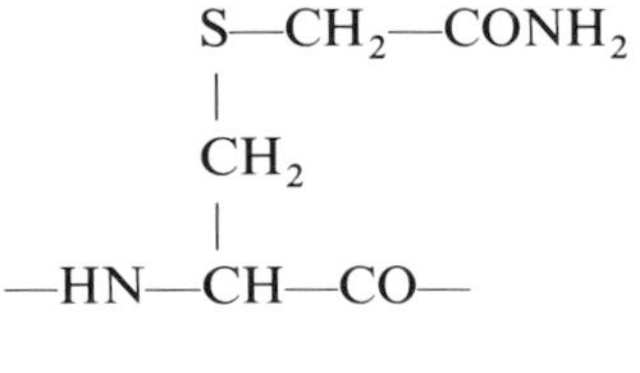

$$S—CH_2—CONH_2$$
$$|$$
$$CH_2$$
$$|$$
$$—HN—CH—CO—$$

yields a derivative which retains almost full growth-promoting activity. Thus, the disulphide bridges are not required for the expression of biological activity (or for the maintenance of three-dimensional structure) though they may increase the stability of the hormone in the circulation.

Oxidation of the four methionine residues of bovine growth hormone with hydrogen peroxide can be carried out without loss of activity. However, carboxymethylation of these residues, with iodoacetate at low pH, giving carboxymethyl methionine:

$$
\begin{array}{c}
CH_3 \\
| \\
S^{\pm}\!-\!CH_2COO^- \\
| \\
CH_2 \\
| \\
CH_2 \\
| \\
-\!NH\!-\!CH\!-\!CO\!-
\end{array}
$$

leads to almost complete inactivation.

The reactivity of the methionines towards carboxymethylation has been studied; methionine-124 is much less reactive than any of the other methionine residues. Interestingly, methionine-124 is the one methionine that is conserved in all of the growth hormone sequences that have been studied so far; its low reactivity suggests that it may be located away from the surrounding solvent.

2.4.2 Active fragments of growth hormone[10]

Many indications suggest that peptides which represent only part of the amino acid sequence of growth hormone may retain significant biological activity. For example, a peptide corresponding to residues 96–134 of bovine growth hormone (derived by partial tryptic digestion) has been shown to retain about 10% of the growth-promoting activity of the parent hormone.

Of considerable interest is work on the partial digestion of growth hormone with the enzyme plasmin. Under controlled conditions this enzyme cleaves human growth hormone on the C-terminal side of residues 134 and 140, releasing a hexapeptide, and leaving two large peptides joined by a disulphide bridge. This derivative retains almost full biological activity. The two component peptides can be separated after reduction and alkylation of the disulphide bridge (Figure 7.4), and the larger of the two retains significant, but low, activity in several bioassays. When the two peptides are allowed to recombine, a derivative can be recovered which has regained almost full biological activity (Figure 7.4). Hybridization experiments based on this, in which a peptide from one member of the hormone family is recombined with a complementary peptide from another member, are considered further in Chapter 8.

Digestion with several other enzymes, and cleavage using specific chemical methods (such as cyanogen bromide) has confirmed the concept that not all of

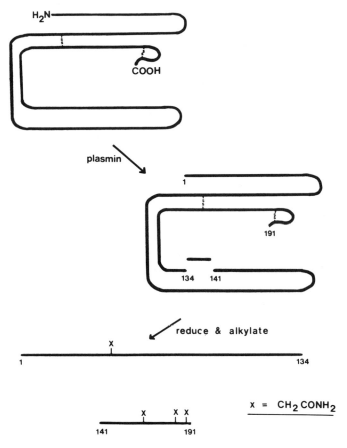

Figure 7.4 Diagram illustrating how human growth hormone can be converted to a two-chain molecule by digestion with the proteolytic enzyme plasmin

the growth hormone molecule is required for retention of at least some biological activity. Retention of activity in relatively small peptides is rather surprising, in view of the fact that growth hormone is a protein with a clear-cut three-dimensional structure. Little, if any, of the tertiary structure can be retained in the 39-residue active peptide isolated from partial tryptic digests, which suggests that the tertiary structure may not be a crucial element in the biological actions of the hormone.

2.5 Chemical synthesis of growth hormone

The complete chemical synthesis of human growth hormone was described in 1970. The synthetic hormone possessed biological activity about 15% of that of the natural hormone. It subsequently became apparent, however, that this synthesis was based on a primary structure which was wrong in several important respects (see Section 2.3), and this, together with the fact that the synthetic protein showed, not surprisingly, considerable heterogeneity, has cast some doubts on the result obtained. Nevertheless, the derivative did show some biological activity and the synthesis still represents a remarkable achievement.

Syntheses of many shorter regions of the growth hormone molecule have also been reported; some of the resultant peptides show low growth-promoting and other biological activities, as mentioned above.

2.6   Naturally-occurring variants of growth hormone[11]

Most preparations of growth hormone display considerable heterogeneity, a good deal of which is due to the occurrence of multiple forms of the hormone in the pituitary glands from which it is prepared. The causes of this heterogeneity are various, and their detailed investigation reveals a good deal about the biochemistry of the hormone. Many of the variants have different biological properties from the main form of the hormone, and may be physiologically important. Heterogeneity at the N-terminus of bovine growth hormone (presence or absence of an additional alanyl residue) appears to be a consequence of ambiguous processing of the precursor of the hormone, pre-growth hormone,[12] while heterogeneity at residue 127 (leucine in some molecules, valine in others) represents an allelic variation in the population from which the hormone is made, some cows having leucine, some valine and others (heterozygotes) both residues at this position.

Preparations of human growth hormone are remarkably heterogeneous.[11,13] Particularly interesting is a variant (the 20K variant) described by U. J. Lewis in which 15 residues (residues 32–46) are deleted,[14] although the molecule remains a single chain. This variant apparently occurs in all human pituitaries, making up about 15% of the total growth hormone content; it has normal growth-promoting activity but its insulin-like activity is much lower than that of normal growth hormone. The origin of this variant is not clear, but it is noteworthy that the beginning of the deleted sequence corresponds exactly to the position of an intervening sequence in the human growth hormone gene; it seems unlikely that this is a coincidence, and probable that the deletion in the protein is a consequence of a lengthening of the intervening sequence in the gene[15] (see also Chapter 18).

Heterogeneity in human growth hormone also reflects the occurrence of forms of the hormone which include one or more enzymic cleavages within their structure. Such 'nicked' forms may be due to modification occurring during preparation of the hormone, but they may also reflect processing of the hormone within the gland *in vivo*.

Finally, size heterogeneity of growth hormone should be mentioned. A substantial proportion of growth hormone in many species occurs as forms of molecular weight greater than that of the predominating monomer. Fairly stable dimers frequently represent 5–20% of the total growth hormone, and occur in the circulation as well as the pituitary gland; they are partly linked covalently by disulphide bridges. Such forms are potentially important, but their true significance remains to be elucidated. It seems unlikely, however, that they represent precursors of growth hormone.

3.   THE BIOLOGY OF GROWTH HORMONE

3.1   Biological actions of growth hormone

3.1.1   *The regulation of somatic growth*

The obvious major action of growth hormone is to promote overall somatic growth. Thus, removal of the pituitary gland from an animal leads to complete cessation of growth, and injection of purified growth hormone can restore growth almost to normal. In some cases injection of growth hormone into non-growing adult animals (such as female rats) will lead to resumption of growth, and may cause giantism. Defective secretion of growth hormone in animals and man can cause dwarfism, which in many cases can be treated successfully by injection of the purified hormone (Figure 7.5). The causes of such dwarfism are discussed in Section 6. Oversecretion of growth hormone by the pituitary (usually due to the presence of a tumour) can cause giantism or, in adult humans in whom growth of the long bones is no longer possible, acromegaly.

The growth-promoting effects of growth hormone apply both to the skeleton and the soft tissue, especially muscle. Actions on skeletal growth involve the stimulation of cartilage growth (which will eventually largely give rise to bone)—administration of growth hormone *in vivo* to a suitable animal stimulates both cell division of chondrocytes and production of cartilage matrix components (principally proteoglycans). The increase in width of the tibial epiphyses caused by growth hormone, already discussed as the basis of the 'tibia test' assay, is a direct consequence of the actions of the hormone on this cartilagenous growth plate.

The growth-promoting effects of growth hormone on cartilage, in mammals at least, appear to be indirect, being mediated by a family of insulin-like peptide hormones, the somatomedins, although direct actions on cartilage have also been proposed.[16] Whether the somatomedins also mediate the actions of growth hormone on other tissues is not yet clear. Their role will be discussed in detail in Section 4.2.

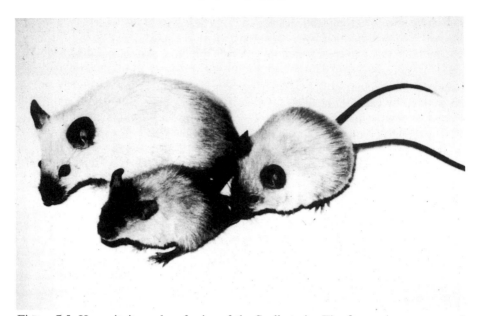

Figure 7.5 Hypopituitary dwarf mice of the Snell strain. The figure shows a normal animal, a hypopituitary dwarf and a dwarf that has received hormone treatment, leading to resumption of growth. (Photograph courtesy of Dr A. T. Holder)

3.1.2   Multiple actions of growth hormone[17,18]

Growth hormone affects many metabolic processes, including nucleic acid and protein synthesis, and the metabolism of carbohydrates and lipids. In many animals the hormone has a marked diabetogenic activity. It is often difficult to decide which of these many effects are primary ones, and which secondary—clearly any hormone that has major effects on protein synthesis in many organs of the body may produce indirect effects on other aspects of metabolism. Quite how the various effects of growth hormone are integrated to produce coordinated growth is not well understood. The biochemical basis of the actions of growth hormone will be discussed in Section 4.

3.1.3   Growth hormone and nutrition

It is clear that in the overall biological context an important function of growth hormone is to regulate the growth of the young animal in relation to the available food supply. Indeed, the nutritional status is presumably the main environmental variable which provides a requirement for hormonal control of growth in the animal. In a completely stable environment, growth could be predetermined by a genetic programme, and day-to-day regulation would be unnecessary. (Environmental variability provides the need for most forms of hormonal control, of course, but usually the link is more obvious.) The effect of a variable food supply is well illustrated by the phenomenon of 'catch-up growth'—an underfed young animal or child shows deficient growth, and a subsequent period of good nutrition leads to greater than normal growth, with recovery of much of the lost ground. The rapid growth occurring in such a 'catch-up' period is, to some extent at least, under hormonal control.

It is notable that in the fasting animal growth hormone levels change markedly. Surprisingly, however, the response in man and the rat is different. In man, fasting leads to an increase in growth hormone levels, whereas in the rat they fall.

3.1.4 Biological specificity; human growth hormone

It is important to re-emphasize here the species specificity of growth hormones, the structural basis of which has already been discussed. All mammalian growth hormones are about equally active in the rat (though that of teleost fish is not)—a factor which has been important for the many experimental studies carried out in this animal. However, non-primate growth hormones are not active in man or monkey. This means that only human (or potentially monkey) growth hormone can be used for clinical purposes. Human growth hormone is in fact used quite extensively for the treatment of hypopituitary dwarfism in man,[19,20] and active schemes have been established in many countries for collecting human pituitary glands at post-mortem for the extraction of growth hormone and other pituitary hormones. Human growth hormone produced by recombinant DNA techniques in bacteria is also active in man.

Not surprisingly, however, human growth hormone remains in short supply, and it is important that only children with dwarfism clearly attributable to lack of growth hormone are treated with the hormone. Absence of detectable circulating growth hormone, as measured by radioimmunoassay, even after treatment designed to stimulate secretion of the hormone (e.g. infusion of arginine) is a good indicator of those patients likely to respond to treatment with the hormone. Among the several types of human dwarf who do not grow when given growth hormone is a group ('Laron dwarfs') which has unusually high circulating growth hormone levels; in such individuals dwarfism is due to defective somatomedin production or inability of the target tissues to respond to the hormone (possibly because of defective receptors).

There is some evidence that human growth hormone may have other clinical applications in addition to treatment of dwarfism. For example, the hormone appears to have beneficial effects on the healing of stress ulcers.[21] However, at present too little hormone is available for widespread use in such treatment; production of the hormone by genetic engineering will improve the situation (see Chapter 18).

Although the lack of action of non-primate growth hormones in man is the most notable example of species specificity, other aspects are also important. Thus, human growth hormone, unlike most mammalian growth hormones (but like prolactin), has substantial lactogenic activity. Furthermore, although most non-primates show a fairly uniform growth response to growth hormone, other responses vary greatly from one species to another. For example, the diabetogenic activity of the hormone is particularly marked in the dog.

3.2   What is the active form of growth hormone *in vivo*?

A certain amount of evidence suggests that growth hormone may exist in an altered form in the circulation, with increased and/or altered activity, compared to the stored form of the hormone. Thus, measurements of the concentration of growth hormone in serum and in pituitary incubation medium are often greater if made by bioassay than radioimmunoassay, and attempts to isolate the bioactive material from such sources suggest a molecule with physicochemical properties quite different from the stored form of growth hormone.[22] Activated forms of the hormone have also been identified as minor components in pituitary extracts.[11,13]

Other workers in the field have proposed that relatively small fragments of growth hormone may be produced *in vivo* and may bring about many of the actions of the hormone on carbohydrate metabolism.[18] Separate fragments with diabetogenic and insulin-like activity have been described, as well as separate peptides with lipolytic activity.

If the views of these various workers are accepted, the stored form of growth hormone must be viewed as a prohormone, which is processed to give one or more different peptides with altered biological properties. The view remains very controversial, however, although it is certainly established, by chemical and enzymatic modification methods, that the multiple actions of growth hormone are associated with different features of the molecule,[17] and presumably are mediated by different receptors.

4.   THE BIOCHEMICAL MODE OF ACTION OF GROWTH HORMONE

The main biological role of growth hormone is to produce and regulate growth. As has been mentioned, at least some of these growth-promoting actions are mediated by somatomedins. Growth hormone also has many short-term effects on metabolism, but it is difficult to relate these in detail to the long-term action of growth promotion. As with most polypeptide hormones, the first step in the action of growth hormone on its target cells involves interaction with a receptor in the plasma membrane.

4.1 Growth hormone receptors[5]

Receptors for growth hormone are found in many different tissues, and are particularly prevalent in the liver. In some species the liver growth hormone receptors appear to be rather complex, several different kinds of receptor being present, as determined by their binding specificity towards growth hormones and related proteins from various species. Growth hormone receptors from rabbit liver have been partially purified, using particularly affinity chromatography on columns containing immobilized growth hormone or monoclonal antibodies. Receptor purified in this way appears to have a molecular weight of about 300 000. Receptors from rat and rabbit liver have also been characterized by affinity labelling (cross-linking) experiments; these suggest a subunit molecular weight of about 40 000–70 000.[23]

4.2 The somatomedins[24–26]

4.2.1 The discovery of somatomedins (sulphation factor)

Salmon and Daughaday in 1957 noticed that although growth hormone administered to hypophysectomized rats caused a rapid and marked increase in the ability of cartilage to take up labelled sulphate (a measure of the rate of synthesis of the proteoglycan of the cartilage matrix), sulphate incorporation by cartilage incubated *in vitro* was not stimulated by the hormone applied *in vitro*. Sulphate incorporation *was* stimulated by serum from normal rats or growth hormone-treated hypophysectomized rats, but much less by serum from untreated hypophysectomized rats or hypopituitary dwarf mice (Figure 7.6). On the basis of these observations, they proposed that the actions of growth hormone on sulphation of cartilage are indirect, and mediated by a factor which they referred to as a sulphation factor. It subsequently became apparent that this, or a related factor, mediates most of the actions of growth hormone on cartilage, including increase of DNA synthesis (measured by ³H-thymidine incorporation) and cell division, RNA synthesis, protein synthesis and proteoglycan synthesis. Sulphation factor was renamed somatomedin in 1972, and it is now clear that there is in fact a whole family of somatomedins.[24,25]

4.2.2 Isolation of somatomedins

The somatomedins are probably produced mainly in the liver, but are not stored there or in any other organ. As a consequence, the only suitable source of the peptides is blood serum or plasma, and this (or protein fractions from serum) has been the starting material for most attempts to purify somatomedins. The concentration of somatomedins in blood serum is low (less than 1 µg/ml) so purification from this source is extremely difficult. Various groups have succeeded in isolating small quantities of material from human serum, using

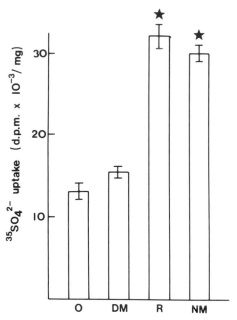

Figure 7.6 Actions of serum on sulphate incorporation into cartilage from young, fasted rats. O, control; DM, 5% serum from hypopituitary mice; R, 5% serum from intact male rats; NM, 5% serum from normal mice. *indicates significant stimulation. (Data from Holder, A. T. (1978) D. Phil. thesis, University of Sussex)

large-scale precipitation techniques followed by higher resolution chromatographic and electrophoretic techniques. In several cases, somatomedin preparations isolated by different groups differ considerably in their chemical and biological properties, and it is now clear that this is due to the presence of more than one somatomedin in serum. Use of different isolation procedures coupled with different assays led to selective enrichment of one or other of the different somatomedins by different groups. Occurrence of at least three different human somatomedins (A, B and C) is now recognized, although it is likely that more than one different form of somatomedin A exists and that somatomedin B is a completely different peptide which would be better not included in the somatomedin family.

4.2.3   Assay

Most somatomedin bioassays are based on the ability of these peptides to stimulate anabolism of cartilage *in vitro*. Stimulation of sulphate incorporation (into mucopolysaccharides) and thymidine incorporation (into DNA) are the most widely used techniques. The nature of the tissue used is important, rat and pig cartilage apparently responding preferentially to somatomedin C and chick chondrocytes to somatomedin A. Somatomedin B was isolated using an assay based on effects on cultured glial cells, but as mentioned above, this peptide is probably wrongly termed a 'somatomedin' (it has no effect in the cartilage assays).

Immunoassays have been described for all of the somatomedins, but doubts remain about the soundness of some of these because of doubts about the purity of the somatomedin preparations used to establish them. In many cases, however, there is good agreement between the values obtained by bioassay and radioimmunoassay. Radioreceptor assays for somatomedins have also been set up, using in most cases receptor preparations made from placenta (a particularly good source of such receptors). In some cases labelled insulin has been used as the labelled ligand for such receptor assays—insulin binds weakly to the somatomedin receptor.

4.2.4   Chemical nature

Chemical studies carried out so far suggest that somatomedins A and C are insulin-like peptides of molecular weight 8000–9000. Each consists of a single polypeptide chain, so they are comparable to proinsulin rather than the two-chain molecule of insulin itself. Somatomedin A is a slightly acidic peptide, whereas somatomedin C is basic. The complete amino acid sequence of somatomedin C has been determined, and is identical to that of insulin-like growth factor I (see below). The complete sequence of somatomedin A has now also been determined, and suggests that this peptide is virtually identical to somatomedin C.

In the plasma, somatomedins A and C occur in tight association with a specific binding protein of molecular weight about 50 000. An important consequence of this is to increase the half-life of somatomedin in the circulation—free somatomedin has a half-life of 10–20 minutes, whereas in association with the binding protein this period is increased to about 3 hours.

4.2.5   The relation of somatomedins to other growth factors; the somatomedin family

With the isolation of the somatomedins it became clear that these peptides are very similar to several other types of 'growth factor'. The first of these is non-suppressible insulin-like activity (NSILA) which has now been resolved into two peptides, renamed insulin-like growth factors (IGF) I and II.[27] NSILA was first recognized as a factor in serum which possessed insulin-like biological activity, but which could not be inhibited by antibodies to insulin. Like the

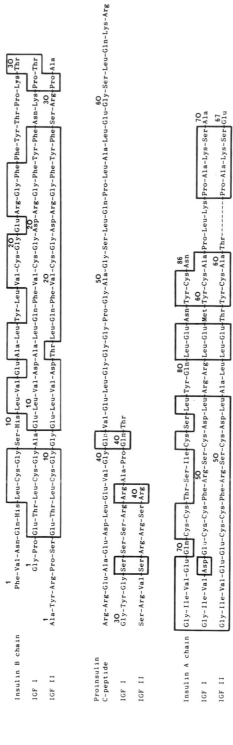

Figure 7.7 Amino acid sequences of human proinsulin, IGF I and IGF II. The sequences of the insulin B chain, proinsulin C-peptide and insulin A chain are shown, and compared with corresponding regions of the sequences of IGF I and IGF II. Identical sequences are enclosed in 'boxes'

somatomedins, NSILA had to be isolated from serum, an enormously difficult task. Eventually, however, significant quantities of these peptides were isolated from human serum, and it was then realized that their insulin-like activity is relatively weak, but that they have potent growth-promoting activity, particularly on cartilage (hence their redesignation as insulin-like growth factors). Their properties are thus very similar to those of the somatomedins. The chemistry of these factors has been studied in detail, their amino acid sequences have been determined (Figure 7.7) and predictions have been made about their three-dimensional structure.[28] The amino acid sequence of IGF I appears to be the same as that of somatomedin C, indicating the identity of these two hormones. The sequences of IGF I and IGF II are homologous with that of proinsulin.

Another growth factor that resembles the somatomedins is the multiplication stimulating activity (MSA) described by Dulak and Temin. This is a factor produced by a few strains of cultured rat liver cells. It has the property of stimulating the proliferation of many types of cell in tissue culture (hence the name) and has been purified and characterized from conditioned medium. Its amino acid sequence is similar to that of human IGF II. It also has biological activity on cartilage similar to that of somatomedins, and it has been suggested that it corresponds to, or is very similar to, rat IGF II. A further somatomedin/IGF that has been partially characterized by Herington and coworkers is an acidic form, which might correspond to a two-chain molecule analogous to insulin.

There is thus a whole family of peptides making up the somatomedins and their relatives. Precisely how many members of the family really exist in their own right is not yet clear, but it seems likely that at least four such peptides occur in human plasma, and can be separated on the basis or differences in isoelectric point: somatomedin C/IGF I, IGF II/MSA, somatomedin A and one or more acidic somatomedins.

It should also be made clear at this point that many other peptide growth factors exist, each of which plays an important part in regulating the growth of one or more cell types. It is possible that at least some of these factors, like somatomedins, are partially under the control of growth hormone. Whether they should be considered as true hormones is not yet clear. A full treatment of this increasingly important topic is not possible in this book. Some of the better-characterized of these peptide growth factors are listed in Table 7.1; a full account may be found in ref. 29 and 29a.

4.2.6 *The production of somatomedins and its regulation*[30,31]

The somatomedins are produced by, but not stored in, the liver, the kidney and several other tissues. Perfused liver and cultures of hepatocytes and fibroblasts have been used to study the production of the peptides; in each of these systems

Table 7.1   Some polypeptide tissue growth factors

Factor	Source	Chemical Nature	Main Actions
Epidermal growth factor (EGF) (urogastrone)	mouse submaxillary gland; human urine; several other sites	peptide; 53 residues	stimulates (1) epidermal growth and keratinization; (2) growth of corneal epithelium; (3) gastric acid secretion; (4) growth of GI tract
Platelet-derived growth factor (PDGF)	blood platelets	protein; mol. wt 28 000–38 000	stimulates proliferation of fibroblasts and many other cell types; wound healing?
Fibroblast growth factor (FGF)	brain; pituitary gland	polypeptide; mol. wt approx. 13 000	stimulates proliferation of many cell types, including fibroblasts, adrenal cells, chondrocytes and endothelial cells
Nerve growth factor (NGF)	salivary gland; snake venom; some cultured cell lines	protein; several subunits	stimulates growth and development of peripheral nerve system
Somatomedins (see text)	liver; kidney; many other tissues	peptides; 50–80 residues	stimulates growth and development of peripheral nervous system
Haemopoietic colony stimulatory factors	several organs and cell lines	glycoproteins	haemopoietic cells
Angiogenesis factors	many tissues	?	induce formation of blood vessels
Chondrocyte growth factor	pituitary gland	peptide	stimulates chondrocyte growth

Note: Other, as yet poorly-characterized, growth factors have been described. Further details may be found in refs. 29 and 29a.

growth hormone stimulates somatomedin production, and prolactin and insulin may have similar effects. Nutritional status also appears to regulate somatomedin production. Since the somatomedins are not stored to any significant extent, stimulation of their secretion presumably involves either increased synthesis, or increased production from a precursor. The somatomedin binding protein is also apparently synthesized by the liver, and its synthesis is increased by growth hormone. It is not clear, however, whether synthesis of binding protein and somatomedins is linked.

Considerable evidence suggests that the somatomedins are produced as precursor molecules of higher molecular weight. Thus, mRNA prepared from MSA-producing cells codes for the production of material immunologically similar to MSA, but of molecular weight about 22 000, and short-term labelling of such cells also provides evidence for precursors.[32] The nature of the precursor of somatomedin C/IGF I has been defined by cloning cDNA corresponding to the mRNA for this precursor.[33] This suggests a precursor containing either 130 or 153 amino acids (Figure 7.8); the details of the processing pathway

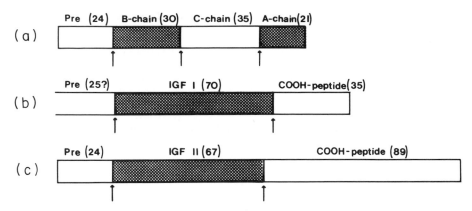

Figure 7.8 Diagram illustrating the structures predicted for the precursors of (a) insulin (preproinsulin), (b) IGF I/Somatomedin C, and (c) IGF II. The arrows indicate sites of proteolytic cleavage occurring during conversion of precursors to native hormones. Shaded regions indicate sequences of insulin, IGF I and IGF II respectively

involved in the maturation of this precursor to the 70-residue peptide of somatomedin C/IGF I are not yet clear. The precursor of IGF II is rather longer than that of IGF I (Figure 7.8).

The levels of somatomedins A and C in normal human serum are of the order of 0.1–1 µg/ml. Levels fluctuate considerably less on an hour-to-hour basis than for growth hormone. Levels are low in hypopituitary dwarfs (human or mouse) and high in acromegaly. They are often low in 'Laron dwarfs' despite very high growth hormone levels. Somatomedin levels increase somewhat at the time of the prepubertal growth spurt in man. It should also be noted that serum contains inhibitors of the action of somatomedins which may also play an important role in the overall regulation of growth.

4.2.7 The biochemical mode of action of somatomedins

Rather little is known about this topic, largely because of lack of pure somatomedins. Perhaps the most immediate question as yet unanswered is: Which of the actions of growth hormone are mediated by somatomedins and which are direct? It is fairly generally accepted that in mammals, at least, the actions of growth hormone on cartilage (and hence skeletal) growth are mainly somatomedin mediated. A role for somatomedins in regulating the metabolism of other tissues is less clear, however. Growth hormone has direct effects on skeletal muscle, liver and other tissues *in vitro* (see below) and the need for somatomedin mediation is less obvious. On the other hand, somatomedins also are reported to have effects on these tissues. The question remains unresolved, despite considerable research effort. It has now been shown, however, that somatomedin C/IGF I can promote growth in hypophysectomized rats.[34]

Receptors for somatomedins are found in many different tissues, the placenta being a particularly rich source,[5,35] although little is known about the action of somatomedins on the placenta. As with most other polypeptide hormones, the receptors are associated with the plasma membrane. The IGF I/somatomedin C receptor is structurally similar to the insulin receptor, and probably contains 4 polypeptide chains, 2α and 2β, joined by disulphide bridges. The IGF II receptor is quite different, and appears to contain only a single polypeptide chain. The IGF I receptor is associated with a protein kinase which is activated when the receptor binds the growth factor, leading to phosphorylation of tyrosine residues on some cellular proteins.[36]

At least some of the actions that somatomedins A and C have on cartilage may be mediated by cyclic AMP.[37] Five per cent serum from normal (but not hypophysectomized) rats increased amino acid transport, chondroitin sulphate synthesis, protein synthesis and RNA synthesis in chick cartilage, and also elevated cyclic AMP levels. Dibutyryl cyclic AMP and theophylline produced similar effects. Adenosine inhibited these effects, however, and it is possible that cartilage metabolism is subject to a dual control by cyclic AMP and adenosine. It is thus possible that in the case of these actions of somatomedins, at least, the formation of a hormone–receptor complex is linked to activation of adenylate cyclase. The details of such linkage are as yet poorly understood in the case of cartilage, and a role for cyclic AMP in the actions of somatomedins is not usually accepted.

In addition to their metabolic effects on cartilage, somatomedins are mitogenic for many cell types, including chondrocytes and fibroblasts. The mitogenic actions require the presence of other growth factors, which probably vary from one cell type to another. In the case of fibroblasts, somatomedins synergize with platelet-derived growth factor and epidermal growth factor to promote cell division.

4.3   Actions of growth hormone on protein biosynthesis in liver and muscle[38]

4.3.1   *In vivo actions on nitrogen metabolism*

When growth hormone is administered to an animal *in vivo*, it causes general nitrogen retention; levels of urinary nitrogen and blood urea fall. These effects may be largely due to a stimulation of protein synthesis which occurs in many tissues, most notably (but certainly not exclusively) in skeletal muscle and liver. This increase in protein synthesis is quite rapid, occurring within 30 minutes of injection of growth hormone into hypophysectomized rats.

4.3.2   *In vitro actions on skeletal muscle*

The rat diaphragm has been a favourite model for studies of the action of hormones on skeletal muscle, partly because its thinness increases viability and accessibility *in vitro*. When the diaphragm from a hypophysectomized rat is incubated *in vitro* with a labelled amino acid and a suitable buffer, it will incorporate amino acids into protein. Growth hormone will stimulate such incorporation, and the stimulation appears to reflect a direct effect on protein biosynthesis (although amino acid transport is also stimulated). Insulin also stimulates such amino acid incorporation (like somatomedins) but there are clear-cut differences between the actions of insulin and growth hormone, and growth hormone is able to bring about its effects in the presence of anti-insulin serum. A lag period of about 30 minutes is characteristic of the action of growth hormone, as is a refractory period after 2–3 hours of stimulation. Another important difference between insulin and growth hormone is that the former may alter breakdown (turnover) of protein as well as synthesis, whereas the latter increases synthesis without a concomitant change in protein breakdown.[39]

The effects of growth hormone on diaphragm are generally rather small (characteristically 30–40% stimulation of protein synthesis, although larger effects have been achieved by some workers) and this has often been interpreted as meaning that the *in vitro* actions may be irrelevant with regard to the large *in vivo* effects of the hormone on growth. This may be misleading, however. Protein turnover is very rapid in muscle (and most other tissues), rapid protein breakdown balancing rapid protein synthesis in a non-growing animal such as the hypophysectomized rat. Under such circumstances, a fairly small percentage stimulation of synthesis, without a corresponding stimulation of breakdown, could in fact give rise to a large increase in growth rate. The percentage stimulation of protein synthesis by growth hormone *in vivo* and *in vitro* is in fact fairly similar.

The biochemical mechanism whereby growth hormone stimulates protein synthesis in skeletal muscle is not understood. A role of cyclic AMP has been proposed by some authors and hotly disputed by others. A role for polyamines in the mechanism has also been proposed. Detailed studies of the step in protein synthesis in skeletal muscle is not understood. A role for cyclic AMP has been ability of the ribosomes to promote peptide bond formation rather than increases the number of ribosomes engaged in protein synthesis. The hormone may enhance ribosomal activity by accelerating the peptidyl transferase reaction or the translocation process (or both).

4.3.3   *Actions of growth hormone on the liver*[40]

Growth hormone stimulates protein synthesis in the liver *in vivo*, and in perfused liver *in vitro*. In the perfused liver it also stimulates amino acid uptake and the synthesis of all forms of RNA, which may explain some, but probably not all, of the effects on protein synthesis. As in the case of effects on diaphragm, for most effects on liver there is a 15–60 minute delay. The classical experiments of

Korner showed that ribosomes from livers of hypophysectomized rats possess a reduced ability to carry out protein synthesis, and that this defect is remedied by administration of growth hormone *in vivo*.[41] As for diaphragm, the effect may be primarily on the rate at which the growing polypeptide chain elongates, rather than on the rate of initiation of synthesis of new polypeptides.

Growth hormone does not stimulate the production of all hepatic proteins equally. The case of somatomedin production has already been considered, but this might be thought exceptional. In fact the hormone also selectively stimulates production of several enzymes, particularly those associated with amino acid metabolism such as tyrosine transaminase, tryptophan pyrrolase and ornithine decarboxylase.[42] Other hormones also affect hepatic levels of these enzymes, and the results are difficult to interpret in terms of a simple induction by growth hormone.

Another hepatic protein which is selectively controlled by growth hormone is α_{2u} globulin.[43] This is synthesized in the liver of the male rat and appears in the urine of the animal. Its biological function is not clear. Hypophysectomy abolishes the synthesis of this protein, and normal production can only be restored by administration of four hormones: a glucocorticoid, an androgen, thyroxine and growth hormone. Messenger RNA for α_{2u} globulin cannot be detected in the liver of hypophysectomized rats; treatment with a glucocorticoid, an androgen and thyroxine has little affect on mRNA levels. Further treatment, with growth hormone, greatly increases the production of mRNA for α_{2u} globulin and the protein itself. Thus, growth hormone appears to regulate transcription of the genes for α_{2u} globulin in the liver (although it must be noted that these experiments were carried out *in vivo*, and actions of somatomedin on the liver cannot yet be ruled out). Possibly all the effects of growth hormone on protein synthesis reflect the summation of selective stimulation of transcription or translation of a large number of specific proteins.

4.4   Actions of growth hormone on lipid and carbohydrate metabolism

The actions of growth hormone discussed in Sections 4.2 and 4.3 can be clearly linked to its growth-promoting actions. However, the hormone also has other biological actions which are not so easily linked to growth promotion.

4.4.1   Lipid metabolism[44]

In vivo, growth hormone causes a drop in the level of plasma non-esterified fatty acids (NEFA), followed by a prolonged increase in this level. This appears to be due to increased utilization of lipids—increased uptake of NEFA by muscle somewhat preceding increased output by adipose tissue. The overall effect is that growth hormone tends to divert energy metabolism from carbohydrate catabolism to lipid catabolism—the hormone acting to counteract the action of insulin. The actions of growth hormone on lipid metabolism are particularly marked in man, where growth hormone levels are elevated on fasting, and the hormone presumably plays an important part in stimulating the increased lipid utilization associated with fasting. In the rat, on the other hand, growth hormone levels *fall* on fasting, and here it is less easy to explain the effects of the hormone on lipid metabolism (which are still apparent, though less marked than in man). Again, the species specificity of growth hormone must be stressed.

In vitro, growth hormone has many effects on adipose tissue, including stimulation of lipolysis and actions on glucose utilization. These effects accord well with *in vivo* actions. Some evidence suggests that the actions of the hormone on lipolysis are mediated by cyclic AMP.

4.4.2 Carbohydrate metabolism[45]

In vivo, growth hormone has both insulin-like and diabetogenic effects.[45a] In suitable animals, including rats, both effects can be demonstrated, a transient hypoglycaemia being followed by a prolonged hyperglycaemia. The latter appears to be a consequence of reduced glucose uptake by peripheral tissues and is often associated with elevated insulin levels (due presumably to increased plasma glucose or possibly a direct effect of growth hormone on the pancreas). A classical view of the mechanism of growth hormone proposed that it acted by increasing insulin levels and blocking all of the consequent actions of insulin except for those on protein synthesis; the view is no longer tenable as an explanation of all the actions of growth hormone, but the mechanism could still play a part, in some animals at least. How growth hormone reduces glucose uptake by peripheral tissues is not clear. Direct antagonism may occur, or alternatively the effect may be a consequence of the elevated levels of non-esterified fatty acids consequent upon the lipolytic actions of the hormone, which may act to inhibit glucose uptake by muscle. Thus, because of the operation of the glucose/fatty acid cycle, action of the hormone at one site may produce several metabolic consequences. Whatever the mechanism of the diabetogenic effects, it is clear that they can in some cases be an aetiological factor in diabetes.

In vitro, particularly in diaphragm, growth hormone stimulates glucose uptake and utilization (i.e. has insulin-like effects); diabetogenic effects are difficult to demonstrate *in vitro*, although some such effects have been reported.

Growth hormone has also been shown to have direct effects on the pancreas, enhancing insulin production and secretion. This too, of course, will have important effects on carbohydrate and lipid metabolism (see also Chapter 10).

4.5 Growth hormone and cellular differentiation

Growth hormone has been shown to be involved in the differentiation of several cell types. Studied in considerable detail has been the conversion of fibroblasts of the 3T3 cell line into adipocytes,[46] which requires the presence of growth hormone. The change involves morphological alterations, enzymic differentiation, increased hormone responsiveness, triglyceride accumulation and a limited mitogenic response. Somatomedin C has no effect on this transformation. This action of growth hormone has been studied mainly *in vitro*, but various *in vivo* actions of the hormone may be related to it.

Growth hormone may also be involved in the differentiation of muscle. It has been shown to stimulate the conversion of myoblasts to myotubes *in vitro*, with associated changes in the levels of creatine phosphokinase. It has been suggested that growth hormone may have direct effects on the differentiation of cartilage, acting synergistically with somatomedins to promote cartilage growth. Such a view would accord with observations that have been made that growth hormone may sometimes have direct effects on cartilage, but it is not universally accepted.

4.6 The actions of growth hormone: further aspects

Growth hormone has direct or indirect effects on many tissues additional to cartilage, liver, muscle and adipose tissue. Actions on pancreas and the lymphoid system may be particularly important. These additional actions serve to confuse further what is already an extremely complex picture. Particularly difficult to understand is the apparent conflict between the observations that (in man) growth hormone is a key hormone at times of both active growth and starvation. Metabolic needs would appear to be almost completely reversed between the two conditions, and yet growth hormone levels are elevated in both. Hopefully future research will resolve the apparent paradox.

5.   BIOSYNTHESIS AND SECRETION OF GROWTH HORMONE

5.1   Biosynthesis[47]

Growth hormone is produced in the pituitary in a group of acidophilic cells termed somatotrophs (Figure 3.3). In male animals these are the predominant cells of the anterior pituitary, and growth hormone makes up about 40% of the total hormonal content of the gland. In female animals (rats at least) somatotrophs and lactotrophs (producing prolactin) occur in approximately equal numbers. Low concentrations of growth hormone have also been found in the brain, and it seems likely that the brain can produce some growth hormone, as well as many other pituitary hormones.

With the discovery that insulin is made in the form of a precursor, the possible existence of precursors was studied for many other protein hormones. Large forms of growth hormone are found in both the pituitary gland and the

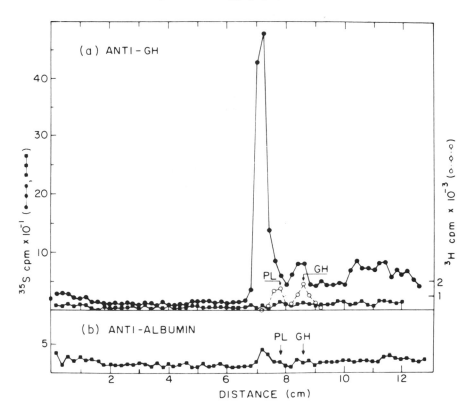

Figure 7.9 An *in vitro* experiment demonstrating the existence of pregrowth hormone. Messenger RNA for rat growth hormone was translated in a wheat germ cell-free system in the presence of [^{35}S] methionine. Labelled proteins were precipitated with antiserum to growth hormone (a) or serum albumin (b) and then subjected to polyacrylamide gel electrophoresis followed by estimation of radioactivity in slices of the gel. The major peak in (a) (——●——) is due to pre-growth hormone; no equivalent peak is seen when antiserum to albumin was used (b) or when mRNA was omitted (——■——). The position of unlabelled growth hormone (GH) and prolactin (PL) markers is indicated (- - -○- - -). (Reproduced with permission from Sussman, P. M., Tushinski, R. J. and Bancroft, F. C, (1976) *Proc. Nat. Acad. Sci. USA* **73**, 29–33. Copyright (1976) the authors)

circulation and the possibility that these might represent precursors of the hormone was considered, but is now thought unlikely. The existence of a short-lived precursor of growth hormone was demonstrated conclusively by Bancroft,[47] who prepared mRNA from cultures of a growth hormone-secreting rat pituitary tumour cell line. This mRNA was translated in a cell-free protein-synthesizing system made from wheat germ (containing ribosomes and other enzymes and components needed for protein synthesis, but lacking endogenous mRNA and proteases). The products of cell-free synthesis were precipitated by

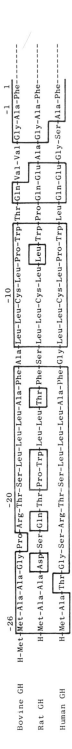

Figure 7.10 The amino acid sequences of the signal peptides of bovine, rat and human growth hormones. Identical regions of sequence are enclosed in 'boxes'. The sequence of the mature hormone starts at residue 1

antiserum to rat growth hormone and characterized by SDS gel electro-phoresis. The main growth-hormone-like product ran with a mobility somewhat slower than that of growth hormone, indicating a rather greater size, by 2000–3000 daltons (Figure 7.9). It was named pre-growth hormone. It is not normally detected in the pituitary gland because it has a very short half-life (about 30 seconds), but small quantities of labelled pre-growth hormone were detected in pituitary cells after biosynthetic experiments in the presence of a protease inhibitor.

The complete amino acid sequences of pre-growth hormones from several species have been deduced from the sequence of the corresponding mRNAs (see Chapter 18). The additional peptide sequence, compared with growth hormone itself, comprises 26 amino acid residues at the N-terminus (Figure 7.10) and is thus a 'classical' signal peptide of the type proposed by Blobel (see Chapter 1) the function of which is to facilitate secretion of the nascent growth hormone chain into the lumen of the endoplasmic reticulum. Once the hormone has entered the endoplasmic reticulum, the signal peptide is rapidly removed by one or more proteases ('signalase'), a process which may occur even before synthesis has been completed. Characterization of this pre-growth hormone structure more or less precludes the possibility of larger covalent precursors of the hormone.

Application of recombinant DNA techniques has provided a good deal of in-formation about the genes for growth hormone. This is considered in Chapter 18.

5.2 Secretion

Within the somatotroph, growth hormone is stored in the form of granules of diameter 300–400 nm. The process of packaging of these, and mechanisms of secretion, appear to be similar to those seen in other polypeptide hor-mone-producing cells.

5.3 Regulation of growth hormone secretion. GHRH and somatostatin

The secretion of growth hormone has been thought classically to be primarily under positive control. If links with the hypothalamus are disrupted (see Chap-ter 3) secretion of the hormone slows markedly, suggesting that hypothalamic factors normally stimulate secretion of the hormone.

5.3.1 Experimental methods

The advent of good radioimmunoassays for growth hormone and availability of various experimental systems has led to intensive study of growth hormone secretion during the past 20 years. As with other pituitary hormones, systems used include intact animals and various *in vitro* preparations including whole pituitary glands, slices of glands, dispersed cells, purified somatotrophs and cultured cells (including cell lines derived from pituitary tumours) (see also

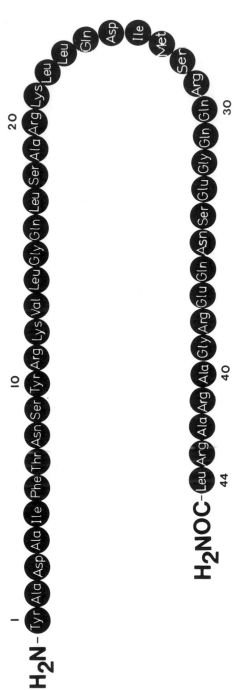

Figure 7.11 The amino acid sequence of human growth hormone releasing factor. The peptide shown is the largest of the naturally occurring GHRH peptides (GHRH$_{1-44amide}$); shorter forms (especially GHRH$_{1-40}$) also occur naturally

Chapter 3). Systems for perifusing various types of preparation have also been devised, allowing rapid removal of the secreted products from the environment of the secreting cells (as would happen *in vivo*). Using these methods together with radioimmunoassay the ability of many factors to stimulate or inhibit growth hormone secretion has been studied, and the mechanism whereby they act has been investigated. It must be borne in mind, however, that there may be some 'processing' of growth hormone before or during secretion, in which case the hormone measured by radioimmunoassay would not necessarily be its most active form (Section 3.2). Indeed, some of the experiments on secretion lend extra weight to this idea.

5.3.2   Growth hormone-releasing hormone (GHRH)

The demonstration that a factor from the hypothalamus can stimulate growth hormone secretion led to several attempts to isolate this molecule. It was only in 1982, however, that isolation and characterization of GHRH was finally achieved. Two groups, led by Guillemin and Vale, respectively, independently isolated two slightly different peptides with GHRH activity, from pancreatic tumours which had given rise to acromegaly via excessive growth hormone secretion.[48,49] One peptide possessed 44 amino acid residues and an amidated C-terminus,[48] the other possessed 40 residues and a free C-terminus[49] (Figure 7.11). Similar peptides have subsequently been isolated from the hypothalamus of several species, and it now seems likely that they represent the true hypothalamic GHRHs. Antibodies to these peptides have been prepared and used to demonstrate their distribution within the brain (the hypothalamus is the richest source, but several other regions also contain GHRH). Administration of antibodies to GHRH *in vivo* has been used to demonstrate the importance of the releasing factor in the physiological control of growth hormone secretion.

The content of GHRH in the hypothalamus is very low (about 1 pmole in the human hypothalamus) which explains the difficulties that were encountered in earlier attempts to isolate the peptide. GHRH has been shown to elevate growth hormone levels in adult humans. It also produces growth hormone release in some human hypopituitary children, and may be useful for treatment of short stature in some such patients.

Analogues of GHRH have been prepared. The C-terminal third of the molecule appears to be relatively unimportant for biological activity, and the peptide comprising residues 1–29 of GHRH with an amidated C-terminus is almost as active as the full 44-residue peptide. GHRH shows sequence homology with some hormone peptides found in the gut.

The precursor of GHRH has been partially characterized, using recombinant DNA techniques.[50] It is substantially larger than GHRH, and its proteolytic processing gives rise to other peptides which might also possess biological activity (Figure 7.12).

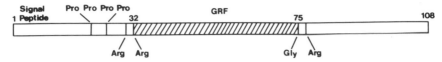

Figure 7.12 The structure predicted for the precursor of growth hormone releasing hormone. Based on data in ref. 50

5.3.3   Somatostatin[51–53]

During the course of one attempt to isolate GHRH, it was noted that some hypothalamic fractions actually inhibited growth hormone release. Attempts were therefore made to isolate this inhibitory factor, and these were successful and resulted in the isolation of small quantities of a highly active inhibitory peptide called growth hormone release inhibiting factor, subsequently renamed somatostatin.[51] This hormone was shown to be a peptide of 14 amino acid residues, the sequence of which was rapidly determined (Figure 7.13). Chemical synthesis of somatostatin was soon achieved (and more recently the peptide has also been made by genetic engineering—see Chapter 18), and a range of analogues has now been synthesized. Availability of substantial amounts of these synthetic peptides has fuelled an intensive research activity over the last few years.

Preparation of antibodies to somatostatin enabled its distribution within the body to be studied by both radioimmunoassay and immunofluorescence techniques. The hormone is found in many different tissues, including many parts of the brain (additional to the hypothalamus), the islets of Langerhans (D cells) and many parts of the gut. Although the concentration in the hypothalamus is much higher than that in most other organs, the total content in the gut or brain substantially exceeds that in the hypothalamus.

This wide distribution of somatostatin brought into question its role as a specific regulator of growth hormone secretion, and this role was further questioned when it was discovered that somatostatin has many other biological actions, including inhibition of secretion of thyrotropin, insulin, glucagon, several gut hormones, pancreatic enzymes and gastric acid. It is not yet clear which of these many actions are of physiological significance, nor whether somatostatin acts as a true hormone, via the circulation, or as a local secretion. It is notable that the peptide has a remarkably short half-life in the circulation—of the order of 2 minutes.

Evidence that circulating somatostatin is of physiological importance in the regulation of growth hormone secretion, at least, was provided by experiments utilizing antiserum to somatostatin *in vivo*[54]. Rats were injected with such antiserum and then fasted. The antiserum prevented the lowering of growth hormone levels usually seen in fasted rats, but had no effect on circulating insulin levels. Presumably increased secretion of somatostatin from either the

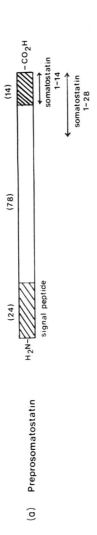

(a) Preprosomatostatin

H₂N— (24) signal peptide (78) (14) —CO₂H

somatostatin 1–14

somatostatin 1–28

(b) Somatostatin 1–28

H_2N-Ser-Ala-Asn-Ser-Asn-Pro-Ala-Met-Ala-Pro-Arg-Glu-Arg-Lys-Ala-Gly-Cys-Lys-Asn-Phe-Phe-Trp-Lys-Thr-Phe-Thr-Ser-Cys-CO_2H

(c) Somatostatin 1–14

H_2N-Ala-Gly-Cys-Lys-Asn-Phe-Phe-Trp-Lys-Thr-Phe-Thr-Ser-Cys-CO_2H

Figure 7.13 The structures of human somatostatin and its precursor. (a) Predicted structure of human preprosomatostatin showing the putative signal peptide (24 residues), the sequence of somatostatin$_{1-14}$ at the C-terminus and sites of cleavage leading to excision of somatostatin$_{1-14}$ and somatostatin$_{1-28}$. (b) Amino acid sequence of somatostatin$_{1-28}$. (c) Amino acid sequence of somatostatin$_{1-14}$. (Based in part on data of Shen, L-P., Pictet, R. L. and Rutter, W. J. (1982) *Proc. Nat. Acad. Sci. USA* **79**, 4575–4579)

hypothalamus or possibly the gut is normally responsible for the reduced growth hormone secretion, and this is blocked by the antiserum.

Many analogues of somatostatin have now been prepared and tested, using the type of approach discussed in Chapter 15. Some of these show preferential retention of one or other of the various activities of the peptide; for example, a derivative in which residue 4 is replaced by phenylalanine retains high activity with respect to inhibition of growth hormone release, but has much reduced activity on the secretion of insulin, etc.—and is thus almost exclusively a 'growth hormone release inhibiting peptide'.

Somatostatin appears to be synthesized as a precursor protein containing about 120 amino acids, the sequence of somatostatin being at the C-terminus.[55] Processing of this precursor can give rise to a longer form of somatostatin, containing 28 residues (extended at the N-terminus), and there is considerable evidence that this form is found in the circulation, has biological activity and is probably of physiological significance (Figure 7.13).

5.3.4 The mechanism of action of GHRH and somatostatin[56,57]

GHRH elevates cyclic AMP levels in pituitary tissue or cells *in vitro*, probably by activating adenylate cyclase. Derivatives of cyclic AMP stimulate growth hormone release, as does theophylline. As in other systems, activation of protein kinase appears to be a consequence of elevated cyclic AMP levels. This in turn presumably stimulates growth hormone secretion by phosphorylating key components of the secretory pathway, but the nature of these has not yet been determined. Prostaglandins also stimulate growth hormone secretion *in vitro* and may play a role in the mechanism of action of GHRH. Somatostatin does not appear to alter levels of cyclic AMP in the pituitary very markedly, although the evidence here is somewhat conflicting. Much of the evidence currently available suggests that it acts at a rather late stage in the secretory pathway, since it will block the stimulatory action of both cyclic AMP and prostaglandins as well as GHRH on growth hormone secretion. Somatostatin alters the flux of calcium into and out of the somatotroph, and its actions on calcium handling may be the key to its mechanism of action. As with most other polypeptide hormones, calcium is an essential requirement for active secretion of growth hormone, and modulation of calcium levels may play a major role in regulating secretion of the hormone. GHRH may act at this level in addition to its effects on cyclic AMP.

5.3.5 Physiological control of growth hormone secretion

The overall control of growth hormone secretion is different from that of most pituitary hormones in that simple feedback loops are less easy to recognize. Growth hormone release is stimulated *in vivo* by insulin-induced hypoglycaemia (although direct actions of insulin cannot be ruled out), which may relate to its diabetogenic actions, and by infusion of amino acids, particularly arginine, which could relate to its effects on protein synthesis. Both of these effects are mediated by the hypothalamus. Starvation stimulates growth hormone secretion in man but inhibits it in the rat. Feedback by circulating somatomedin levels probably also regulates growth hormone secretion.

As with other hypothalamic hormones, the control of release of GHRH and somatostatin is at least partly under the control of neural pathways in the brain, and the role of specific neurotransmitters in such pathways has been studied. The bulk of the available evidence suggests that α-adrenergic receptors stimulate growth hormone secretion and β-adrenergic receptors inhibit this process. Dopamine and serotonin may also be involved in the pathways in the CNS which regulate hypothalamic control of growth hormone secretion.

5.4   Regulation of growth hormone synthesis

A variety of agents may act directly on growth hormone synthesis, although indirect effects, via depletion of stored hormone, are sometimes difficult to exclude. GHRH has been shown to act directly to increase the rate of transcription of growth hormone genes.[58] Corticosteroids and thyroxine markedly stimulate the transcription of growth hormone genes in cultured pituitary tumour cells (GH cells),[59] but the relationship of these actions to effects occurring in the normal gland is not yet clear.

6.   THE MOLECULAR BASIS OF GROWTH HORMONE DEFICIENCY[60,61]

Lack of growth hormone in man and animals leads to hypopituitary dwarfism. It is now clear that this can be a consequence of several different defects at the molecular level, some of which are quite well understood. Lack of the genes for growth hormone is relatively rare, but a Swiss family showing hereditary growth hormone deficiency has been shown to lack at least 7.5 kilobase pairs of DNA which includes the normal growth hormone gene.[61] More usually, growth hormone genes are present, but synthesis or secretion of the hormone is defective; in some cases of this kind GHRH can stimulate growth hormone release, suggesting that the defect lies at the hypothalamic rather than at the pituitary level. Short stature in some individuals may also be due to the production of a defective growth hormone, sometimes one which retains immunological but not biological activity.[11]

In other cases defects may lie at the level of the actions of growth hormone due to defective receptors, somatomedin production or responsiveness to somatomedin actions. 'Laron' dwarfism and short stature in African pygmies appear to be consequences of such defects.[19]

7.   GROWTH HORMONE LEVELS IN THE CIRCULATION

Basal growth hormone levels in man are low—about 2 ng/ml. They increase episodically, every few hours, reaching levels of 10–30 ng/ml and then falling fairly quickly back to the basal level. Secretion appears to be increased by exercise, emotional stress and high protein meals. However, the bulk of growth hormone secretion in children and young adults occurs during sleep, especially during the first hour or so of onset.

Growth hormone in the circulation is not thought to be associated with a binding protein. A considerable proportion of the hormone in man circulates as a dimer.

REFERENCES

1. Wallis, M. (1978) In: *Chemistry and Biochemistry of Amino Acids, Peptides and Proteins,* Vol. 5 (Ed. Weinstein, B.). Dekker, New York, pp. 213–320.
2. Reichert, L. E. (1975) *Methods in Enzymology,* **37**, 360–380.
3. Li, C. H. (1977) In: *Hormonal Proteins and Peptides,* Vol. IV (Ed. Li, C. H.). Academic Press, New York, pp. 1–41.
4. Hunter, W. M. (1976) In: *Hormone Assays and their Clinical Application,* 4th edn (Eds. Loraine, J. A. and Bell, E. T.). Churchill Livingstone, Edinburgh, pp. 221–284.
5. Wallis, M. (1980) In: *Cellular Receptors for Hormones and Neurotransmitters* (Eds. Schulster, D. and Levitzki, A.). Wiley, London, pp. 163–183.
6. Bewley, T. A. and Li, C. H. (1975) *Adv. Enzymol. Mol. Biol.,* **42**, 73–165.
7. Li, C. H., Liu, W.-K. and Dixon, J. S. (1966) *J. Amer. Chem. Soc.,* **88**, 2050–2051.
8. Martial, J. A., Hallewell, R. A., Baxter, J. D. and Goodman, H. M. (1979) *Science,* **205**, 602–607.
9. Wallis, M. (1981) *J. Molecular Evolution,* **17**, 10–18.
10. Wilhelmi, A. E. (1982) In: *Hormone Drugs, Proc. FDA/USP Workshop on Drug Ref. Stand. for Insulins, Somatotropins and Thyroid Hormones.* US Pharmacopeial Convention, Rockville MD, pp. 369–381.
11. Chawla, R. K., Parks, J. S. and Rudman, D. (1983) *Ann. Rev. Med.,* **34**, 519–547.
12. Wallis, M. and Davies, R. V. (1976). In: *Growth Hormone and Related Peptides* (Eds. Pecile, A. and Müller, E. E.). Excerpta Medica, Amsterdam, pp. 1–13.
13. Lewis, U. J., Singh, R. N. P., Tutwiler, G. F., Sigel, M. B., VanderLaan, E. F. and VanderLaan, W. P. (1980) *Recent Progr. Hormone Res.,* **36**, 477–508.
14. Lewis, U. J., Bonewald, L. F. and Lewis, L. J. (1980) *Biochem. Biophys. Res. Commun.,* **92**, 511–516.
15. Wallis, M. (1980) *Nature,* **284**, 512.
16. Isaksson, O. G. P., Jansson, J. O. and Gause, I. A. M. (1982) *Science,* **216**, 1237–1239.
17. Paladini, A. C., Pena, C. and Retegui, L. A. (1979) *Trends in Biochem. Sci.,* **4**, 256–260.
18. Bornstein, J. (1978) *Trends in Biochem. Sci.,* **3**, 83–86.
19. Laron, Z. (1981) In: *Endocrine Control of Growth* (Ed. Daughaday, W. H.). Elsevier, New York, pp. 175–205.
20. Tanner, J. M. (1972) *Nature,* **237**, 433–439.
21. Rudman, D. (1981) In: *Insulins, Growth Hormone and Recombinant DNA Technology* (Ed. Gueriguian, J. L.) Raven, New York, pp. 161–175.
22. Ellis, S., Vodian, M. A. and Grindeland, R. E. (1978) *Recent Progr. Hormone Res.,* **34**, 213–238.

23. Haeuptle, M. T., Aubert, M. L., Djiane, J. and Kraehenbuhl, J.-P. (1983) *J. Biol. Chem.*, **258**, 305–314.
24. Preece, M. A. and Holder, A. T. (1982) In: *Recent Advances in Endocrinology and Metabolism* (Ed. O'Riordan, J. L. H.). Churchill-Livingstone, New York, pp. 47–73.
25. Phillips, L. S. and Vassilopoulou-Sellin, R. (1980) *New Eng. J. Med.*, **302**, 371–380.
26. Phillips, L. S. and Vassilopoulou-Sellin, R. (1980) *New Eng. J. Med.*, **302**, 438–445.
27. Bradshaw, R. A. and Niall, H. D. (1978) *Trends in Biochem. Sci.*, **3**, 274–278.
28. Blundell, T. L. and Humbel, R. E. (1980) *Nature*, **287**, 781–787.
29. Baserga, R. (Ed.) (1981) *Tissue Growth Factors*. Springer-Verlag, Berlin.
29a. James, R. and Bradshaw, R. A. (1984) *Ann. Rev. Biochem.* **53**, 259–292.
30. Daughaday, W. H., Phillips, L. S. and Herington, A. C. (1976) In: *Growth Hormone and Related Peptides* (Eds. Pecile, A. and Müller, E. E.). Excerpta Medica, Amsterdam, pp. 169–177.
31. Wallis, M. (1981) In: *Proceedings of the Bristol Somatomedins Colloquium* (Ed. Spencer, G. S. G.), U.K. Somatomedin Club, Bristol, pp. 20–26.
32. Yang, Y. W.-H., Acquaviva, A. M., Bruni, C. B., Romanus, J. A., Nissley, S. P. and Rechler, M. M. (1983) In: *Insulin-like Growth Factors/Somatomedins* (Ed. Spencer, E. M.). deGruyter, Berlin, pp. 603–610.
33. Jansen, M., van Schaik, F. M. A., Ricker, A. T., Bullock, B., Woods, D. E., Gabbay, K. H., Nussbaum, A. L., Sussenbach, J. S. and Van den Brande, J. L. (1983) *Nature*, **306**, 609–611.
34. Schoenle, E., Zapf, J., Humbel, R. E. and Froesch, E. R. (1982) *Nature*, **296**, 252–253.
35. Rechler, M. M., Kasuga, M., Sasaki, N., De Vroede, M. A., Romanus, J. A. and Nissley, S. P. (1983) In: *Insulin-like Growth Factors/Somatomedins* (Ed. Spencer, E. M.). deGruyter, Berlin, pp. 459–490.
36. Rubin, J. B., Shia, M. A. and Pilch, P. F. (1983) *Nature*, **305**, 438–440.
37. Lebovitz, H. E., Drezner, M. K. and Neelon, F. A. (1976) In: *Growth Hormone and Related Peptides* (Eds. Pecile, A. and Müller, E. E.). Excerpta Medica, Amsterdam, pp. 202–215.
38. Kostyo, J. L. and Nutting, D. F. (1974) In: *Handbook of Physiology, Section 7: Endocrinology*, Vol. 4, Part 2 (Eds. Knobil, E. and Sawyer, W. H.). American Physiological Society, Washington D.C., pp. 187–210.
39. Goldberg, A. L., Tischler, M., DeMartino, G. and Griffin, G. (1980) *Fed. Proc.*, **39**, 31–36.
40. Jefferson, L. S., Robertson, J. W. and Tolman, E. L. (1972) In: *Growth and Growth Hormone* (Eds. Pecile, A. and Müller, E. E.). Excerpta Medica, Amsterdam, pp. 106–123.
41. Korner, A. (1970) In: *Ciba Foundation Symposium, Control Processes in Multicellular Organisms* (Eds. Wolstenholme, G. E. W. and Knight, J.). Churchill, London, pp. 86–107.
42. Korner, A. and Hogan, B. L. M. (1972) In: *Growth and Growth Hormone* (Eds. Pecile, A. and Müller, E. E.). Excerpta Medica, Amsterdam, pp. 98–105.
43. Roy, A. K., Chatterjee, B., Demyan, W. F., Milin, B. S., Motwani, N. M., Nath, T. S. and Schiop, M. J. (1983) *Recent Progr. Hormone Res.*, **39**, 425–461.
44. Goodman, H. M. and Schwartz, J. (1974) In: *Handbook of Physiology, Section 7: Endocrinology*, Vol. 4, Part 2 (Eds. Knobil, E. and Sawyer, W. H.). American Physiological Society, Washington D.C., pp. 211–231.
45. Altszuler, N. (1974) In: *Handbook of Physiology, Section 7: Endocrinology*, Vol. 4, Part 2 (Eds. Knobil, E. and Sawyer, W. H.). American Physiological Society, Washington D.C. pp. 233–252.
45a. Young, F. G. (1937) *Lancet* **ii**, 372–374.

46. Morikawa, M., Nixon, T. and Green, H. (1982) *Cell,* **29**, 783–789.
47. Bancroft, F. C., Sussman, P. M. and Tushinski, R. J. (1976) In: *Growth Hormone and Related Peptides* (Eds. Pecile, A. and Müller, E. E.). Excerpta Medica, Amsterdam, pp. 84–93.
48. Guillemin, R., Brazeau, P., Böhlen, P., Esch, F., Ling, N. and Wehrenberg, W. B. (1982) *Science,* **218**, 585–587.
49. Rivier, J., Spiess, J., Thorner, M. and Vale, W. (1982) *Nature,* **300**, 276–278.
50. Gubler, U., Monahan, J. J., Lomedico, P. T., Bhatt, R. S., Collier, K. J., Hoffman, B. J., Bohlen, P., Esch, F., Ling, N., Zeytin, F., Brazeau, P., Poonian, M. S. and Gage, L. P. (1983) *Proc. Nat. Acad. Sci. USA,* **80**, 4311–4314.
51. Brazeau, P., Vale, W., Burgus, R., Ling, N., Butcher, M., Rivier, J. and Guillemin, R. (1973) *Science,* **179**, 77–79.
52. Vale, W., Brazeau, P., Rivier, C., Brown, M., Boss, B., Rivier, J., Burgus, R., Ling, N. and Guillemin, R. (1975) *Recent Progr. Hormone Res.,* **31**, 365–397.
53. Brazeau, P., Epelbaum, J., Tannenbaum, G. S., Rorstad, O. and Martin, J. B. (1978) *Metabolism,* **27**, 1133–1137.
54. Tannenbaum, G. S., Epelbaum, J., Colle, E., Brazeau, P. and Martin, J. B. (1978) *Metabolism,* **27**, 1263–1267.
55. Hobart, P., Crawford, R., Shen, L., Pictet, R. and Rutter, W. J. (1980) *Nature,* **288**, 137–141.
56. Bilezikjian, L. M. and Vale, W. W. (1983) *Endocrinology,* **113**, 1726–1731.
57. Law, G. J., Ray, K. P. and Wallis, M. (1984) *FEBS Lett.,* **166**, 189–193.
58. Barinaga, M., Yamonoto, G., Rivier, C., Vale, W., Evans, R. and Rosenfeld, M. G. (1983) *Nature,* **306**, 84–85.
59. Rosenfeld, M. G., Amara, S. G., Birnberg, N. C., Mermod, J.-J., Murdoch, G. H. and Evans, R. M. (1983) *Recent Progr. Hormone Res.,* **39**, 305–351.
60. Friesen, H. G. (1980) *Endocrine Rev.,* **1**, 309–318.
61. Phillips, J. A., Hjelle, B. L., Seeburg, P. H. and Zachmann, M. (1981) *Proc. Nat. Acad. Sci. USA,* **78**, 6372–6375.

Chapter 8

Hormones of the adenohypophysis: Prolactin with placental lactogen and the molecular evolution of the growth hormone/prolactin family

1. THE CHEMISTRY OF PROLACTIN[1-3]

1.1 Isolation

Recognition of the occurrence of a factor in pituitary extracts which could stimulate mammary growth and lactation led to identification and purification of the protein hormone prolactin in the 1930s and 1940s. Early purification methods used mainly selective precipitation, but more recently chromatographic and electrophoretic methods have been developed which give better yields and purity. Prolactin is relatively difficult to solubilize when in the stored form in the pituitary (i.e. in subcellular granules) and other hormones can be largely removed by extraction at neutral or slightly acidic pH before prolactin is solubilized under alkaline conditions. There is generally much more prolactin in pituitary glands from female animals than in those from males, and the former therefore represent a preferable source for isolation. Prolactin concentrations in human pituitaries are usually particularly low.

1.2 Assay[4]

Bioassays for prolactin mainly exploit the actions of the hormone on the mammalian mammary gland or the crop sac of the pigeon, although wide interest in the comparative biology of prolactin has led to the development of alternative bioassays in various lower vertebrates. Assays using effects on the mammary gland involve induction of milk or milk-protein synthesis *in vivo*, or the induction of milk proteins (casein or α-lactalbumin) or specific enzymes (e.g. N-acetyl-lactosamine synthetase) in mammary gland explants or dispersed mammary gland cells *in vitro*. The commonly used pigeon crop sac assay exploits the fact that prolactin stimulates formation of pigeon 'milk' by promoting cell division in the walls of the crop sac—cells rich in protein and fat are sloughed off the crop wall and regurgitated as a milk-like fluid. The bioassay measures crop-sac weight or thickening after injection of prolactin *in vivo*. Another, more recent, bioassay for prolactin utilizes the ability of the hormone to stimulate cell division in a rat lymphoid cell line. The bioassays for prolactin are for the most part highly specific, sensitive and in at least some cases have a clear-cut biochemical end-point.

Radioimmunoassays have been developed for prolactin and are very widely used, particularly for studies on secretion and to measure blood levels of the hormone (including clinical uses). As in the case of growth hormone, there is a good deal of species specificity in the immunological properties of prolactin, and specific assays have to be used for different species. Radioreceptor assays have also been described for prolactin, using membrane-bound or solubilized receptors from mammary gland or liver.[5] As in the case of growth hormone, there is some discrepancy between measurements obtained using bioassay and radioimmunoassay, and some authors have suggested that prolactin occurs in the circulation in an activated form.

1.3   Structure and chemistry[1,2,6]

1.3.1   Primary structure

The amino acid sequence of sheep prolactin was determined in 1970, although a few small alterations have now been made to this structure. The hormone comprises a single polypeptide chain of 199 amino acid residues. The sequences of bovine (Figure 8.1), pig, human, rat, mouse, horse and (partial) whale prolactins have now been reported also. There is considerable species specificity in the sequences, that of the rat being particularly different from the others, at about 40% of all amino acid residues (see also Section 6). In all cases except the horse the protein contains three internal disulphide bridges, one linking distant parts of the sequence and the other two forming small loops near the amino and carboxyl ends of the chain. Sequences of cDNA/mRNA for rat, bovine and human prolactins are now available.

Prolactin has not been crystallized, so the detailed three-dimensional structure of the protein has not been determined. Use of circular dichroism and optical rotatory dispersion reveals a considerable amount of organized secondary structure within the three-dimensional structure; 50–60% of the polypeptide chain is organized into α-helix, a very similar proportion to that seen in growth hormone. Other structural resemblances between growth hormone and prolactin are discussed in Section 6.

Prolactin preparations display much less evidence for microheterogeneity (or naturally-occurring variants) than growth hormone, although this may be a reflection of less intensive study. Microheterogeneity due to partial deamidation of asparagine and glutamine residues is apparent in almost all preparations, but microheterogeneity due to other causes is rare, although glycosylated forms of prolactin may occur. Size heterogeneity (existence of stable aggregates,

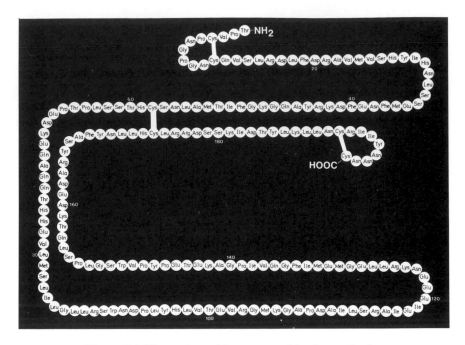

Figure 8.1 The amino acid sequence of bovine prolactin

especially dimers) on the other hand is seen in many prolactin preparations, and occurs in the pituitary and circulation in man. As in the case of growth hormone, its significance is not clear, although it seems unlikely that these fairly stable large forms of prolactin represent biosynthetic precursors of the hormone.

1.3.2 Structure–function relationships

Lack of a detailed three-dimensional structure is a major drawback in understanding the relation of structure to function in prolactin. Relatively little work has been carried out on chemical and enzymic modification of prolactin and its effect on biological activity. Detailed studies on the tyrosine residues of ovine and bovine prolactin suggest that these can be extensively modified by nitration or iodination without major loss of activity, and various other studies also suggest that the structural requirements for biological activity in prolactin are not extremely stringent.

There has been little work on the chemical synthesis of fragments of prolactin.

1.4 Human prolactin[7,8]

Until the early 1970s it was not clear whether a real human prolactin existed. The presence of large quantities of human growth hormone which, as has been discussed in Chapter 7, possesses marked lactogenic activity, made detection of a separate prolactin difficult, and the prolactin-like activity of the growth hormone made the need for a separate prolactin less obvious. In the period 1970–1972, however, evidence from several sources proved that human prolactin does in fact exist. Thus, it was found that much of the lactogenic activity in serum cannot be suppressed by antibodies to human growth hormone, and that human or monkey pituitaries (or pituitary tumours) when maintained in culture secrete increased amounts of lactogenic hormone but decreased amounts of growth hormone. Furthermore, it had long been recognized that human pituitary dwarfs with completely defective growth-hormone production are able to lactate normally.

Subsequently small amounts of prolactin were isolated from human pituitaries and shown to be similar in chemical, immunological and biological properties to other mammalian prolactins. The amino acid sequence of the hormone has now been determined, and shown to be homologous with other prolactins; indeed, human prolactin is less dissimilar to non-primate prolactins than human growth hormone is to non-primate growth hormones. It is now clear that the prolactin content of the human pituitary is rather low except in late pregnancy and during lactation.

With the recognition and isolation of human prolactin, radioimmunoassays were established, and the occurrence of the hormone in the human circulation was investigated. Levels are generally low in men and non-pregnant, non-lactating women. They increase steadily (though not dramatically) during pregnancy, and reach high values in lactating women, with peaks after suckling of 1–4 µg/ml. Very high levels of prolactin are found in human amniotic fluid; the significance of this is not clear.

Pathologically high levels of prolactin in men and women are associated with various clinical conditions, and usually are the consequence of a prolactin-secreting pituitary tumour. They often, but not always, give rise to galactorrhea (inappropriate lactation), and are frequently associated with sexual disorders. Treatment with drugs (such as bromocryptine) which inhibit prolactin secretion can frequently rapidly cure these consequences of oversecretion.

1.5 What is the active form of prolactin?

As in the case of growth hormone, there have been various suggestions that prolactin may exist in the circulation, at least partly as an activated form. The stored form of the hormone would be activated either in the pituitary gland, during secretion or in the circulation, presumably by limited proteolysis.

Evidence for this idea comes from observations that when rat anterior pituitary glands are incubated *in vitro* the medium appears to contain a great deal more prolactin as assayed by bioassay than can be measured by immunoassay. Furthermore, exogenously administered prolactin has been reported to be relatively inactive in stimulating mammary growth and milk formation, whereas treatments stimulating endogenous production of prolactin are highly active in this respect. A fragment of prolactin corresponding to about two-thirds of the molecule has been reported to occur in rat pituitary, and to possess much enhanced mitogenic activity on the mammary gland.[9]

2.   THE BIOLOGICAL ACTIONS OF PROLACTIN[10,11]

The main actions of prolactin in mammals are generally considered to be those concerned with the promotion of mammary growth and lactation.[10-13] The hormone acts on the mammary gland along with several other hormones. It is essential for production of the various components of milk (proteins, fat, lactose, etc.) and appears to regulate the final differentiation of the secretory cells and production of these components. The hormone also plays a role in mammary growth, but its precise function here is less well understood. There has been much interest in the possibility that prolactin plays a role in the aetiology of mammary cancer. It certainly sustains and promotes growth of some rodent mammary tumours, but a link with human breast cancer has not been clearly established.[11,14] Further details of the biochemical mode of action of prolactin on the mammary gland will be considered in Section 3.

Various other functions for prolactin in mammals have been proposed. It certainly has luteotrophic actions in rodents and some other mammalian groups, and probably plays a part in maintaining the corpus luteum in early pregnancy. On the other hand, prolactin has an anti-gonadal action in different circumstances (possibly higher concentrations) in female mammals, and may be at least partly responsible for the relative infertility seen in lactating animals and women. Prolactin may also be involved in the control of spermatogenesis in male mammals.

Many other actions have been proposed for prolactin in mammals, but their physiological importance remains unclear. The very wide range of effects clearly established in lower vertebrates (see next section) does suggest, however, that actions additional to those on the mammary gland and gonads (including, possibly, behavioural effects) cannot be discounted.

2.1   The comparative endocrinology of prolactin[15-18]

A prolactin-like hormone has been demonstrated in most vertebrate groups, although that occurring in some fish groups differs markedly from that found in

Table 8.1   Some actions of prolactin

Cyclostomes

Electrolyte metabolism in hagfish

Teleosts

Various osmoregulatory actions
Dispersion of pigment xanthophores in some species
Parental behaviour (nest-building, buccal incubation of eggs) in some species
Maintenance of brood pouch in male seahorse

Amphibians

Larval growth; thyroxine antagonism in metamorphosis
Eft water drive in newts
Limb regeneration
Osmoregulatory actions

Reptiles

Tail regeneration
Epidermal sloughing
Anti-gonadotropic
Osmoregulation

Birds

Crop milk production (pigeons and doves)
Formation of brood patch
Stimulation of salt gland secretion
Anti-gonadal
Parental behaviour

Mammals

Stimulation of mammary growth and development of milk secretion
Actions on male reproductive organs
Luteotrophic in many species
Parental behaviour

Note: Many other actions have been described (see, for example, ref. 58).

mammals and has been termed 'paralactin' by some authors. Clearly, since non-mammalian vertebrates do not produce milk, the hormone must have different actions in the lower vertebrates. In fact an extraordinary range of such actions has been found, some of which are listed in Table 8.1. A detailed discussion of these various actions is not possible here, but it should be noted that a high proportion of the actions described are concerned with regulation of salt and water relationships or secondary sexual features. Although a great deal of work has been carried out on the comparative zoology and physiology of these actions of prolactin, little is known about their biochemistry.

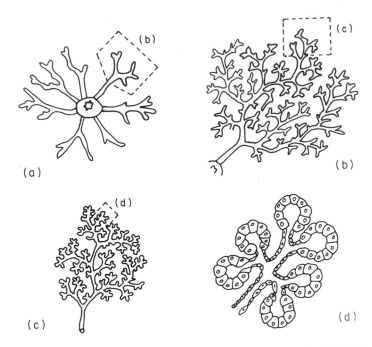

Figure 8.2 The development and organization of the mammary gland. (a) The gland of an immature animal consists of simple ducts radiating from the nipple. (b) Extensive branching of the ductal system occurs in the adult female. (c) During pregnancy the branching increases in complexity and the branch termini give rise to secretory alveoli. (d) The cellular structure of the alveoli; during lactation milk globules accumulate within the lumen of the alveolus. (Based in part on Turner, R. W. and Bagnara, J. T. (1971) *General Endocrinology*, 5th Edn.)

3.   THE BIOCHEMICAL MODE OF ACTION OF PROLACTIN ON THE MAMMARY GLAND[19-22]

3.1   The hormonal control of milk secretion and mammary development[23]

The organization of the mammary gland is illustrated diagrammatically in Figure 8.2. It consists of three main types of tissue: secretory tissue, organized into alveoli, which actually produce the milk; ducts which carry the milk from the alveoli through a hierarchical system, eventually to the nipple; and adipose and connective tissue in which the alveoli are distributed. Smooth muscles associated with the alveoli serve to force the milk from the alveoli through the ductal

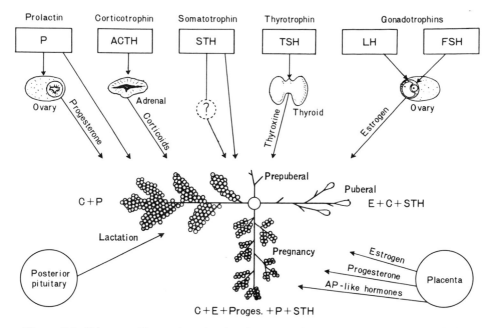

Figure 8.3 Diagram illustrating the development of the mammary gland, and the role of hormones in this process. Note that somatotrophin (STH) = growth hormone. P: prolactin, E: oestrogen, C: corticosteroid. (Reproduced with permission from Lyons, W. R., Li, C. H. and Johnson, R. E. (1958) *Recent Prog. Hormone Res.* **14**, 219–254. Copyright (1958) Academic Press Inc.)

system, and are largely controlled by the posterior pituitary hormone oxytocin (see Chapter 4).

The development of the ductal and secretory tissue is illustrated in Figures 8.2 and 8.3. The early part of the developmental scheme occurs in all maturing female animals. The subsequent stages occur in pregnancy and early lactation. At the end of lactation the gland involutes, reverting to a state similar to that of the mature, nulliparous animal, ready for the cycle of development to begin again with the next pregnancy. This scheme is, of course, a generalized one, and details vary considerably from one species to another.

The development of the mammary gland is under hormonal control, as is the production of milk by the fully developed gland. The growth and development of the gland at puberty is largely under the control of oestrogens (and possibly progesterone), although growth hormone or prolactin (possibly through the mediation of somatomedins) also appears to be necessary at this stage. This stage involves mainly the development of the extensive ductal system, few mature alveoli being present prior to the first pregnancy. Further development during pregnancy involves some additional branching of the ductal system, and development of the alveoli, which actually produce the milk components. This stage is regulated by oestrogens, progesterone, glucocorticoids and prolactin; in some species thyroid hormones and growth hormone may also play a part. During late pregnancy in some species the role of prolactin may be partly taken over by placental lactogen.

Actual secretion of milk is inhibited by progesterone, and possibly oestrogens, and the high levels of these hormones during pregnancy probably serve to inhibit lactation before parturition; they also probably inhibit prolactin secretion. At parturition the levels of these steroid hormones fall rapidly, and this plus the associated increase in prolactin secretion serves to initiate lactation. The importance of prolactin in this respect is indicated by the fact that in a pregnant or pseudo-pregnant rabbit, injection of prolactin alone, intraductally (via the teat), induces local secretion of milk. However, adrenal corticoids are also required for this effect—prolactin is unable to initiate normal milk secretion in an adrenalectomized animal. Once milk secretion has been initiated, prolactin appears to be the main hormone involved in maintaining milk production, although adrenal corticoids, thyroid hormones and growth hormone may also play a role. In lactating dairy cows, growth hormone appears to be more important than prolactin—possibly because maintenance of the very high milk production requires diversion of the entire 'metabolic effort' of the animal towards milk synthesis.

3.2　Cellular aspects[21,24,25]

Elucidation of the sequence of events which occurs at the cellular level as the mammary gland develops and differentiates has been helped enormously by *in vitro* studies on explants of mammary tissue from the mouse and other animals, carried out by various workers, particularly Topper and his colleagues. When explants about $1-2\,mm^3$ are removed from a mammary gland of a mid-pregnant mouse (before production of milk proteins has started) and placed in culture, they soon die unless hormones are present. In the presence of insulin, long-term survival is maintained, and DNA synthesis and cell division occur, but there is no production of milk proteins such as casein. Further development of the explants *in vitro* requires the presence of a corticosteroid. The precise role of this is not clear; it is thought to induce a state of 'covert differentiation', prerequisite for the action of prolactin, but specific changes are not clear. Explants which have been cultured for 2–3 days in the presence of insulin and corticosteroid will respond to the presence of prolactin by starting to produce substantial quantities of milk proteins.

These stages are summarized in Figure 8.4. It should be emphasized, of course, that insulin, corticosteroids and prolactin are not the *only* hormones required for mammary gland development; others are involved, as discussed in the previous section, but have already brought about their effects *in vivo* by the stage of mid-pregnancy when the explants are removed. Nevertheless, the

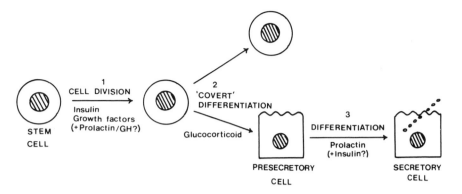

Figure 8.4 Diagram illustrating the possible roles of various hormones in promoting proliferation and differentiation of mouse mammary epithelial cells *in vitro*

sequence of events observed *in vitro* clearly parallels quite closely that occurring *in vivo* and the *in vitro* system provides a powerful model for study of the biochemical events involved. It allows a distinction to be drawn between three major aspects of the developmental sequence.

(1) First is a phase of proliferation and growth, which is induced by insulin (*in vitro* at least). A cycle of cell division, induced by insulin, seems to be required before the mammary alveolar (epithelial) cells can differentiate and start to produce milk. *In vitro*, insulin induces DNA synthesis, RNA synthesis, histone synthesis, phosphorylation of histones and non-histone nuclear protein and RNA and DNA polymerases. There is some doubt, however, as to whether insulin is the hormone responsible for these events *in vivo*; levels of insulin required *in vitro* are much higher than physiological levels, and it is likely that the 'natural' factor responsible is another growth factor, possibly a somatomedin.

(2) After the phase of proliferation induced by insulin, mammary cells must undergo a 'covert differentiation', as referred to above. This is induced by corticosteroids, which may in part stimulate production of the secretory apparatus (rough endoplasmic reticulum, Golgi body, etc.) needed by a milk-secreting cell.

(3) The final phase of development is the terminal differentiation of the cell to one able to produce and secrete milk proteins (casein, lactalbumin, etc.) and other milk products. This is induced by prolactin.

3.3 Receptors for prolactin[5]

There is a good deal of evidence that prolactin, like other polypeptide hormones, acts on its target cells by binding to receptors on the plasma membrane. Thus, prolactin linked covalently to beads of cross-linked dextran or agarose is able to stimulate production of milk protein in mouse mammary explants *in vitro*.

Membrane-associated receptors for prolactin have been identified in mammary tissue and several other organs, and have been used as the basis for a radioreceptor assay for the hormone. Such receptors can be solubilized, using detergents, and considerable progress has been made with the purification of the receptors, using, particularly, affinity chromatography. The partially purified receptors have been utilized to prepare anti-receptor antibodies, and these will block not only binding of labelled prolactin to receptors, but also the biological actions of prolactin *in vitro*; a clear link has thus been established in this case between the binding of the hormone to its putative receptor and its ability to bring about biochemical effects in its target cell.[26]

The nature of partially purified prolactin receptors has been investigated using cross-linking techniques.[27] A polypeptide chain of molecular weight about 35 000 was shown to be labelled, but whether this comprises the entire receptor has not yet been established.

3.4 Internalization of prolactin

Although considerable evidence shows that prolactin can exert at least some of its actions without entering its target cells, other work indicates that a considerable proportion of the hormone is internalized after binding to the receptors on the plasma membrane.[28] Internalization occurs in a similar fashion to that of insulin and many other peptide hormones—typical receptor-mediated endocytosis. After internalization, the hormone is located within the target cell in membrane-bound endocytotic vesicles (Figure 8.5). The fate of such internalized prolactin is not yet clear. It may be that fusion with lysosomes always follows rapidly, leading to degradation of hormone and/or receptor. Alternatively, the internalized hormone may move to sites of action within the cell, such as the nucleus or Golgi body, where it may have direct actions. Another possibility is that internalization serves to prolong the effective life of the hormone–receptor complex and the actions that it may have on the cell. Whatever the full significance of internalization, it is clear that this process can explain the downregulation of receptors that is often observed: the number of receptors available decreases following initial interaction of hormone with target cell.

At least some evidence suggests that prolactin may have direct actions within the cell; actions of the hormone on transcription in isolated nuclei have been reported.[29]

3.5 Actions of prolactin on milk-protein synthesis[19,30]

In stimulating the final stages of differentiation of the mammary epithelial cells to fully-functioning, milk-secreting cells, prolactin brings about a wide range of

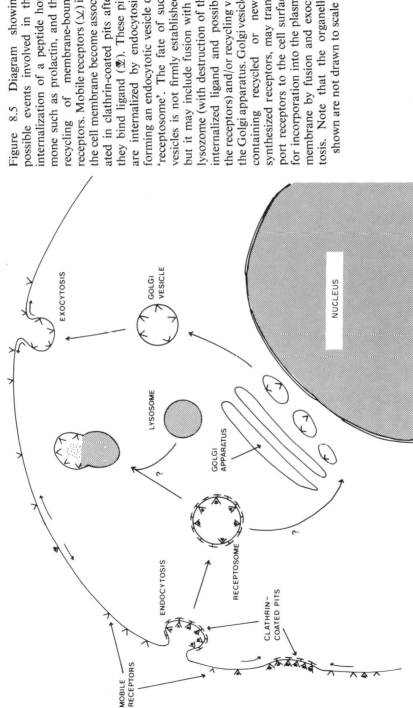

Figure 8.5 Diagram showing possible events involved in the internalization of a peptide hormone such as prolactin, and the recycling of membrane-bound receptors. Mobile receptors (⋎) in the cell membrane become associated in clathrin-coated pits after they bind ligand (⊻). These pits are internalized by endocytosis, forming an endocytotic vesicle or 'receptosome'. The fate of such vesicles is not firmly established, but it may include fusion with a lysozome (with destruction of the internalized ligand and possibly the receptors) and/or recycling via the Golgi apparatus. Golgi vesicles containing recycled or newly synthesized receptors, may transport receptors to the cell surface for incorporation into the plasma membrane by fusion and exocytosis. Note that the organelles shown are not drawn to scale

EXOCYTOSIS

GOLGI VESICLE

LYSOSOME

GOLGI APPARATUS

NUCLEUS

RECEPTOSOME

ENDOCYTOSIS

CLATHRIN–COATED PITS

MOBILE RECEPTORS

effects. The actions on induction of the milk proteins casein and lactalbumin have perhaps been most intensively studied. Mouse mammary explants incubated *in vitro* with insulin and a corticosteroid produce little of these proteins. Addition of prolactin to the incubation medium induces their production, giving a 10–20 fold stimulation during a period of about 24 hours. Enzymes concerned with the production of other milk components, such as lactose synthetase and enzymes of lipid metabolism including acetyl-CoA carboxylase and lipoprotein lipase, are also induced by prolactin, in a similar way. The hormone also stimulates the general secretory apparatus of the cell (endoplasmic reticulum and associated ribosomes, Golgi body, etc.), although the production of these components may have already started under the influence of corticosteroids.

Induction of milk proteins such as casein appears to involve increased transcription of the appropriate genes (i.e. gene derepression). Levels of mRNA for casein increase substantially (Figure 8.6), although not by as great a factor as for the increase in casein production, suggesting that increased translation of mRNA may also be involved. The increased levels of mRNA may be partly a consequence of decreased degradation as well as increased transcription. The

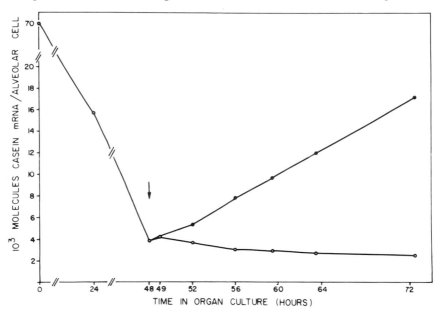

Figure 8.6 Analysis of casein mRNA accumulation in mammary gland tissue from pregnant rats incubated *in vitro* in the presence or absence of prolactin. —○—: insulin and cortisol were the only hormonal additions. —●—: prolactin, insulin and cortisol were added to the medium at the point indicated by the arrow. (Reprinted by permission from Rosen, J. M. (1981). In *Prolactin* (ed. Jaffe, R. B.), pp. 85–126. Copyright (1981) Elsevier Science Publishing Co., Inc.)

evidence for increased levels of casein mRNA derives from experiments in which these levels are measured directly, in *in vitro* translation systems, or using hybridization to labelled cDNA. They accord well with observations that an early effect of prolactin is stimulation of the synthesis of heterogeneous nuclear RNA, now recognized to be the precursor of cytoplasmic mRNA.

3.6 Second messengers and the actions of prolactin[19,20]

If prolactin binds to a receptor on the plasma membrane of the mammary gland epithelial cell, some mechanism is required to couple the hormone–receptor binding to the subsequent events within the cell. Internalization could provide this, but it seems unlikely that it represents the whole story. It has been proposed that activation of the sodium/potassium ATPase (which is of course associated with the plasma membrane) causes ionic changes within the cell which then cause the subsequent changes.[31] However, a variety of alternative 'second messengers' have also been proposed, the functions of which would include (directly or indirectly) switching on specific genes. A role for cyclic AMP has been proposed, but the bulk of the available evidence suggests that this nucleotide is not directly involved as a second messenger in the mammary gland cells. It has been suggested that levels of cyclic AMP-dependent protein kinase are limiting, rather than cyclic AMP itself, and that these can be increased by prolactin by induction of the appropriate gene(s) (Figure 8.7).[32] Other proposed second messengers include prostaglandins (prolactin appears to increase mammary levels of prostaglandin $F_{2\alpha}$, possibly by activating phospholipases[33] which in turn hydrolyse phospholipids containing arachidonic acid—the precursor of the prostaglandins), polyamines (prolactin stimulates ornithine decarboxylase and other enzymes concerned with polyamine production), products of phosphatidyl inositol metabolism and cyclic GMP. In no case is the evidence for one of these compounds as a second messenger conclusive, however, and the absence of a well-established second messenger remains a major stumbling block in understanding the mechanism of action of prolactin.

It is possible that the prolactin receptor acts as a protein kinase to activate various proteins in the cell by phosphorylation, in the way that insulin and some growth factors may act. No evidence exists for this in the case of prolactin, as yet, but it is perhaps worth noting that if such a mechanism did apply, internalization in endocytotic vesicles could provide a mechanism whereby the activated receptor-kinase could be moved to internal sites within the cell at which phosphorylation could occur.

3.7 Prolactin and the activation of sodium/potassium ATPase[31]

Another well-established effect of prolactin on the mammary gland is stimulation of sodium/potassium ATPase, with consequent increase of uptake of potassium and extrusion of sodium from the epithelial cells. Indeed, it is likely that

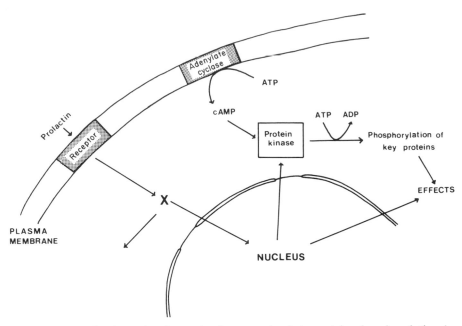

Figure 8.7 Mechanism whereby prolactin may stimulate protein phosphorylation by cAMP-dependent protein kinase. X represents a putative second messenger that may mediate the effects of prolactin at the nuclear and other levels. (Based in part on Turkington, R. W., Majumder, G. C., Kadohama, N., MacIndoe, J. H. and Frantz, W. L. (1973) *Recent Prog. Hormone Res.* **29**, 417–449)

the hormone plays an important part in controlling salt and water relationships in the mammary gland. In the lactating animal, milk secretion involves a considerable loss of both water and salts, and control of these materials is of crucial importance for both the mother and the offspring.

4.　BIOSYNTHESIS AND SECRETION OF PROLACTIN

4.1　Biosynthesis[34]

Prolactin is produced in acidophilic cells which are termed lactotrophs, and which are clearly distinguishable from somatotrophs. In most adult female mammals (although not women) lactotrophs are approximately equal in number to somatotrophs; during pregnancy and lactation their numbers increase considerably. In males there are generally fewer lactotrophs than somatotrophs. As with several other pituitary hormones, prolactin occurs also in the brain, and it seems likely that it is formed there.

Like growth hormone, prolactin is synthesized as a precursor, pre-prolactin, which has a 'signal peptide' 25–30 amino acids long attached to the N-terminus

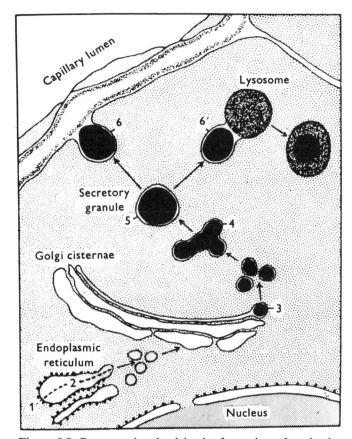

Figure 8.8 Processes involved in the formation of prolactin granules (1–5), their secretion by exocytosis (6) and their degradation by crinophagy (6'). (Reproduced with permission from Farquhar, M. G. (1971). In *Subcellular Organization and Function in Endocrine Tissues* (eds. Heller, H. and Lederis, K.), pp. 79–122. Copyright (1971) Cambridge University Press)

of the protein.[35,36] This was detected by translating pituitary mRNA in wheat germ and other cell-free systems, immunoprecipitating and characterizing the prolactin-like product as a protein 2000–3000 daltons heavier than prolactin itself. The existence and nature of the signal peptide in rat, human, and bovine pre-prolactins has also been determined by cloning cDNA corresponding to the prolactin mRNA and determining the DNA sequence (see Chapter 18).

As in the case of other pre-hormones, pre-prolactin is very rapidly processed to the normal hormone in the lactotroph. The function of the signal peptide is presumably to commit the protein hormone to the secretory pathway by ensuring its sequestration into the endoplasmic reticulum.

4.2 Secretion[37]

Prolactin is stored in the lactotroph in the form of large, rather irregular, membrane-bound granules, about 600 nm in diameter. It has been proposed that these large granules are formed by the fusion of several smaller, immature ones,[37] a process which has not been invoked for the formation of granules of most other protein hormones (Figure 8.8). Another process which has been noticed particularly in the lactotroph is crinophagy—fusion of a granule with a lysosome, resulting in destruction of the granule.[37] Crinophagy has been observed using electron microscopy, and accords well with biochemical studies which suggest a rapid degradation of prolactin in the pituitary.[38] It is possible that crinophagy occurs in most secretory cells, but is particularly marked in lactotrophs because at weaning secretion of prolactin must be rapidly turned off, and overproduction of the hormone may be a consequence. It should also be recalled, however, that there is some evidence for conversion of prolactin to a form with altered biological activity (Section 1.5); distinction between such 'processing' and degradation may not be very clear experimentally.

Prolactin granules are secreted from the lactotrophs by a process of exocytosis which is similar to that seen in other secretory cells.

4.3 Regulation of prolactin secretion[39−41]

As with other hormones of the anterior pituitary, secretion of prolactin is controlled by factors from the hypothalamus. The regulation of prolactin secretion has been studied using the same type of *in vitro* and *in vivo* experimental systems that have been applied to growth hormone (Chapter 7). The availability of good specific bioassays helped such studies, although these have now been largely replaced by radioimmunoassays.

4.3.1 Hypothalamic factors

Unlike the other anterior pituitary hormones, control of prolactin secretion by the hypothalamus is largely inhibitory. If the pituitary is removed from the influence of the hypothalamus, either by culturing *in vitro*, or by transplanting to a different site in the body, or by sectioning the pituitary stalk, prolactin secretion increases markedly, whereas release of other pituitary hormones (except melanotropin) decreases. As a consequence of such experiments it was proposed that hypothalamic factors capable of inhibiting prolactin release are conveyed from the hypothalamus to the anterior pituitary via the hypothalamo–hypophysial portal system. In accordance with this idea it has been shown on many occasions that hypothalamic extracts can inhibit prolactin release both *in vivo* and *in vitro*.

The nature of the hypothalamic prolactin release inhibiting factor has been studied for many years. Partial purification of the factor (or factors) from hypothalami has been achieved, but complete characterization has not. However, it was recognized some years ago that one factor which does inhibit prolactin release both *in vivo* and *in vitro* is dopamine; other catecholamines also have effects, but are less active. Dopamine agonists and antagonists have characteristic inhibitory and stimulatory actions on prolactin secretion *in vivo*, and are indeed used to regulate prolactin levels clinically (bromoergocryptine is particularly effective). Although some of their effects may be mediated via the hypothalamus, it seems likely that at least some are direct, at the pituitary level. Pituitary cells have receptors for dopamine, and the catecholamine can inhibit prolactin secretion dramatically in a variety of *in vitro* pituitary preparations. Furthermore, dopamine levels in the portal vessels leading from hypothalamus to pituitary gland are high (and at a level which could inhibit prolactin secretion *in vitro*) at times of low prolactin secretion, and decrease at times when prolactin secretion increases.

There is thus considerable evidence that dopamine itself is a physiological prolactin release inhibiting factor. Whether it is the *only* such factor, or whether there are other hypothalamic factors (PIFs) of this kind (possibly themselves regulated at the hypothalamic level by dopamine), is not yet clear. Some authors have claimed that all the prolactin release inhibiting activity of hypothalamic extracts can be explained in terms of their dopamine content, but others have strongly disputed this.

Although the influence of the hypothalamus on prolactin secretion appears to be largely inhibitory, there is also evidence for hypothalamic factors which can stimulate prolactin release. These have not yet been fully characterized, but it is known that three such factors originally recognized for other properties are also able to stimulate prolactin release. These are thyrotropin releasing hormone (TRH), vasoactive intestinal peptide (VIP) and the opiate peptides, enkephalins and β-endorphin. All of these are able to stimulate prolactin release under certain circumstances, both *in vivo* and *in vitro*, although their physiological significance remains in some doubt.

4.3.2   Mechanism of action of hypothalamic factors on prolactin secretion

The mechanism whereby dopamine inhibits prolactin secretion is poorly understood. Although conventional catecholamine receptors are usually considered to exist on the plasma membrane of their target cells, there is some evidence that at least some dopamine occurs in lactotrophs, associated with secretory granules.[42] A role for cyclic AMP as a mediator of the effects of dopamine on prolactin secretion has been proposed (dopamine appears to lower the level of cyclic AMP in pituitary glands, presumably by inhibiting adenylate cyclase)[43] but a considerable amount of evidence militates against such a role as far as basal secretion is concerned.[44] Factors which increase levels of cyclic AMP within the cell are unable to reverse the effects of dopamine on prolactin secretion, although they sometimes stimulate such secretion in the absence of dopamine. Some evidence suggests that dopamine may inhibit prolactin secretion by lowering levels of Ca^{2+} within the cell.[45] It seems probable, in fact, that the catecholamine can act both via lowered Ca^{2+} or lowered cyclic AMP levels, but that the latter action only operates when levels of the cyclic nucleotide have been elevated by another agent.

Stimulation of prolactin release by TRH also appears not to be mediated by elevation of cyclic AMP levels,[46] although, again, some evidence does support a role for this nucleotide. A role for Ca^{2+} in mediating the effects of TRH seems likely,[46,47] possibly involving release of Ca^{2+} from intracellular stores. VIP, on the other hand, stimulates production of pituitary cyclic AMP markedly, and it seems likely that its actions are mediated by the nucleotide.

The stimulation of prolactin release that is produced by opiate peptides is seen only in the presence of dopamine. Enkephalins can reverse the inhibitory effect of dopamine *in vitro*, although by themselves they are without a direct effect on prolactin secretion.[48] The mechanism whereby they bring about this reversal of the action of dopamine is not understood. It seems likely that the opiate peptides also act at the hypothalamic level to lower dopamine release.

4.3.3   Physiological control of prolactin secretion

Variation of prolactin levels according to physiological status is rather species specific. The situation in the rat has been studied in considerable detail, and will be considered here. In the immature female rat, prolactin levels are low. They increase somewhat as sexual maturity is achieved, and then a distinct cyclical variation begins associated with the oestrous cycle (Figure 8.9); prolactin levels show a peak at pro-oestrus, coinciding more or less with the peaks of luteinizing and follicle stimulating hormones which occur at this time. The increased prolactin secretion at this period is probably due to elevated levels of oestrogen —which can stimulate prolactin production both directly and indirectly (via the hypothalamus); it probably relates to the luteotrophic function of prolactin in the rat. In the pregnant rat, levels of prolactin increase only slightly during pregnancy. Shortly before, and more markedly after, parturition, however, they increase rapidly and as lactation commences dramatic peaks of prolactin secretion are seen, which are associated with suckling—levels increasing several fold within 1–3 minutes of the commencement of suckling and falling off quite rapidly after suckling. The stimulation of the release of the hormone by suckling appears to be the consequence of a neural loop, nervous signals from the mammary gland passing to the brain and thence to the hypothalamus where they cause decreased secretion of dopamine and possibly increased secretion of prolactin releasing factor(s). Electrical stimulation of some parts of the brain can

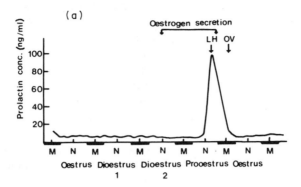

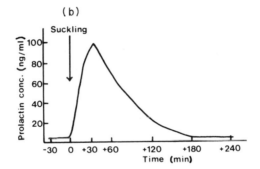

Figure 8.9 Blood prolactin levels during the oestrous cycle (a) and lactation (b) in the rat. In (a) prolactin levels on the 4 days of the oestrous cycle (oestrus, dioestrus 1, dioestrus 2, pro-oestrus; M = midnight, N = noon; solid bars indicate dark periods) are shown. Note the peak of prolactin secreted on the afternoon of pro-oestrus, associated with the peak of LH secretion, and shortly before ovulation (OV). In (b) prolactin levels in a nursing rat are shown; suckling started at time O (after a period of 8 h without suckling). (Based on Neill, J. D. (1974) In *Handbook of Physiology, Section 7, Endocrinology, Vol. IV-2* (eds. Knobil, E. and Sawyer, W. H.), pp. 469–488)

stimulate the secretion of prolactin and this has been used to trace the neural pathway which controls prolactin secretion. A similar pathway controls the release of oxytocin which is also required, of course, for successful suckling (see Chapter 4).

The control of prolactin secretion at the physiological level thus involves both humoral signals (such as steroids) and nervous ones. As already mentioned, the detailed pattern of control varies considerably from one species to another (for example, in humans there is no apparent variation of prolactin levels during the menstrual cycle), but the reflex arc described for suckling and prolactin release in the rat does seem to apply also in most mammalian species.

4.3.4   Prolactin secretion in lower vertebrates

What has been described so far for prolactin secretion has referred mainly to mammals. The situation differs considerably in other vertebrate groups. Obviously, in lower vertebrates prolactin release is not controlled by suckling, and other physiological loops apply, related to the actions of the hormone in each particular group. More important from a biochemical point of view is the fact that in birds and reptiles prolactin is under primarily positive regulatory control of the hypothalamus, in contrast to the negative influence seen in mammals. The nature of the hypothalamic factors regulating prolactin secretion in lower vertebrates has not been established. It is clear that the situation may differ substantially from that in mammals, although a different balance between stimulatory and inhibitory signals (such as TRH and dopamine) may apply, rather than completely different signals.

4.4   Regulation of prolactin synthesis

Dopamine may have direct effects on prolactin synthesis as well as secretion, but these are observed only after several hours of exposure of the gland to the catecholamine or its agonists. TRH stimulates prolactin synthesis, an effect that has been most extensively studied using a rat pituitary tumour cell line (GH cells). Oestrogens also stimulate prolactin synthesis, an effect that is largely due to enhanced transcription of the prolactin gene,[34] as is shown by a marked increase in mRNA level. The effect of oestrogen is seen *in vitro* and *in vivo*, and is probably responsible for the differences in prolactin content and production seen between male and female animals, as well as differences seen through the oestrous cycle. The effects of oestrogen on prolactin synthesis can be reversed by progesterone.

5.   PLACENTAL LACTOGENS[49]

5.1   Discovery and characterization of human placental lactogen

The presence of a lactogenic factor in the human placenta had been suspected for many years, but it was not until the early 1960s that Josimovich and Mac-Claren demonstrated that the placental activity is associated with a protein which is rather similar to pituitary growth hormone. They named the protein human placental lactogen and showed that it cross-reacts with human growth

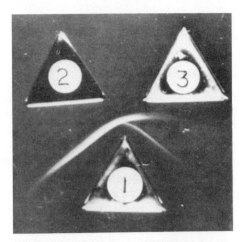

Figure 8.10 Demonstration of the immuno-logical cross-reaction between human growth hormone and human placental lactogen by immunodiffusion. Well 1 contained antiserum against human growth hormone; well 2 contained human growth hormone and well 3 serum containing human placental lactogen. The 'spur' on the immunoprecipitin lines indicates partial immunological cross-reaction between the growth hormone and placental lactogen. (Reproduced with permission from Josimovich, J. B. and MacLaren, J. A. (1962) *Endocrinology* **71**, 209–220. Copyright (1962) the Endocrine Society)

hormone immunologically (Figure 8.10). This lactogenic protein was rapidly isolated and shown to occur in the placenta at very high concentrations (several hundred mg/placenta during the last trimester of pregnancy).

Human placental lactogen is a protein hormone of molecular weight about 21 000, which exists as a dimer under many conditions. Its amino acid sequence was determined by several groups.[1] Like human growth hormone the protein contains 191 amino acid residues, including two internal disulphide bridges. The sequence is very similar to that of human growth hormone, the two hormones differing at only about 15% of all amino acid residues. The evolutionary rela-tionships of this sequence to other members of the growth hormone/prolactin family will be discussed in Section 6.

The tertiary structure of human placental lactogen has not been determined; although crystals of the hormone have been obtained they were not sufficiently well ordered for detailed crystallographic analysis. However, like the growth hormones and prolactins, placental lactogen possesses about 50% α-helix (as determined by circular dichroism), suggesting that it has a tertiary structure similar to that of the other protein hormones in this family.

5.2 Structure–function relationships[1]

In many assays (but not those using vertebrates other than mammals) human placental lactogen has lactogenic activity about equivalent to that of mammalian prolactins. However, despite its resemblance to human growth hormone, its growth-promoting activity is relatively low (only about 1–10% that of growth hormone). A good deal of chemical modification and enzymic modification work has been carried out on placental lactogen and suggests that the lactogenic activity of the hormone is relatively resistant to such modification. However, the derivatives have not been characterized very fully, and firm conclusions about the relation of structure to function are difficult to draw.

An interesting series of experiments on fragments of human placental lactogen produced by plasmin digestion has been described, which involved formation of hybrids between one fragment produced from human growth hormone (see Chapter 7) and a second obtained from the placental hormone. When an N-terminal fragment (residues 1–134) from growth hormone was recombined with a C-terminal fragment (residues 141–191) from placental lactogen, the resultant hybrid retained growth-promoting activity equivalent to that of growth hormone. When the corresponding N-terminal peptide came from placental lactogen and the C-terminal one came from growth hormone, the hybrid had very low growth-promoting activity. Thus, the growth-promoting activity appears to be associated largely with the N-terminal two-thirds of growth hormone.[50]

5.3 The biological actions of human placental lactogen[51]

The biological role of placental lactogen remains rather unclear. The best-established activity, promotion of lactation, presents something of an enigma since the hormone disappears at the very time (parturition) that lactation is most needed. The actions proposed for the hormone relate to three main areas: (a) control of mammary growth and lactation; (b) control of foetal growth; and (c) control of maternal metabolism, although these by no means cover all the effects that have been observed. That placental lactogen does not play a very crucial role in the human is suggested by the evidence of individuals who lack the protein completely, due to a gene deletion.[52] Pregnancy and parturition is reported to be normal in such circumstances, and the children are normal and healthy.

A role for the hormone in controlling mammary growth and lactation is hard to dispute. However, although the hormone stimulates milk formation and secretion in animal assays, both *in vivo* and *in vitro*, it has been suggested that (unlike prolactin) it has no such activity on cultured human mammary tissue. A possible explanation is that the hormone promotes mammary growth, but not the final stages of milk formation and secretion. Indeed, it may even inhibit the actions of prolactin on these late stages (a role which would accord with the very high plasma concentration of the hormone reached during late pregnancy). The hormone would thus act during pregnancy to prepare the mammary glands for lactation, but would prevent the actual onset of lactation. At parturition rapidly falling placental lactogen levels would remove the inhibitory influence, and lactation would commence, under the control of prolactin.

A direct role for placental lactogen in foetal growth is unlikely since the hormone probably does not cross the placenta. However, it seems probable that placental lactogen can increase somatomedin levels, and these may have a direct, or possibly indirect, effect on the growth of the foetus.

There is a good deal of evidence that placental lactogen alters maternal metabolism so that there is a diversion of nutrients from the mother to the foetus. In the last trimester of pregnancy there is a diminished responsiveness to insulin, impaired glucose tolerance, and increased mobilization of lipid stores. Infusion of placental lactogen into non-pregnant human subjects produces similar effects, which occur despite an increase in plasma insulin levels. Which of these various responses represent direct and significant effects of placental lactogen remains in doubt, however. They may of course be particularly important in individuals suffering from poor nutrition.

Little has been done on the biochemical mode of action of human placental lactogen, partly because its actions at the physiological level are so poorly understood. A major problem in understanding the full role that the hormone plays arises from the inadequacy of the animal models. As discussed below, it seems likely that placental lactogens from non-primate mammals are not strictly homologous with the human hormone, and as a consequence results obtained when the human hormone is applied to animal tissues are of questionable significance.

5.4 Biosynthesis and secretion of human placental lactogen

Studies on the biosynthesis of the human hormone, especially experiments in which placental mRNA is translated in *in vitro* systems, indicate that, like growth hormone and prolactin, it is synthesized as a pre-hormone with a signal peptide attached to the N-terminus. The amino acid sequence of this N-terminal extension has been determined. Processing of this precursor occurs very rapidly, the signal peptide being removed immediately it has served its purpose—to transfer the nascent polypeptide chain to the lumen of the endoplasmic reticulum.

The total content of placental lactogen in the placenta is substantial, and daily production is of the order of several hundred milligrams per day. However, the placenta is of course a fairly large organ, so the overall *concentration* of the lactogen is relatively low. The hormone is produced mainly by the syncytiotrophoblast, where it has been located by immunofluorescent techniques. The hormone seems to be secreted soon after biosynthesis, probably by exocytosis of cytoplasmic granules. There is considerable debate about the factors that control the secretion of placental lactogen, but little agreement. It may be in fact that there is little specific control of secretion, the hormone simply being exported into the blood almost as fast as it is synthesized. Certainly there is little evidence of fluctuation of plasma levels of the hormone—these increase steadily during pregnancy, correlating roughly with the size of the placenta.

The plasma concentration of human placental lactogen reaches very high levels before term (5–6 μg/ml), one to two orders of magnitude greater than peak values for growth hormone or prolactin. The level of placental lactogen gives quite a good indication of the condition of the placenta. Low levels suggest placental insufficiency.

5.5 Non-primate placental lactogens[49,53]

Placental lactogens have now been detected in many non-primate mammalian groups, although none has been fully characterized yet. Placental lactogens have been isolated from the cow, sheep and rat, and their occurrence in hamster, guinea pig, mouse and goat, among other species, has been established. Preliminary data on the amino acid sequences of the cow and sheep hormones suggest that they are fairly different from human placental lactogen. As will be discussed in the next section, it is likely that they arose independently from the human placental lactogen, but the lack of detailed structural or biological information precludes any firmer conclusions at this stage.

6. THE MOLECULAR EVOLUTION OF THE GROWTH HORMONE/PROLACTIN FAMILY[54,55]

6.1 Structural relationships

Complete or partial amino acid sequences are now available for eight growth hormones and six prolactins, as well as human placental lactogen. Comparison between these various proteins has allowed a picture of the probable evolutionary relationships in the family to be developed.

Among the growth hormones, the sequences of the pig, rat, horse and whale (each of which belongs to a different mammalian order) hormones are very similar, differing at only 2–5% of all residues. The sheep and ox hormones are also very similar (not surprising since these are closely related artiodactyls) but their sequences differ considerably from any of those in the pig/rat/horse/whale group, at about 9% of all residues. Human growth hormone differs very markedly from any of the non-primate growth hormones, at about 35% of all residues. Artiodactyls (pig, sheep, ox), rodents (rat), perissodactyls (horse), cetaceans (whale) and primates (man) are mammalian orders which are generally considered to have diverged at about the same time in evolutionary history (about 75 million years ago), so the conclusion has to be drawn that rates of evolution in this protein family have been remarkably variable, a basic rather slow rate having accelerated on at least two occasions—once, modestly, in the evolution of the ruminants, and once, rather dramatically, during evolution of the primates. This is further emphasized when the partial sequence of a shark growth hormone is considered.[56] This appears to be structurally more similar to the sequences of non-primate mammalian growth hormones than is human growth hormone, despite the fact that the lines leading to sharks and mammals diverged over 400 million years ago.

As has been mentioned, the amino acid sequence of human placental lactogen is rather similar to that of human growth hormone. The two differ at only 15% of all amino acid residues. Human growth hormone is thus more similar to human placental lactogen than it is to any of the non-primate growth hormones. This indicates that the gene duplication which gave rise to the placental lactogens in man (and presumably other mammals) must have occurred after the evolutionary divergence of the primates from the other mammalian orders. This in turn brings into question the relationship between the human placental lactogen and the placental lactogens of non-primate mammals; these must have arisen as a consequence of one or more independent gene duplications. The human and non-primate placental proteins are thus not strictly speaking homologous proteins, and extrapolation of results obtained in laboratory animals to the human situation will be even more questionable in the case of these hormones than it is for others.

Comparisons of the sequences of prolactins show that here too the rate of evolution has been variable. Pig and whale prolactins are very similar, as are the sheep and ox hormones. However, comparison of the sequences of pig (or whale) prolactin with that of the sheep (or ox) hormone or with human prolactin shows a considerable number of differences—about 20% of all residues differing in each case. Rat prolactin is very different from any of the others so far investigated—the sequence differing from those of prolactins from each of the other species at at least 40% of all residues. Thus, the evolutionary rate for prolactin seems to have accelerated on at least three occasions in mammalian evolution—in the evolution of the primates, ruminants and, most markedly, that of the rodents. It should be noted that in most families of proteins the rate of evolution is remarkably constant, so the situation in the growth hormone/prolactin family is rather unusual.

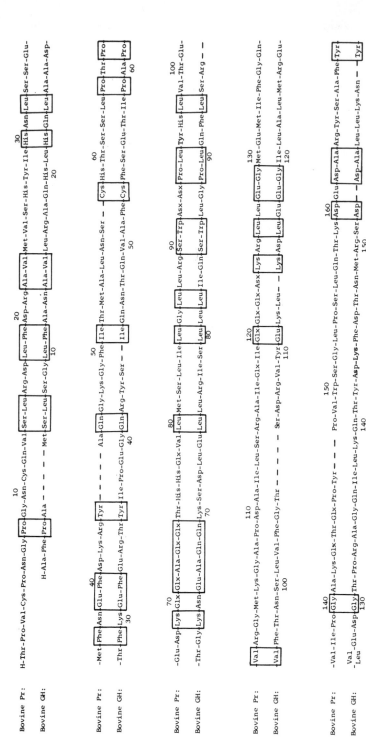

Figure 8.11 Comparison of the amino acid sequences of bovine growth hormone and prolactin. Identical sequences are enclosed in 'boxes'. (Reproduced with permission from Wallis, M. (1975) *Biol. Revs.* **50**, 35–98. Copyright (1975) Cambridge University Press)

Nucleotide sequences are available for the cDNAs and genomic DNA corresponding to several members of the growth hormone–prolactin–placental lactogen family. Comparison of these leads to similar evolutionary conclusions to those based on comparison of protein sequences. Of particular interest is the observation that the growth hormone/placental lactogen genes in man are members of a complex family of linked genes, only two of which are normally expressed (see also Chapter 18).

When the sequence of any growth hormone is compared with that of any prolactin, considerable homology is recognizable (Figure 8.11), and this is the basis for including the two hormones in a single protein family. The sequences are identical at about 25% of all residues, although several 'gaps' have to be introduced into them to maintain this degree of homology. This degree of resemblance is similar to that between haemoglobin and myoglobin—two proteins for which the tertiary structures show very close resemblance.

6.2 An evolutionary tree for the growth hormone-prolactin family

The detailed sequence comparisons discussed in the previous section allow the construction of an evolutionary tree for the growth hormone–prolactin family (Figure 8.12). This simply summarizes the relationships within the group. The two main branches of the tree must have arisen as the result of a gene duplication which occurred early in the evolution of the vertebrates, or possibly in an invertebrate ancestor—growth hormone and prolactin are known to be present as independent hormones in nearly all groups of lower vertebrates. Divergent evolution of the duplicate genes then gave rise to the various prolactins and growth hormones that we know today. Further duplication of the gene for growth hormone gave rise to the primate placental lactogen, as indicated, and presumably one or more corresponding duplications gave rise to the non-primate placental lactogens (although it is not yet clear whether these arose from the growth hormones or prolactins). The lengths of the arms shown in Figure 8.12 reflect the number of point mutations accepted in the evolution of the various hormones; the great variability in the lengths is a consequence of the variable rates of evolution already referred to. Of course, Figure 8.12 emphasizes also the paucity of the information currently available; as more sequences are determined it is likely that the tree will become increasingly complex.

It should also be noted that it has been suggested that there are internal homologies within the sequences of the growth hormones and prolactins. These would indicate an increase in the size of the gene for these proteins by a process of partial gene duplication. However, the degree of internal homology is small, and its significance is not universally accepted.

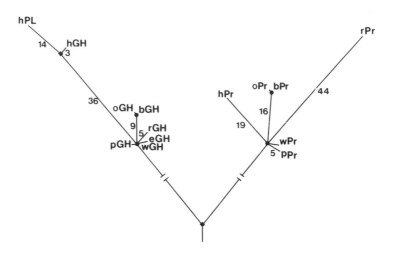

Figure 8.12 An evolutionary tree for the growth hormone–prolactin family. Species are indicated by lower-case letters: h, human; o, ovine; b, bovine; p, pig; r, rat; e, equine; w, whale. GH, growth hormone; Pr, prolactin; PL, placental lactogen. ◆ indicates gene duplication, ● indicates specific divergence without duplication. Branch lengths are related to the number of point mutations/100 residues (shown in several cases on the diagram) calculated to have been accepted in the evolution of each specific molecule from a common ancestor. (Reproduced with permission from Wallis, M. (1981) *J. Mol. Evolution* **17**, 10–18. Copyright (1981) Springer Verlag)

6.3   Biological and immunological relationships

The picture of molecular evolution which has been developed on the basis of structural studies accords well with various observations on the immunological and biological relationships within this hormone family. Thus, the marked structural differences between human growth hormone and the non-primate growth hormones (interpreted here as a consequence of an enhanced rate of evolution in the primate line) are associated with the marked differences in biological actions between the human and non-primate hormones that have already been discussed. Similarly, there is little immunological cross-reactivity between human and non-primate growth hormones, although clear cross-reactivity has been detected between mammalian, avian, reptilian, amphibian and fish (but not teleost) growth hormones, the degree of cross-reactivity diminishing as the evolutionary distance increases (Figure 8.13).[57]

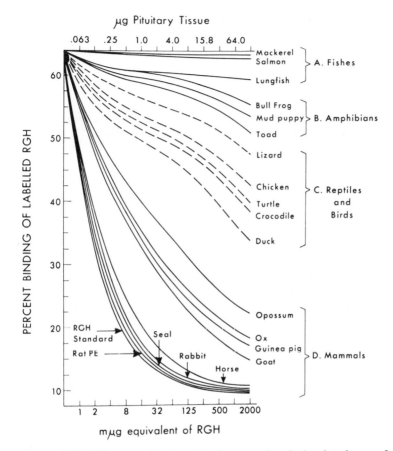

Figure 8.13 Diagram showing the immunochemical relatedness of growth hormones in pituitary extracts from various vertebrates. Based on a radioimmunoassay using antiserum to rat growth hormone. (Reproduced with permission from Hayashida, T. (1970) *Gen. Comp. Endocr.* **15**. 432–452. Copyright (1970) Academic Press Inc.)

6.4 Proliferin: a new member of the growth hormone-prolactin family

Recently a new protein, related to prolactin, has been identified and named proliferin.[59] This is produced by cultured mouse fibroblasts in which proliferation has been stimulated by serum or growth factors. Its sequence is more similar to that of prolactin than growth hormone, but the homology is less than that between the mammalian prolactins. The function of proliferin is not clear, but it seems likely that it arose as the consequence of a duplication of the prolactin gene before the radiation of the placental mammals.

REFERENCES

1. Wallis, M. (1978) In: *Chemistry and Biochemistry of Amino Acids, Peptides and Proteins,* Vol. 5 (Ed. Weinstein, B.). Dekker, New York, pp. 213–320.
2. Lewis, U. J. (1975) *Pharmac. Therap. B.,* **1**, 423–435.
3. Niall, H. D. (1981) In: *Prolactin* (Ed. Jaffe, R. B.). Elsevier, New York, pp. 1–17.
4. Franks, S. (1979) In: *Hormones in Blood,* 3rd. Edn., Vol. 1 (Eds. Gray, C. H. and James, V. H. T.). Academic Press, New York, pp. 279–331.
5. Wallis, M. (1980) In: *Cellular Receptors for Hormones and Neurotransmitters* (Eds. Schulster, D. and Levitzki, A.) Wiley, Chichester, pp. 163–183.
6. Li, C. H. (1980) In: *Hormonal Proteins and Peptides,* Vol. 8 (Ed. Li, C. H.). Academic Press, New York, pp. 1–36.
7. Pasteels, J. L. and Robyn, C. (Eds.) (1973) *Human Prolactin.* Excerpta Medica, Amsterdam.
8. Frantz, A. G. (1978) *New England J. Med.,* **298**, 201–207.
9. Mittra, I. (1980) *Biochem. Biophys. Res. Commun.,* **95**, 1750–1759, 1760–1767.
10. Jaffe, R. B. (Ed.) (1981) *Prolactin.* Elsevier, New York.
11. Horrobin, D. F. (1973–83) *Prolactin,* Vols. 1–8. Churchill Livingstone, Edinburgh.
12. Fulkerson, W. J. (1979) *Hormonal Control of Lactation,* Vol. 1 Churchill Livingstone, Edinburgh.
13. Jacobs, L. S. (1977) *Adv. Exp. Biol. Med.,* **80**, 173–191.
14. Friesen, H. G. (1976) *Recent Results Cancer Res.,* **57**, 143–149.
15. Dellman, H.-D., Johnson, J. A. and Klachko, D. M. (1977) *Comparative Endocrinology of Prolactin.* Plenum Press, New York.
16. Ensor, D. M. (1978) *Comparative Endocrinology of Prolactin.* Chapman and Hall, London.
17. Nicoll, C. S. (1981) In: *Prolactin* (Ed. Jaffe, R. B.). Elsevier, New York, pp. 127–166.
18. Nicoll, C. S. (1982) *Perspectives in Biology and Medicine,* **25**, 369–381.
19. Rosen, J. M. (1981) In: *Prolactin* (Ed. Jaffe, R. B.). Elsevier, New York, pp. 85–126.
20. Rillema, J. A. (1980) *Fed. Proc.,* **39**, 2593–2598.
21. Topper, Y. J. and Freeman, C. S. (1980) *Physiol. Rev.,* **60**, 1049–1106.
22. Shiu, R. P. C. and Friesen, H. G. (1980) *Ann. Rev. Physiol.,* **42**, 83–96.
23. Cowie, A. T., Forsyth, I. A. and Hart, I. C. (1980) *The Hormonal Control of Lactation.* Springer-Verlag, Berlin.
24. Topper, Y. J. (1970) *Rec. Progr. Hormone Res.,* **26**, 287–302.
25. Turkington, R. W. (1972) In: *Biochemical Actions of Hormones,* Vol. 2 (Ed. Litwack, G.). Academic Press, New York, pp. 55–80.
26. Shiu, R. P. C. and Friesen, H. G. (1976) *Science,* **192**, 259–261.
27. Haeuptle, M.-T., Aubert, M. L., Djiane, J. and Kraehenbuhl, J.-P. (1983) *J. Biol. Chem.,* **258**, 305–314.
28. Posner, B. I., Bergeron, J. J. M., Josefsberg, Z., Khan, M. H., Khan, R. J., Patel, B. A., Sikstrom, R. A. and Verma, A. K. (1981) *Recent Progr. Hormone Res.,* **37**, 539–582.
29. Chomczynski, P. and Topper, Y. J. (1974) *Biochem. Biophys. Res. Commun.,* **60**, 56–63.
30. Rosen, J. M., Matusik, R. J., Richards, D. A., Gupta, P. and Rodgers, J. R. (1980) *Recent Progr. Hormone Res.,* **36**, 157–193.
31. Falconer, I. R. and Rowe, J. M. (1977) *Endocrinology,* **101**, 181–186.
32. Turkington, R. W., Majumder, G. C., Kadohama, N., MacIndoe, J. H. and Frantz, W. L. (1973) *Recent Progr. Hormone Res,* **29**, 417–449.
33. Rillema, J. A., Wing, L.-Y. C. and Foley, K. A. (1983) *Endocrinology,* **113**, 2024–2028.
34. Gorski, J. (1981) In: *Prolactin* (Ed. Jaffe, R. B.). Elsevier, New York, pp. 57–83.

35. Maurer, R. A., Stone, R. and Gorski, J. (1976) *J. Biol. Chem.*, **251**, 2801–2807.
36. Austin, S. A. and Wallis, M. (1979) *Biochim. Biophys. Acta*, **564**, 534–545.
37. Farquhar, M. G. (1971) *Mem. Soc. Endocrinol.*, **19**, 79–124.
38. Shenai, R. and Wallis, M. (1979) *Biochem. J.*, **182**, 735–743.
39. Weiner, R. I. and Bethea, C. L. (1981) In: *Prolactin* (Ed. Jaffe, R. B.). Elsevier, New York, pp. 19–55.
40. Leong, D. A., Frawley, S. and Neill, J. D. (1983) *Ann. Rev. Physiol.*, **45**, 109–127.
41. Macleod, R. M. (1976) In: *Frontiers in Neuroendocrinology*, Vol. 4 (Eds. Martini, L. and Ganong, W. F.). Raven Press, New York, pp 169–194.
42. Nansel, D. D., Gudelsky, G. A. and Porter, J. C. (1979) *Endocrinology*, **105**, 1073–1077.
43. Labrie, F., Godbout, M., Beaulieu, M., Borgeat, P. and Barden, N. (1979) *Trends in Biochem. Sci.*, **4**, 158–160.
44. Ray, K. P. and Wallis, M. (1982) *Mol. Cell. Endocrinol.*, **27**, 139–155.
45. Ray, K. P. and Wallis, M. (1982) *Mol. Cell. Endocrinol.*, **28**, 691–703.
46. Ray, K. P. and Wallis, M. (1984) *Mol. Cell. Endocrinol.*, **36**, 131–139.
47. Gershengorn, M. C. (1982) *Mol. Cell. Biochem.*, **45**, 163–179.
48. Enjalbert, A., Ruberg, M., Arancibia, S., Priam, M. and Kordon, C. (1979) *Nature*, **280**, 595–597.
49. Talamantes, F., Ogren, L., Markoff, E., Woodard, S. and Madrid, J. (1980) *Fed. Proc.*, **39**, 2582–2587.
50. Burstein, S., Grumbach, M. M., Kaplan, S. L. and Li, C. H. (1978) *Proc. Nat. Acad. Sci. USA*, **75**, 5391–5394.
51. Chard, T. (1979) In: *Hormones in Blood*, 3rd edn, Vol. 1 (Eds. Gray, C. H. and James, V. H. T.). Academic Press, London, pp. 333–361.
52. Wurzel, J. M., Parks, J. S., Herd, J. E. and Nielsen, P. V. (1982) *DNA*, **1**, 251–257.
53. Fellows, R. E., Bolander, F. F., Hurley, T. W. and Handwerger, S. (1976) In: *Growth Hormone and Related Peptides* (Eds. Pecile, A. and Müller, E. E.). Excerpta Medica, Amsterdam, pp. 315–326.
54. Wallis, M. (1981) *J. Molecular Evolution*, **17**, 10–18.
55. Miller, W. L. and Eberhardt, N. L. (1983) *Endocrine Rev.*, **4**, 97–130.
56. Lewis, U. J., Singh, R. N. P., Lindsey, T. T., Seavey, B. K. and Lambert, T. H. (1974) In: *Advances in Human Growth Hormone Research* (Ed. Raiti, S.). DHEW, Washington D.C., pp. 349–371.
57. Hayashida, T. (1972) In: *Growth and Growth Hormone* (Eds. Pecile, A. and Müller, E. E.). Excerpta Medica, Amsterdam, pp. 25–37.
58. Nicoll, C. S. and Bern, H. A. (1972) In: *Lactogenic Hormones* (Eds. Wolstenholme, G. E. W. and Knight, J.). Churchill Livingstone, Edinburgh, pp. 299–324.
59. Linzer, D. I. H. and Nathans, D. (1984) *Proc. Nat. Acad. Sci. USA*, **81**, 4255–4259.

Chapter 9

Insulin and the islets of Langerhans

1. STRUCTURE AND PROPERTIES OF INSULIN

1.1 Discovery of insulin

The discovery of insulin in 1922[1] and its rapid application to the successful treatment of diabetes mellitus is one of the major medical triumphs of this century. Diabetes is a condition in which blood glucose is permanently elevated and the urine also contains large amounts of glucose as an abnormal constituent. Although diabetes had been known since classical antiquity, and is described in early Chinese medical texts, the association between the disease and the pancreas was not finally established until von Mering and Minkowski in 1889[2] showed that pancreatectomy in animals resulted in a severe and fatal form of diabetes closely resembling human diabetes seen in the young. This discovery was made at a time when there was a growing interest in endocrine glands and their function. It had already been noted that extirpation of the thyroid gland in sheep resulted in a disease resembling human myxoedema, and it was found natural therefore to investigate in detail the factor in human pancreas in whose absence diabetes apparently developed.

Not long before the discovery that pancreatectomy produced a permanent diabetes, Paul Langerhans discovered small aggregations of tissue which were histologically distinct from the enzyme-producing parts of the pancreas. Subsequently, these were seen to show damage in some cases of diabetes,[3] and it was assumed that they might be associated with an anti-diabetic principle.

Although complete success was achieved in 1922 when Banting and Best extracted insulin from dog pancreas, there were many earlier attempts to extract anti-diabetic substances and some came near to success. Thus, in 1907[4] Zuelzer produced an extract of pancreas, by using aqueous ethanol, which successfully reduced the urine glucose of some diabetic patients, although it caused a pyrogenic response and was too toxic for general use. At about this time, a number of other workers showed that after blocking the pancreatic duct, the acinar or enzyme-producing part of the pancreas degenerated in a few weeks, while leaving the islets intact. By this means, an organ could be produced in which extracts of the endocrine pancreas would be less likely to undergo enzymatic destruction. Scott in 1912[5] obtained a partial atrophy of the pancreas by duct ligation, and showed that saline extracts from endocrine pancreas reduced the urinary glucose in depancreatized dogs. However, because such extracts also produced a toxic response on injection, the work was discontinued. Banting and Best used pancreas in which the pancreatic duct had been ligated 7–10 weeks earlier and which was extracted by Ringers solution in the cold. This material, after filtration, consistently reduced the blood glucose of depancreatized dogs. It was soon discovered that such material could also be extracted with ethanol. There were a number of reasons for the success of Banting and Best; thus they used foetal pancreas in their initial experiments in which the acinar enzyme content was greatly reduced. They also used chilled pancreas, which inactivated enzymes, and they were also among the first to use reliable methods of glucose estimation in blood and were therefore accurately able to assess the biological effects of their product. They showed that such pancreatic extracts lowered the blood glucose of depancreatized dogs. Such extracts were also capable of lowering the blood glucose of diabetic patients, who, for the first time, became effectively treatable.

It should also be noted that Banting and Best were not completely alone in their discovery of insulin. Paulesco[6] had also published in 1921 an account of the properties of a blood sugar-lowering material which he obtained by saline extraction of dog pancreas. This reduced blood sugar and ketones of depancreatized dogs, but was too toxic for injection into man, owing to the presence of many impurities. The real achievement of Banting and Best and of their colleagues was to produce a material which could safely and effectively be injected into man, and restore diabetics to virtually normal health.

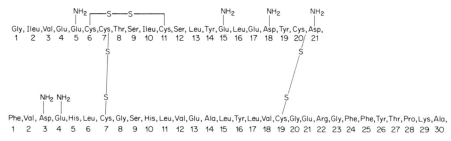

Figure 9.1   The amino acid sequence of pig insulin

1.2   Sequence of beef insulin

Insulin was one of the first proteins to be obtained in a state of considerable purity for pharmaceutical purposes, and therefore was soon the object of considerable attention as a convenient protein to analyse. The amino acid composition of insulin from cattle was established at an early date and it was shown to possess a relatively high cystine content, although tryptophan and methionine were absent. It was later shown that insulin possessed a high content of free α-amino groups, indicating that it was a protein composed of relatively short polypeptide chains. Later, using beef insulin, it was shown by Sanger, who reacted the free amino groups with the reagent fluorodinitrobenzene (FDNB), that there were two distinct types of chain, a glycyl chain and a phenylalanyl chain (see Figure 9.1). These chains could be separated after treating insulin with performic acid which oxidized the cystine residues to cysteic acid. The amino acid sequence of the two chains was then established. In the case of the B chain this was achieved by breaking up the chain into smaller fragments by trypsin and chymotrypsin, end groups again being determined by FDNB. The A chain proved rather more resistant to enzyme attack, and was finally broken up

Table 9.1 Some sequence differences in insulins of various species

Species	A chain 1, 8, 9, 10	B chain —1, 1, 29, 30
Man	Gly. Thr. Ser. Ile	Phe. Lys. Thr.
Horse	Gly. Thr. Gly. Ile.	Phe. Lys. Ala.
Pig	Gly. Thr. Ser. Ile.	Phe. Lys. Ala.
Sheep	Gly. Ala. Gly. Val.	Phe. Lys. Ala.
Beef	Gly. Ala. Ser. Val.	Phe. Lys. Ala.
Rabbit	Gly. Thr. Ser. Ile.	Phe. Lys. Ser.
Rat I*	Gly. Thr. Ser. Ile.	Phe. Lys. Ser.
Rat II*	Gly. Thr. Ser. Ile.	Phe. Met. Ser.
Chicken	Gly. His. Asn. Thr.	Ala. Lys. Ala.
Cod	Gly. His. Arg. Pro.	Met. Ala. Lys.---

* Asp substitution for Glu at position 4, on A chain.

by partial acid hydrolysis into smaller peptides whose structures could be more easily established. By this means it was possible to establish the complete sequence of chains, and in later experiments to determine how they were joined by disulphide bridges.[7] Insulin was the first protein for which a complete amino acid sequence was determined, work for which Sanger received a Nobel Prize in 1958.

1.3 Sequence of insulins from other species

Since Sanger's delineation of the sequence of beef insulin, the insulins of many other species have been examined, and their structures determined.[8] Details of these are shown in Table 9.1. Several features emerge from this table. Thus, there are frequent changes in the sequence within the intrachain disulphide ring in certain mammals (positions 8, 9, 10). Pig, sheep, beef and horse insulins differ because of changes here. Some insulins, such as that from the guinea pig, show very marked changes in sequence, in comparison with that found in cattle insulin. Knowledge of such species differences makes it possible to decide which residues are invariant in insulin, and necessary therefore for biological activity.

In some animals, there are two different molecular species of insulin present in the same pancreas. This is so for the rat and mouse (whose insulins have the same structure), as well as for certain fish (tuna and toadfish).

1.4 X-ray crystallography of insulin

Progress on determining the domains of insulin which combine with receptors had to await the complete resolution of insulin by X-ray crystallography. This

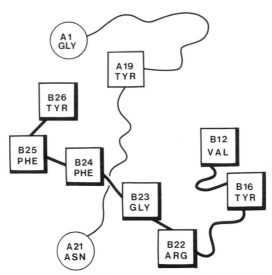

Figure 9.2   Residues probably involved in insulin–receptor interactions

was first achieved by Hodgkin and her coworkers[9,10] using pig insulin, in which six molecules of the insulin monomer were combined with two atoms of zinc. This structure consists of three dimers in which (except in the case of guinea pig insulin) the zinc is coordinated with the histidine residue at B10. The dimer exists in solution, but at the very high dilution at which insulin circulates (10^{-10} M), insulin will be present as a monomer. It is likely that many of the residues which involve dimer formation from the monomers are also concerned with the insulin–receptor interaction. This region includes a hydrophobic sequence near the C-terminus of the B chain (B24 phenylalanine, B25 phenylalanine, B26 tyrosine, B12 valine, as well as residue B16 tyrosine) and more-polar groups from the A chain which are directed to the exterior of the molecule.[10] This arrangement is shown diagrammatically in Figure 9.2.

1.5   Relationship between structure and activity (see also Chapter 15)

Insulins and insulin secretory cells have been described amongst many members of the animal kingdom. Thus, molecules resembling insulin appear to exist as far down the evolutionary scale as echinoderms. Moreover, quite recently similar substances closely resembling insulin have been found in many insects, including the honeybee and the blowfly.

Information about the structure which is necessary for the biological activity of insulin has been obtained from work on chemically modified insulins, and more recently from studies on the X-ray crystallography of insulin as well as from a careful study of the invariant amino acids in different animal species.

Early work showed that esterification of the carboxyl groups or reduction of the disulphide linkages caused complete inactivation of insulin. In later work a wide variety of chemical agents have been used to modify insulins. When amino groups are modified, loss of biological activity usually occurs, particularly when this affects the N-terminal glycine. Removal of the N-terminal phenylalanine (B1) does not affect activity. Enzymatic removal of the C-terminal asparagine by carboxypeptidase abolishes activity although this is not so if the C-terminal alanine (B30) is removed. Complete iodination of insulin, by substitution in the tyrosine residues, also abolishes the activity of insulin, although this is not true for less-substituted derivatives.

Many of these observations are supported by work on synthetic insulins in which the chains have been chemically modified. From such studies it seems that the N-terminal glycine is essential for activity, whereas if the intrachain disulphide link on the A chain is replaced by a CH_2–S bridge, activity is retained. In general, studies on such insulins have suggested that the chemical modification often results in disturbance of conformation, rather than the identification of specific functional groups involved in biological activity.

1.6   Insulin antibodies

Despite its low molecular weight, insulin is antigenic. This was first convincingly demonstrated in 1955, when it was shown by Moloney and Coval[11] that antibodies could be induced in guinea pigs by injection of insulin in Freund's adjuvant. Such antibodies were able to neutralize the hypoglycaemic effect of insulin injected into animals.

The sites on the insulin molecule for antibody combination are quite different from those which are thought to be involved in its biological activity. They have been discussed in detail elsewhere.[12] Insulin antibodies are sometimes of importance clinically, since they are produced in diabetics treated with animal insulins. Such antibodies can bind large quantities of injected insulin, and this property was the basis of the first immunoassay devised by Berson and Yalow.[13]

1.7   Assay of insulin

The earliest methods used to assay insulin depended on the effects of the hormone in lowering the blood glucose of rabbits and mice. Later, effects of insulin on isolated tissues were used—for example, the effects of insulin on accelerating glucose uptake and glycogen synthesis in rat diaphragm muscle preparations. Alternatively, insulin increases the oxidation of ^{14}C-labelled glucose to $^{14}CO_2$ in adipose tissue, and this has been the basis of a reliable assay procedure. These biological methods are neither very precise nor very sensitive, and they are now used only rarely. A more recent type of assay (a radioreceptor assay) which measures the binding of insulin to liver membranes has been used to measure the biological activity of insulin. The most widely used types of assays for insulin are immunoassays, which utilize a variety of techniques to separate free and bound insulin (see Chapter 2). The procedures now most generally used are double antibody techniques, or methods dependent on precipitating the insulin antibody complex with ethanol; alternatively, free insulin may be removed with charcoal. It has to be remembered that such methods measure the *immuno-logical* reactivity of insulin, which may differ from its *biological* activity under some circumstances. Circulating concentrations of insulin lie between 0.2 ng/ml (fasting) and 2 ng/ml following food.

1.8   General properties of insulin

The insulin monomer has a molecular weight of 5700. It has an isoelectric point of 5.3. Although insulin is stable under neutral and strongly acid conditions, it is relatively easily inactivated by alkali. When treated with cysteine or glutathione, it also loses biological activity due to reduction of disulphide linkages.

2.   CELLULAR SOURCE OF INSULIN

It has already been mentioned that Paul Langerhans in 1869 discovered tissue scattered throughout the pancreas which appeared to him to be of different structure from that of acinar cells. The development of microscopic techniques by Lane[14] and others clearly showed that at least two kinds of cell were present, designated A (or α) and B (or β) cells. Later, by the use of silver impregnation methods it became possible to differentiate the A cells into two types, the A_2 cells responsible for the production of glucagon, and the A_1 or D cells. It was at first thought that the A_1 cells might produce gastrin. It is now thought, however, that these cells manufacture the small polypeptide hormone somatostatin. In addition, there are a number of other cells in the islets which are associated with the putative hormone, pancreatic polypeptide, whose function is at present

Table 9.2   Cellular composition of pancreatic islets

Cell type	Frequency (%)	Hormone present
B (or β)	60–80	Insulin
A (α, A_2)	20–40	Glucagon
D (δ, A_1)	6–15	Somatostatin
PP	1	Pancreatic polypeptide

Data from Gepts and Lecompte.[15]

unknown. The relative proportions of each cell type vary somewhat from species to species and from one part of the pancreas to another. Representative values for human pancreas islets are shown in Table 9.2.

Insulin appears to be associated solely with the islets of Langerhans in vertebrates. Nevertheless, chemically related materials such as various growth factors are elaborated in the liver and elsewhere.

3. THE BIOSYNTHESIS OF INSULIN

The general cellular mechanism for the biosynthesis of polypeptide hormones through various precursors has already been described in Chapter 1. Insulin is made by such a system.

3.1 Proinsulin

The conclusion that insulin was made as a precursor was derived from pulse-chase experiments in which labelled leucine was incubated with slices of an insulin-producing tumour and the pattern of labelling of the early products followed. In these experiments it was shown that material with a molecular weight of 9000 was first labelled rather than insulin itself.[16] In later experiments, this material was shown to be converted by tryptic digestion to material with the general properties of insulin. This precursor was designated proinsulin (Figure 9.3). Later experiments by Steiner and his colleagues demonstrated that it was possible to isolate proinsulin from a number of commercial insulins, and

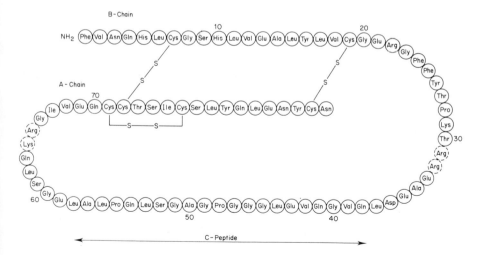

Figure 9.3 Structure of proinsulin in man

material so isolated was used to ascertain its amino acid sequence.[17] Trypsin treatment results in the formation of insulin and a smaller peptide termed the C or connecting peptide, which joined the chains together.

In all species so far examined, insulin appears to be made through a proinsulin, perhaps to enable sulphydryl groups to be brought into a position to facilitate formation of the correct covalent, S–S linkages. The two chains may be joined chemically although the yields are frequently low.

3.2   Nature of the C peptide

The C peptide shows considerable variation in chain length and amino acid composition[18] when compared between one species and another. Its chain length is 31 residues in man, although it is less by 5 residues in the sheep and by 8 in the dog. In those C peptides so far investigated, phenylalanine and tyrosine and sulphur-containing amino acids have been consistently absent. In the rat, it is of considerable interest that there are two proinsulins, showing variation in both the insulin sequence as well as in the C peptide. The variation in sequence suggests that its structural requirements are not very exacting.

3.3   Intracellular conversion of proinsulin to insulin

Some details of the conversion process are still obscure. However, it appears that trypsin is not the enzyme involved, since the product of its activity would be insulin lacking a C-terminal alanine. The conversion process is thought to take place at an early stage in the formation of the granule. It has been suggested[19] that two different kinds of enzyme might be involved in the process. One enzyme, acting as an endopeptidase, resembles trypsin in its specificity and cleaves between glycine of the A chain and arginine, as well as between arginine and the beginning of the C peptide. The other acts like carboxypeptidase B and removes the arginyl and lysyl residue from insulin and the C peptide (see Figure 9.3). This enzyme, which has a molecular weight of 31 000, functions as a thiol protease.

3.4   Release of proinsulin and C peptide—properties of proinsulin

Although historically it was thought that only insulin would be released into the circulation, it has been conclusively shown that C peptide is released in equimolar quantities with insulin when B-cell granules are discharged. In addition, small quantities of proinsulin also appear to be released from the B cells. The mechanism for the release of this is unclear. In general, proinsulin possesses only a small fraction of the biological activity of insulin of the same species. Its immunological activity is also less with respect to the corresponding insulin, although enough to ensure considerable cross-reactivity with antisera made against insulin. The levels of proinsulin are not normally increased when the islets are stimulated by glucose.

3.5   Preproinsulin

Some time after the discovery of proinsulin, it became clear that an even earlier precursor of insulin was detectable in the islets of Langerhans. This discovery followed the identification of similar precursors for immunoglobulins by Milstein and his collaborators.[20] The discovery of these prepeptides resulted from work on the translation of insulin messenger in various types of cell-free systems. Since the absolute amounts of preproinsulin at any time in the pancreas are exceptionally small, partial sequences were obtained by incubating labelled amino acids with insulin messenger and determining the positions of the label. More recently, it has become easier to delineate the sequence of these precursors by direct reference to the nucleotide sequence of the insulin gene itself. In general, these polypeptides ('signal peptides') possess an additional 23 to 30 amino acids attached to the N-terminus of the proinsulin (Figure 9.4).[21] Their function (which has been explained in Chapter 1) is to assist the translocation of the precursor into the lumen of the endoplasmic reticulum. They therefore have a short half-life.

$$NH_2 \underset{24\,aa}{\overline{\text{Pre-region}}} - Glu\Lambda la \| Phe \underset{30\,aa}{\overline{\text{B-chain}}} ArgArg \underset{31\,aa}{\overline{\text{C-peptide}}} LysArg \underset{21\,aa}{\overline{\text{A-chain}}} COOH$$

Figure 9.4   The sequence of rat preproinsulins I and II

In the case of insulin, the signal peptide structure shows quite considerable variation from species to species, but importantly small neutral amino acids, such as alanine, cysteine or serine, are present at the N-terminal side of the junction with the phenylalanine of the B chain. The endopeptidase which splits off the pre-hormone fragment seems to show broad specificity in contrast to the very limited specificity exhibited by the enzymes splitting off C peptide from proinsulin.

3.6   The regulation of insulin biosynthesis

There has been much interest in the way in which the control of insulin biosynthesis may be regulated, since this may be altered under a variety of physiological circumstances, as well as in disease. The subject has been recently reviewed.[22] It now seems certain that the concentration of glucose reaching the islets is the most critical factor in regulating the rate of biosynthesis.[23,24] Several other sugars and sugar metabolites, however, share with glucose this stimulatory effect, such as mannose. However, the non-utilizable sugar, galactose, is without effect. Mannoheptulose, an inhibitor of glucose phosphorylation inhibits insulin biosynthesis, suggesting that utilization of carbohydrate sugars is essential for the effect.

Provision of energy seems not to be the major factor which promotes biosynthesis since pyruvate (which is freely metabolized by islets), does not influence the rate of insulin synthesis. Perhaps some factor produced early in glycolysis is the responsible agent. Glucose itself cannot be excluded, since it appears to be effective in increasing synthesis in non-metabolizing diluted homogenates of islets.

In this connection, interest has very naturally focused on whether or not the effect is transcriptional, translational or both. Prolonged incubation of islets with glucose in tissue culture leads to an increase in insulin mRNA, suggesting a transcriptional effect.[25] In fact, it seems likely that the regulation of transcription is the major way in which in the long term the control of insulin synthesis is effected.

In the short term (under an hour), glucose is mainly functioning to increase the translation of existing messenger[26] and mRNA levels are unchanged over such short periods of time. Perhaps glucose (or a glucose metabolite) acts to increase rates of initiation of insulin synthesis.

As far as transcription is concerned, one suggestion is that cyclic AMP in the B cell of the islet is involved in the mechanism. This may rise in the islet after prolonged glucose feeding, when insulin synthesis increases. The same is true in pregnancy, where cyclic AMP levels and insulin synthesis are increased.[27] However, a direct connection between cyclic AMP and the transcription of the insulin gene has yet to be reported.

3.7 The insulin gene

The nucleotide sequence of the gene has recently been established for a number of species, including man and the two differing rat insulins.[28] Some surprising differences have emerged. Thus, the human insulin gene has a large intervening sequence splitting in two the code for the C peptide, which is not apparent in rat insulin I (see Figure 9.5).

Differences in the DNA sequence have recently been reported for the human gene for those regions near the beginning of the sequence (the 'flanking genes') in some types of human diabetes.[29] It is too early to say whether these have any real significance in terms of disease.

3.8 Other hormones and insulin synthesis

It has been known for some time that growth hormone may directly increase the rate of insulin biosynthesis when incubated *in vitro* with isolated islets of Langerhans over prolonged periods of time.[30] This is also the case for placental lactogen. *Excessive* amounts of growth hormone may affect the islets of Langerhans directly, leading to B cell destruction, especially in carnivores.[31] Adrenaline, by contrast, appears to inhibit insulin biosynthesis at low concentration, and this may be physiologically important.

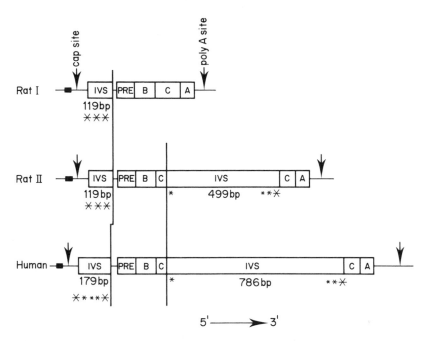

Figure 9.5 Human and rat insulin genes (reprinted by permission from *Nature*, **284**, 26–32. Copyright © 1980 Macmillan Journals Limited)

4. THE SECRETION OF INSULIN

4.1 Historical background

It only became possible to study the release of insulin in any detail when methods for its accurate assay became available, although earlier work by Anderson and Long[32] in which rat pancreas was perfused with a solution containing glucose showed that this sugar promoted insulin release, as indicated by effects on the lowering of the blood sugar in mice. With the introduction of immunoassay techniques, the release of small quantities of insulin could easily be measured—first from slices of pancreas and later by many other workers from isolated islets of Langerhans. As a result a systematic study could be made of the factors likely to influence insulin release physiologically. Just as importantly, the use of isolated islets of Langerhans made it possible to study the metabolism of islets and relate this to secretion.

4.2 The major substances influencing insulin release physiologically

In most animals, glucose and amino acids are the major physiological determinants of insulin release, although in ruminants, short-chain fatty acids

Table 9.3   Insulin secretagogues

		Remarks
Sugars (and derivatives)	Glucose Mannose	
	Xylitol ⎱ Ribitol ⎬ Ribose ⎰	Effective in some species
	Glyceraldehyde Inosine Glucosamine	
Amino acids	esp. Arginine Lysine Leucine Phenylalanine	
Ketone bodies	β-hydroxybutyrate Acetoacetate	Effect variable in differing species
Drugs	Sulphonylureas Dibutyryl cyclic AMP Theophylline Caffeine	

may also be important. In addition, a very large number of other substances induce insulin release, acting as pharmacological agents, often by affecting some aspect of islet metabolism. A list of such substances is shown in Table 9.3. Some of these secretagogues act directly by altering concentrations of cyclic AMP.

4.3   Effects of glucose on the islets of Langerhans

Quite soon after the introduction of immunoassay as a method of estimating insulin, it became possible to analyse accurately the blood insulin changes following glucose administration in man or other animals. The changes in blood insulin levels following 50 g of glucose given by mouth are shown in Figure 9.6. Blood insulin values increase up to five-fold from fasting levels, returning to the fasting values in 2 hours. The corresponding values for blood glucose are also shown in Figure 9.6. It will be seen that blood glucose changes very little in the course of the tolerance test (from about 4.5 to 6.5 mM). It is in fact kept within rather narrow limits by the extra insulin secreted.

Similar results have been obtained when the release of insulin has been studied in preparations of pancreas incubated *in vitro*. For example, when isolated islets of Langerhans are incubated with various glucose concentrations, the release of insulin follows a sigmoidal pattern (Figure 9.7). This pattern of insulin release has been seen in several species of mammal. It indicates that at a certain glucose concentration the islets suddenly begin to secrete and that the

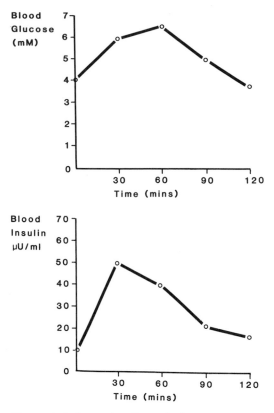

Figure 9.6 Serum insulin and glucose responses
to glucose by mouth in man

response is most sensitive over a comparatively narrow range of glucose concentrations. This mechanism is clearly designed to ensure that glucose values after food return as soon as possible to the fasting values.

Stimulation of perfused pancreas with glucose leads to an initial spike of insulin after 1–2 minutes followed by an abrupt fall. There is a second rise which continues up to 1–2 hours. It may be that insulin released in the first phase is that present in granules which are in very close proximity to the plasma membrane. A similar rather rapid rise in blood insulin is seen when glucose is administered intravenously in man (Figure 9.8).

More insulin appears to be secreted when glucose is administered orally in comparison with the results of infusing the same quantity of glucose intravenously. This has led most investigators to conclude that a factor produced by the direct interaction of sugars with cells in the gastrointestinal tract evokes the release of another polypeptide hormone. This hormone may directly modify the response of the islets. There have been several candidates for this

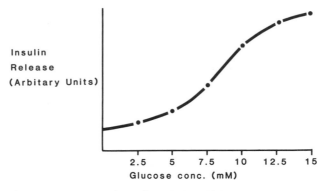

Figure 9.7 Insulin release from isolated islets of Langerhans

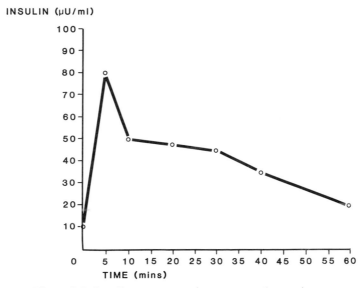

Figure 9.8 Insulin response to intravenous glucose in man

role. Among the most likely is gastric inhibitory peptide, GIP (see Chapter 12). Such a mechanism for stimulating insulin release ensures the maximal response of the islet after feeding.

4.4   Other substances which promote insulin release

As will be seen from Table 9.3, a wide variety of substances increases insulin secretion. Although most attention has been focused on sugars and amino acids, in some animals fatty acids or ketone bodies may be relatively important as physiological secretagogues.

The stimulation of insulin release by leucine was first demonstrated by Cochrane in 1956[33] in children with a hypoglycaemic syndrome. Since then, many other amino acids have been shown to be effective, particularly arginine, lysine, valine and phenylalanine.[34] In fact, mixtures of ten essential amino acids were much more effective than individual amino acids in raising plasma insulin in man. In pancreas preparations *in vitro*, leucine and arginine are effective in the absence of glucose.

The obvious role of amino acid-induced insulin release is to dispose of amino acids produced from a protein meal. Insulin has a stimulatory effect on protein synthesis in many tissues (see Chapter 10). The mechanism of amino acid-stimulated insulin release is, however, under dispute. It may involve the production of reducing equivalents and ATP.[35]

It was first shown in dogs that infusion of ketone bodies affects insulin release.[36] However, in many other species the effects of these substances were less evident. In the rat insulin release due to ketone bodies appears to be a potentiation of the secretory response to glucose.[37] Since this effect, at least in the rat, is only seen with high concentrations of glucose, its real physiological significance is doubtful. There is evidence in some animals, particularly ruminants, that short-chain fatty acids are also secretagogues.

Many drugs will affect insulin release by inducing changes in islet metabolism. The best known group is the sulphonylurea drugs which are clinically important in the treatment of mild diabetes. These may perhaps be effective by raising B cell cyclic AMP concentration. Adrenaline (which lowers cyclic AMP levels), however, inhibits both insulin release and its biosynthesis, and this may be physiologically important.

Substances which provoke insulin release are frequently divided into two major groups—initiators and potentiators. The initiators will promote release in their own right, whereas potentiators do so only in the presence of limiting concentrations of another substance. Glucose is the most important initiator. Other initiators are leucine, glyceraldehyde and the sugar mannose. Fructose and pyruvate act as potentiators. Presumably the function of the initiator is to provide a certain minimum flux through metabolism to sustain exocytosis.

4.5 The regulation of insulin release by sugars

The way in which glucose stimulates the release of insulin is still the subject of controversy. Some of the difficulties in this area are conceptual in that sometimes a simplistic biochemical view has been put forward to explain a process of great morphological complexity, which takes place in distinct steps. The subject has been reviewed in detail elsewhere.[38]

The process of exocytosis is energy dependent since ATP is needed for granule translocation. However, it soon became evident that B cell ATP was not in itself rate limiting, although glucose must be metabolized in order to provide it. Many other factors will be needed to facilitate the complex process of granule extrusion, apart from ATP. Among these the more important are Ca^{2+} ions, NADP(H) and NAD(H), as well as cyclic AMP. Such factors may be needed at various stages in the secretory cycle and their concentrations will depend on glucose metabolism.

A second problem to be considered is the role of glucose as a trigger to set in motion exocytosis. It has frequently been suggested that there are two ways in which the trigger might operate. On the one hand, glucose might interact with a receptor and increase the concentration of an intracellular second messenger (such as cyclic AMP). Alternatively, glucose metabolism might lead to the elaboration of one or more factors which initiate release. Neither of these views is mutually exclusive, although at present the balance of opinion favours the production of some metabolic factor related to glucose metabolism. (This theory has been termed by Ashcroft[39] 'the substrate site hypothesis' of insulin release).

Among the factors which have been identified as perhaps affecting exocytosis are glucose 6-phosphate, intermediates in the pentose phosphate pathway, and phosphoenolpyruvate.

No doubt several metabolic pathways are stimulatory and may be necessary for granule release to take place. In fact, until the precise biochemical mechanisms which involve translocation of granules are properly understood it will not be possible to delineate precisely how the initial trigger operates. In what follows a provisional hypothesis to explain glucose-mediated secretion is put forward.

That there is a close connection between insulin secretion and B cell metabolism became evident at a very early stage in the study of islets.[40] Thus, the inhibition of metabolism by classical inhibitors such as azide, cyanide or 2,4-dinitrophenol, also results in a total inhibition of insulin secretion. Moreover, only those sugars which are metabolized rapidly by islets, such as glucose or mannose, evoke secretion, whereas sugars like fructose, which are only poorly metabolized, and galactose, which is not metabolized at all, are not secretagogues. Again, when the rates of glucose oxidation in islets are plotted against glucose concentration and compared with the effects of glucose concentration on insulin release, both curves are sigmoidal and almost superimposable.

A role for NAD(P)H has been suggested in secretion. It has been shown, for example, that a lowering of islet NAD inhibits secretion. Reducing equivalents will become available from the operation of several pathways in metabolism which are known to be important for secretion. The exact connection between such substances and the process of exocytosis, however, is far from clear.

4.6 Cyclic AMP and secretion

Early studies on insulin secretion also showed that caffeine and theophylline (two phosphodiesterase inhibitors) increased insulin release from pancreas slices. Similar results have been obtained when whole islets are incubated with the cyclic AMP derivative, dibutyryl cyclic AMP. It seemed likely, therefore, that cyclic AMP was involved in the release process, as it is in the release of many other polypeptide hormones. Nevertheless, attempts to show an elevation of cyclic AMP in islets following incubation with glucose were not successful initially, perhaps due to the use of starved animals which normally show markedly depressed intracellular concentrations of B cell cyclic AMP. It is now established that glucose produces a rise in cyclic AMP in most species, although such rises may be relatively transitory and maximal after only a few minutes exposure to glucose.

Secretagogues other than glucose generally also produce an elevation of cyclic AMP concentration in the islet. It seems therefore that increased intra-islet concentrations of cyclic AMP may be associated in some way with the secretory process. The relationship between a rise in cyclic AMP concentration and release is, however, by no means a simple one. Thus, theophylline can raise cyclic AMP levels in islets in the absence of glucose without stimulating secretion.

In general, glucose metabolism is necessary for the elevation of cyclic AMP concentration in whole islets, although this may not necessarily be the case for islet homogenates. The manner in which metabolism increases cyclic AMP concentration in islets is uncertain. However, over some hours glucose appears to increase the activity of adenylate cyclase.[41] Such an effect may be important in the long-term regulation of cyclic AMP concentration in islets. Alternatively, adenylate cyclase may be activated by a rise in cytosolic Ca^{2+}. This would mean that the primary trigger for insulin release would be changes in Ca^{2+} concentrations within the B cells of the islets.

The primary purpose of the elevation of cyclic AMP in most tissues is to increase phosphorylation of specific proteins through protein kinase enzymes. In islets, such proteins may be associated with the microtubules as well as other systems.

4.7 Calcium and insulin release

The release of insulin is dependent upon the presence of calcium, as was shown in early studies on the perfused pancreas.[42] Later studies by Malaisse[43] using Ca^{45} showed that there was a diminution in the efflux of calcium when the islets were stimulated by many secretagogues such as glucose, leucine, mannose and sulphonylureas. Other workers have suggested that glucose may increase cytosolic calcium levels, perhaps from the mitochondrial pool. It is also possible that secretagogues may directly increase calcium uptake into the B cells.

Further supportive evidence for a fundamental role for calcium in the release process came from the use of the ionophore A 23187.[44] This substance complexes with Ca^{2+} and carries it across the plasma membrane. Importantly, this substance in the presence of calcium induces insulin release, even in the absence of glucose.

It is now thought that the protein, calmodulin, mediates many of the effects of calcium within the islets. Thus, the calmodulin inhibitor, trifluoroperazine, also inhibits glucose-stimulated insulin release. In addition, many inhibitors of calcium transport, e.g. Co^{2+}, Mg^{2+} and Mn^{2+}, as well as ruthenium red and verapamil, also depress glucose-stimulated insulin release.

In general, the effects on calcium influx appear to be closely related to glucose metabolism. Thus, if calcium is the trigger for release, then it will be glucose metabolism which activates the mechanism. It is uncertain how glucose metabolites may affect calcium movement between organelles. However, fructose 1,6-diphosphate inhibits Ca^{2+} uptake by mitochondria[45] and calcium uptake into microsomes is also inhibited by cyclic AMP.

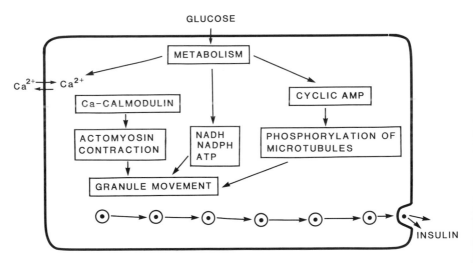

Figure 9.9 General scheme for insulin secretion due to glucose

What is the role of calcium in the secretory process? A provisional hypothesis[46] suggests that glucose metabolism increases cytosolic Ca^{2+} concentrations in the B cells of the islets, acting either on uptake or efflux. The increase in cytosolic calcium is followed by its close association with calmodulin. In turn the Ca^{2+}–calmodulin complex activates both microfilament contraction (involving actomyosin within islets), and promotes the polymerization of microtubules. The action of actomyosin propels the granules along pathways determined by the microtubules. This contractile mechanism shows analogies with contraction in muscle tissue. As in muscle tissue, calcium would then play a leading role in a contractile process. The precise mode of action of cyclic AMP in this system is uncertain. It is, however, likely to be involved in increasing cytosolic calcium from intracellular pools, as well as promoting the phosphorylation of microtubular proteins. The effects of calcium and cyclic AMP will therefore complement one another to produce exocytosis. These events are summarized in Figure 9.9, in which a simplified overview of the mechanisms of insulin release is depicted. The role of calcium in the secretion of insulin has been reviewed by Wollheim and Sharp.[47] Very recently, the turnover of phosphatidylinositol and the breakdown of polyphoshoinositides have been linked to the release of insulin. One substance so produced is inositol 1,4,5-trisphosphate. This has been shown to produce a rapid mobilization of calcium from islet microsomes.[48]

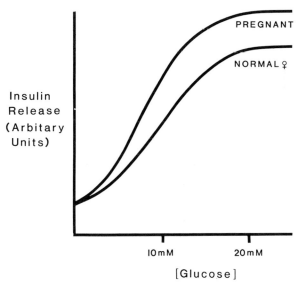

Figure 9.10 Effects of glucose on insulin release from islets in pregnancy

4.8 The long-term regulation of insulin release

The adjustment of insulin release to changes in the physiological state of an animal is very important for the maintenance of proper metabolic homeostasis;

for example, in pregnancy, or after excessive carbohydrate intake, the release of insulin may be considerably increased. Such long-term regulation may well be both cyclic AMP dependent and dependent also on the activation of enzymes important in glucose metabolism within the B cells.

The need for insulin is substantially increased in pregnancy and islets in pregnancy increase in size and cellularity in the rat.[49] They are also much more sensitive to the effects of glucose so that a given concentration of glucose induces greater response in terms of insulin release (Figure 9.10). This appears to be associated both with increased cyclic AMP concentrations as well as with increased glucose oxidation in islets. There is similarly an increase in islet sensitivity to glucose following treatment with growth hormone. This effect is not seen, for example, in tissue culture until some hours after exposure of islets to growth hormone. The mechanism of this effect is not known.

Much earlier evidence had pointed to an effect of growth hormone on insulin secretion in whole animals, since after its injection blood insulin is often significantly increased. It is thought that the placental hormone, placental lactogen, acts similarly. This may well be one of the factors operating in pregnancy to increase blood insulin.

The islet response to changes in dietary carbohydrate

Blood insulin concentrations are greatly diminished in starvation. In isolated islets the response to glucose is also significantly diminished.[50] The reverse situation is seen in obesity. For example, in animals made obese with the appetite-stimulating drug gold thioglucose, blood insulin concentrations are very considerably increased. Again, oxidation rates for glucose are increased in such islets and there appear to be transitory rises in cyclic AMP levels.[51]

It will be seen, therefore, that the B cells of the islet may respond both immediately and in a long-term sense to large numbers of physiological stimuli. On account of the importance of insulin in the overall regulation of metabolism, it is imperative for the islet cells to show this very wide range of responsiveness.

REFERENCES

1. Banting, F. G. and Best, C. H. (1922) *J. Lab. Clin. Med.,* **7**, 251–266.
2. Von Mering, J. and Minkowski, O. (1889) *Arch. Exp. Path.,* **26**, 371–373.
3. Opie, E. L. (1901) *J. Exp. Med.,* **5**, 527–540.
4. Zuelzer, G. L. (1908) *Ztschr. Exper. Path. u Therap Berlin,* **5**, 307–318.
5. Scott, E. L. (1911) *Amer. J. Physiol.,* **29**, 306–310.
6. Paulesco, N. C. (1921) *Comp. Rend. Soc. Biol.,* **73**, 555–559.
7. Ryle, A. P., Sanger, F., Smith, L. F. and Kitai, R. (1955) *Biochem. J.,* **60**, 541–556.
8. Humbel, R. E., Bosshard, H. R. and Zahn, H. (1972) In: *Handbook of Physiology* Section 7, *Endocrinology,* Vol. 1, (Eds. Steiner, D. F. and Freinkel, N.) American Physiological Society, Washington, D.C., pp. 111–132.

9. Blundell, T. L., Dodson, G. G., Dodson, E., Hodgkin, D. C. and Vijayan, M. (1971) *Recent Progr. Hormone Res., 27,* 1–40.
10. Pullen, R. A., Lindsay, D. G., Wood, S. P., Tickle, I. J., Blundell, T. L., Wollmer, A., Krail, G., Brandenburg, D., Zahn, H., Gliemann, J. and Gammeltoft, S. (1976) *Nature,* **259,** 369–373.
11. Moloney, P. J. and Coval, M. (1955) *Biochem. J.,* **59,** 179–185.
12. Arquilla, E. R., Miles, P. V. and Morris, J. W. (1972) *Handbook of Physiology* Section 7, *Endocrinology,* Vol. 1, (Eds. Steiner, D. F. and Freinkel, N.) American Physiological Society, Washington, D.C., pp. 159–173.
13. Berson, S. A. and Yalow, R. S. (1959) *J. Clin. Invest.,* **38,** 1996–2016.
14. Lane, M. A. (1907) *Amer. J. Anat.,* **7,** 409–422.
15. Gepts, W. and LeCompte, P. (1981) *Amer. J. Med.,* **70,**105–115.
16. Steiner, D. F., Clark, J. L., Nolan, C., Rubenstein, A. H., Margoliash, E., Aten, B. and Oyer, P. E. (1969) *Recent Progr. Hormone Res.,* **25,** 207–282.
17. Chance, R. E., Ellis, R. M. and Bromer, W. W. (1968) *Science,* **161,** 165–167.
18. Grant, P. T. and Coombs, T. L. (1970) In: *Essays in Biochemistry,* Vol. 6 (Eds. Campbell, P. N. and Dickens, F.). Academic Press, London, pp. 69–92.
19. Docherty, K., Carroll, R. J. and Steiner, D. F. (1982) *Proc. Nat. Acad. Sci. USA,* **79,** 4613–4617.
20. Milstein, C.; Brownlee, G. G., Harrison, T. M. and Matthews, M. B. (1972) *Nature, New Biology,* **239,** 117–120.
21. Chan, S. J., Keim, P. and Steiner, D. F. (1976) *Proc. Nat. Acad. Sci. USA,* **73,** 1964–1968.
22. Campbell, I. L., Hellqvist, L. N. B. and Taylor, K. W. (1982) *Clin. Sci.,* **62,** 449–455.
23. Taylor, K. W. (1964) *Ciba Foundation Colloquia on Endocrinology,* **15,** 89–94.
24. Lin, B. J. and Haist, R. E. (1973) *Endocrinology,* **92,** 735–742.
25. Brunstedt, J. and Chan, S. J. (1982) *Biochem. Biophys. Res. Commun.,* **106,** 1383–1389.
26. Itoh, N., Sei, T., Nose, K. and Okamoto, H. (1978) *FEBS Lett.,* **93,** 343–347.
27. Bone, A. J. and Howell, S. L. (1977) *Biochem. J.,* **166,** 501–507.
28. Bell, G. I., Pictet, R. L., Rutter, W. J., Cordell, B., Tischer, E. and Goodman, H. M. (1980) *Nature,* **284,** 26–32.
29. Owerbach, D. and Nerup, J. (1982) *Diabetes,* **31,** 275–277.
30. Whittaker, P. G. and Taylor, K. W. (1981) *Diabetologia,* **18,** 323–328.
31. Young, F. G. (1939) *Brit. Med. J.,* II, 393–396.
32. Anderson, E. and Long, J. A. (1948) *Recent Progr. Hormone Res.,* **2,** 209–227.
33. Cochrane, W. A., Payne, W. W., Simpkiss, M. J. and Woolf, L. L. (1956) *J. Clin. Invest.,* **35,** 411–472.
34. Fajans, S. S. and Floyd, J. C. (1972) In: *Handbook of Physiology* Section 7, *Endocrinology,* Vol. 1, (Eds. Steiner, D. F. and Freinkel, N.) American Physiological Society, Washington, D.C., pp. 473–493.
35. Hutton, J. C., Sener, A. and Malaisse, W. J. (1979) *Biochem. J.,* **184,** 303–311.
36. Madison, L. L., Mebane, D. and Lochner, A. (1963) *J. Clin. Invest.,* **42,** 955.
37. Biden, T. J. and Taylor, K. W. (1983) *Biochem. J.,* **212,** 371–377.
38. Hedeskov, C. J. (1980) *Physiol. Rev.,* **60,** 442–509.
39. Ashcroft, S. J. H. (1980) *Diabetologia,* **18,** 5–15.
40. Coore, H. G. and Randle, P. J. (1964) *Biochem. J.,* **93,** 66–78.
41. Howell, S. L., Green, I. C. and Monatgue, W. (1973) *Biochem. J.,* **136,** 343–349.
42. Grodsky, G. M. and Bennett, L. L. (1966) *Diabetes,* **15,** 910–913.
43. Malaisse, W. J. (1973) *Diabetologia,* **9,** 167–173.
44. Wollheim, C. B., Blondel, B., Trueheart, P. A., Renold, A. E. and Sharp, G. W. G. (1975) *J. Biol. Chem.,* **250,** 1354–1360.
45. Sugden, M. C. and Ashcroft, S. J. H. (1978) *Diabetologia,* **15,** 173–180.

46. Howell, S. L. and Tyhurst, M. (1982) *Diabetologia,* **22**, 301–308.
47. Wollheim, C. B. and Sharp, G. W. G. (1981) *Physiol. Rev.,* **61**, 914–961.
48. Prentki, M., Biden, T. J., Janjic, D., Irvine, R. F., Berridge, M. J. and Wollheim, C. B. (1984) *Nature,* **309**, 562–564.
49. Green, I. C. and Taylor, K. W. (1972) *J. Endocrinol.,* **54**, 317–325.
50. Hedeskov, C. J. and Capito, K. (1974) *Biochem. J.,* **140**, 423–433.
51. Caterson, I. D. and Taylor, K. W. (1982) *Diabetologia,* **23**, 119–123.

Chapter 10

Mechanism of action of insulin

1. INTRODUCTION

Long before the discovery of insulin in 1922,[1] it was suggested that a material able to lower blood sugar and reverse the changes seen in diabetes might be present in the pancreas of normal animals. In fact, as early as 1912 Knowlton and Starling[2] showed that very crude extracts of pancreas appeared to promote the utilization of glucose by perfused rat hearts. However, as soon as insulin was discovered in 1922 and purified, an intensive search began for its mode of action. Initially, attention was focused on its effects on carbohydrate metabolism, but later it became very evident that it functioned just as importantly as an anabolic hormone which could regulate fat and protein synthesis.

Not all tissues in mammals are equally responsive to insulin. Thus, although insulin may produce immediate and rapid effects on liver, muscle and adipose tissue, some tissues, such as kidney, are much less responsive, while others such as brain or red blood cells appear not to respond at all. In fact the major immediate effects of insulin are mainly manifest on muscle, adipose tissue and liver. Insulin is also necessary for the growth and functioning of the mammary glands.

2. INSULIN AND CARBOHYDRATE METABOLISM

The administration of insulin to animals leads to a marked fall in blood sugar, evident a few minutes after intravenous injection, although not seen for up to half an hour following a subcutaneous injection due to the much slower rate of insulin absorption. The fate of glucose disappearing under the influence of insulin soon became of great interest. Some of the earliest experiments in this connection were carried out by Best and his associates in 1926,[3] when it was shown that the glucose disappearing could be accounted for both by extra glycogen laid down in tissues and extra carbon dioxide produced by its oxidation. There were thus two major pathways for the extra glucose metabolized in the presence of insulin—it could either be oxidized or stored as glycogen. The precise details of the fate of glucose under these circumstances were only ascertained when the effects of insulin were examined on isolated tissues.

2.1 Insulin and muscle

The principal function of carbohydrate metabolism in muscle is to metabolise glucose with the provision of energy which may be used in muscular contraction. The details of how ATP is used in that process need not be further considered in detail here and are dealt with in a number of reviews.[4,5] The major pathway for glucose metabolism by muscle is shown in Figure 10.1. It will be seen that glucose after phosphorylation may either be converted to glycogen or alternatively be converted to pyruvic acid. Some other metabolic pathways,

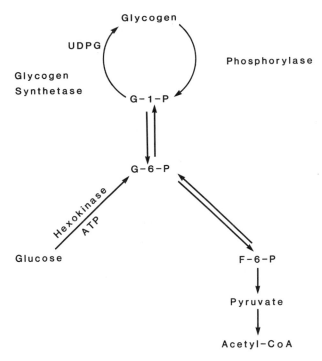

Figure 10.1 Glycogen metabolism in muscle

such as the pentose phosphate pathway, which are important in liver and adipose tissue, show a very low activity in muscle, and in the absence of glucose 6-phosphatase, gluconeogenesis does not take place.

Work with isolated rat diaphragm carried out in the early 1940s[6] showed that insulin greatly increased the uptake of glucose by this muscle tissue and that a large part of the glucose taken up from the medium became converted to glycogen. In such tissues it was also shown that insulin increased the oxidation of pyruvate to carbon dioxide. An increased production of lactate from glucose was also observed under these circumstances. Similar experiments were carried out with ^{14}C-labelled glucose, and it was evident that insulin could increase the uptake and oxidation of glucose by muscle tissue.[7] Studies of this kind showed that insulin could regulate the flow of glucose carbon into glycogen or eventually through glycolysis into carbon dioxide.

2.2 Control of glycogen breakdown in muscle

Although the pathways for glycogen synthesis and breakdown have now been investigated in considerable detail, their exact control by hormones is still a matter of some debate. In muscle this process is controlled, particularly in skeletal muscle, by the hormones adrenaline (which breaks down glycogen) and insulin (which accelerates the synthesis of glycogen), and by Ca^{2+} ions. The nucleotide, cyclic AMP, plays a major role in these processes. Thus, under the influence of adrenaline interacting with β receptors on the membrane, cyclic AMP is generated by a stimulation of adenylate cyclase.

In turn, cyclic AMP activates a phosphate-transferring enzyme—a protein kinase—so that phosphate groups become attached to a serine hydroxyl. It is now clear that the protein kinase consists of regulatory and catalytic subunits, and that the reaction proceeds as:

$$R_2C_2 + 4cAMP \rightleftharpoons R_2(cAMP)_4 + 2C \text{ (catalytic)},$$

where R is a regulatory and C a catalytic unit. The kinase is inactive when the regulatory and catalytic subunits are combined (see Chapter 16).

The protein kinase then activates a second enzyme (a phosphorylase kinase) which in turn regulates the phosphorylation of a third enzyme, phosphorylase. The whole process, termed a 'cascade' reaction, in which one catalytic system activates another, is shown in Figure 10.2.

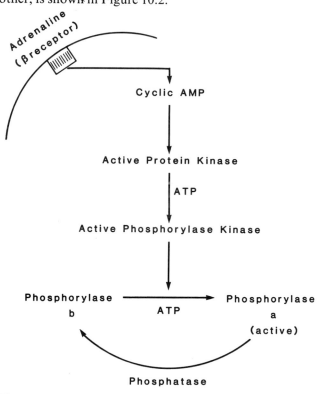

Figure 10.2 Activation of phosphorylase by cascade
mechanism

The final effect is the conversion of phosphorylase b to phosphorylase a, its active form, and the consequent breakdown of glycogen to glucose 1-phosphate. Phosphorylase itself consists of two subunits, an α and a β, which are separately phosphorylated. Details of these processes are given in reviews elsewhere.[8,9] The inactivation of phosphorylase a and its reconversion to phosphorylase b is through phosphatase action.

2.3 Control of glycogen synthesis by insulin[8]

It was first shown by Villar-Palassi and Larner[10] that treatment of rat diaphragm with insulin increased the activity of the enzyme, glycogen synthetase. Later experiments showed that this enzyme exists in two forms which are easily interconvertible (Figure 10.3).

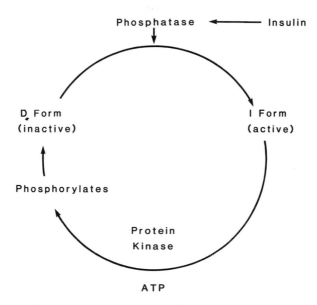

Figure 10.3 Activation of glycogen synthetase

One form which is inactive in the absence of glucose 6-phosphate was termed the D or glucose 6-phosphate dependent form, and the other, which is active in the absence of glucose 6-phosphate, was termed the I form or glucose 6-phosphate independent form. Under normal physiological circumstances the D form, which is also strongly inhibited by ATP and ADP, is inactive.

As with many other forms of regulatory enzymes, one form is a phosphorylated one (the D form) and the other (the I form) is dephosphorylated. The conversion of the D form to the I form is catalysed by a phosphatase. The reverse reaction, by which the I form is reconverted to the D form, is through the action of a protein kinase. The amount of enzyme in the I form, i.e. the active form, is increased by insulin, which activates the phosphatase (Figure 10.3).

It is now thought that the D form is really a mixture of forms which are phosphorylated at several sites. The enzyme may be phosphorylated at up to 12 sites per subunit by a series of kinase enzymes. At least four such enzymes have been described. These include a cyclic AMP-dependent kinase, one which is $Ca^{2+}-$calmodulin dependent, the enzyme phosphorylase kinase and a cyclic AMP-independent kinase.

One important physiological regulator of synthetase phosphorylation is the cyclic AMP-dependent protein kinase. In liver tissue this will be activated for example by glucagon which increases cyclic AMP, and induces phosphorylation on two distinct sites on the synthetase enzyme. Hence, in liver, glucagon inactivates glycogen synthetase. By contrast, in some tissues insulin can induce a fall in cyclic AMP concentrations. It might be supposed, therefore, that a fall in cyclic AMP due to insulin would activate the synthetase. However, despite numerous investigations, no effect of insulin on cyclic AMP concentrations has been observed in skeletal muscle. Moreover, insulin is able to activate glycogen synthetase in the muscles of a strain of mice which genetically lack the enzyme phosphorylase kinase. Therefore insulin cannot operate through a cyclic AMP-dependent mechanism.

It now seems possible that insulin is effective by stimulating the activity of a

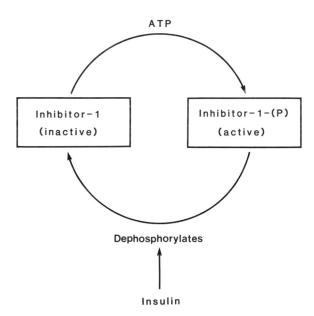

Figure 10.4 Activation of phosphatase-1 inhibitor

rather generally active phosphatase, termed protein phosphatase-1, which dephosphorylates the enzyme. A way in which this might be achieved is by decreasing the effect of a specific inhibitor of the phosphatase, termed inhibitor-1. This is a small molecular weight peptide which is activated by the addition of a phosphate group to a serine residue. It has been suggested that the effects of insulin in dephosphorylating a number of regulatory enzymes are all attributable to removing this phosphate group on the inhibitor, thereby enhancing the activity of the phosphatase (Figure 10.4).

2.4 Control of glycogen phosphorylase in muscle

Adrenaline is known to stimulate the glycogenolytic cascade in muscle through the generation of cyclic AMP. By contrast, insulin inhibits glycogen phosphorylase by methods which do not involve cyclic AMP. The most likely method by which this takes place is through the activation of a protein phosphatase by a lowering of the effective concentration of inhibitor-1. The same inhibitor and the same phosphatase (protein phosphatase-1), seem to be involved as for the dephosphorylation of glycogen synthetase. This time, however, the result is to increase the net conversion of phosphorylase a to phosphorylase b, and hence to prevent glycogen breakdown.

2.5 Insulin and the liver

In the past this subject has been bedevilled with controversy, mainly because it was difficult to obtain viable liver preparations in which effects of insulin could be shown. Early on, however, it was shown that liver slices responded directly to insulin with an increased deposition of glycogen.[11] These observations supported earlier studies in which labelled glucose was administered to dogs and in which it could be shown that hepatic glucose production was greatly reduced in the presence of insulin. This effect may in part be due to a direct lowering of cyclic AMP which has been measured directly in rat liver.[12] Lowering of cyclic AMP results in a decreased activity of phosphorylase leading to diminished glycogen breakdown and an increased activity of glycogen synthetase. In consequence, the liver puts out much less glucose into the portal vein and blood glucose is lowered.

In addition, insulin has an important effect in modifying gluconeogenesis. This has been shown in direct perfusion experiments in which the rate of conversion of precursors of radioactive glucose (such as lactate) to glucose have been measured.[13] Conversion of such substances to glucose is quickly impaired in the presence of insulin, implying a direct and rapid control mechanism, again possibly operative through diminished levels of cyclic AMP.

Three enzymes seem to be involved in the switch from gluconeogenesis to glycolysis. The first of these is phosphofructokinase, which is stimulated by insulin through an increase in the liver concentration of the recently-discovered modulator, fructose 2,6-bisphosphate.[14] With a lowering of cyclic AMP, the concentration of fructose 2,6-bisphosphate increases. This substance increases the affinity of phosphofructokinase for fructose 6-phosphate, and stimulates glycolytic activity.

A second enzyme, pyruvate kinase, is also concerned with increasing the flux through glycolysis and is controlled by insulin. It also exists in an active (dephosphorylated) form and an inactive (phosphorylated) form. Phosphorylation is activated through a cyclic AMP-dependent protein kinase and this is the assumed mechanism of its inactivation by glucagon. As with phosphofructokinase, it has an important role in the transition from a glycolytic to a primarily gluconeogenic pathway (see Figure 10.5).

The third enzyme concerned with turning on gluconeogenesis is the enzyme phosphoenolpyruvate carboxykinase. There is evidence that insulin may diminish the activity of this enzyme by converting an active form of the enzyme to an inactive one. According to Lardy,[15] inactivation involves converting an iron-containing co-factor (a ferroactivator) into a latent form. This may constitute a rapid form of control for this enzyme as opposed to insulin's long-term effects via the repression of protein synthesis (see below).

Insulin and long-term effects on liver enzymes

Insulin has a number of important effects in activating enzymes which are concerned with glycolysis and its reversal. In the long term, it may alter the rate of synthesis of several key enzymes by acting specifically on enzyme biosynthesis.

(a) Glucokinase. This enzyme, which is responsible for phosphorylating glucose to glucose 6-phosphate and which has a high K_m with respect to glucose, is inducible following insulin treatment.[16] In diabetes, its activity is significantly reduced, but may be restored by previous insulin administration. This increase in activity due to insulin is depressed by inhibitors of protein synthesis such as cycloheximide, indicating that in the presence of insulin, *de novo* enzyme synthesis takes place. Under conditions when insulin levels are low, therefore, as in starvation or diabetes, the absolute amounts of the enzyme will be diminished.

(b) Regulatory enzymes in gluconeogenesis. A number of enzymes control the pathway for the conversion of glucose precursors into glucose itself. This process is shown diagrammatically in Figure 10.5. The key enzymes shown here are pyruvate carboxylase, phosphoenolpyruvate carboxykinase, fructose 1,6-bisphosphatase and glucose 6-phosphatase. Insulin appears to act as a long-term regulator of these enzymes and may repress their rate of synthesis. This effect may well be a consequence of the lowering of cyclic AMP levels. Adrenal steroids also seem to be necessary for this change to take place.[17]. When insulin levels are low, as in diabetes or in starvation, gluconeogenesis will therefore be

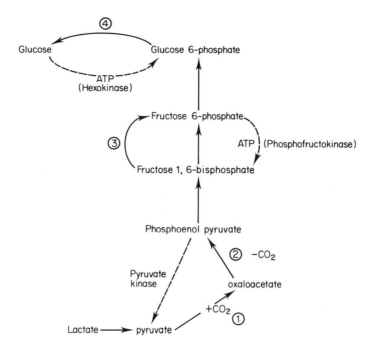

1. Pyruvate carboxylase
2. Phosphoenol pyruvate carboxykinase
3. Fructose 1:6 bisphosphatase
4. Glucose-6-phosphatase

Figure 10.5 Pathways of gluconeogenesis

favoured. Because of the effect of insulin in repressing enzymes concerned with protein breakdown, muscle protein will break down more easily during a lack of insulin, so that the amino acids which will be converted to glucose will flow to the liver with greater facility.

3. INSULIN AND FAT METABOLISM

Insulin administration is known to favour the deposition of body fat at the expense of glucose. This effect is achieved in adipose tissue in which fat is stored as triglyceride. The immediate breakdown of triglyceride into free fatty acids and glycerol is achieved through the action of a hormone-sensitive lipase. Once again this enzyme exists in an active phosphorylated form and an inactive dephosphorylated form, and phosphorylation (i.e. activation) is facilitated by increased levels of cyclic AMP.

Although adipose tissue was formerly thought of as inert tissue, there is now no doubt that it rapidly and efficiently metabolizes glucose, especially under the influence of metabolic hormones. The most important experiments which defined the role of this tissue were carried out by Renold and his co-workers in Boston in the early 1950s.[18] These workers clearly established that insulin increases the oxidation of glucose to carbon dioxide and that glucose, carbon-labelled in postion 1, was also oxidized at a greater rate by adipose tissue in its presence. It seemed evident from this work that insulin was able to increase glucose oxidation both by the glycolytic pathway as well as through the pentose phosphate pathway which is extremely active in adipose tissue. Insulin also increases the incorporation of glucose carbon into the small amounts of glycogen that exist in adipose tissue as well as into newly synthesized fatty acids.

Another effect of insulin is to diminish the output of free fatty acids from adipose tissue into the surrounding medium. The effect is quite independent of the effects of insulin on glucose oxidation, since it can take place in the absence of glucose.[19] It is thought to operate by a lowering of cyclic AMP levels which in turn diminish the activity of the adipose tissue hormone-sensitive lipase.

The oxidation of pyruvate in adipose tissue is also increased by insulin. An important effect here is through a direct action on the enzyme pyruvate dehydrogenase. Much evidence suggests that this enzyme is controlled by insulin.[20] As with many other regulatory enzymes, it exists in two forms: an inactive (phosphorylated) form and an active (dephosphorylated) form. Insulin is able to increase the dephosphorylation of the enzyme by increasing the activity of a specific phosphatase (pyruvate dehydrogenase phosphate phosphatase). The mechanism is a complicated one, and independent of cyclic AMP.

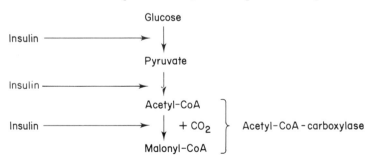

Overall reaction

$$CH_3COCoA + 7HOOCCH_2COCoA + 14NADPH + 14H^+ \longrightarrow$$

$$CH_3(CH_2)_{14}COOH + 7CO_2 + 8CoA(SH) + 14NADP^+ + 6H_2O$$

Figure 10.6 Synthesis of long chain fatty acids from acetyl and malonyl CoA

It is now thought that the α subunit of the complex is inactivated by phosphorylation on three separate sites. Although the full physiological role of these sites remains to be established,[21,22] it seems that phosphorylation on one particular site (site 1) accounts for most of the inactivating effect. According to Randle, phosphorylation at sites 2 and 3 is more concerned with preventing reactivation by the phosphatase.

One important effect of insulin on adipose tissue is therefore to increase the flow of carbon into neutral fat. When glycolysis is increased, the amount of glycerol available for triglyceride formation will be increased. This is because glycerol is formed by a direct reduction of dihydroxy-acetone phosphate by NADH. In addition, the flow of carbon through pyruvate oxidation to acetyl-CoA will also be greater. An increase in the availablity of 2-carbon fragments will favour long-chain fatty acid synthesis, through acetyl-CoA. The pathway of fatty acid biosynthesis is shown briefly in Figure 10.6. It will also be seen from this figure that there is a requirement for reduced NADP. This material is in turn produced by the oxidative limb of the pentose phosphate pathway, whose activity is greatly increased by insulin.

Insulin can also dramatically increase the activity of the enzyme acetyl-CoA–carboxylase in contrast to adrenaline and glucagon which inactivate it. It is now clear that both insulin and adrenaline phosphorylate the enzyme, although they do so at different sites.[23] Phosphorylation through adrenaline involves a cyclic AMP-dependent protein kinase, in contrast to the cyclic AMP-independent kinase associated with insulin activity.

4. INSULIN AND PROTEIN METABOLISM

Clinicians have long been aware that severe diabetes is associated with muscle wasting and a significant loss of body protein. In addition, it was noticed by Von Mering and Minkowski that depancreatized animals excreted large amounts of nitrogen in the urine. Moreover, such changes could be reversed by insulin in man or experimental diabetic animals.

Experiments in the 1920s[24] also showed that insulin could produce a pronounced lowering of plasma amino acid levels. There are two possible reasons for this type of change. The tissue could be breaking down protein less readily under the influence of insulin. Alternatively, amino acid uptake by cells and the synthesis of protein could be increased under the influence of insulin. There is in fact, as will be shown below, evidence that both mechanisms are involved.

4.1 Effects of insulin on protein synthesis in isolated tissue

The experiments of a large number of workers have now demonstrated that insulin increases the incorporation of labelled amino acids into tissues *in vitro*. Thus, Sinex, McMullen and Hastings[25] showed, using isolated rat diaphragm, that insulin increased the incorporation of amino acids into the diaphragm protein. Later, Manchester and Young confirmed and extended these experiments to a number of the naturally-occurring amino acids and showed, importantly, that these effects were independent of the presence of glucose.[26] Insulin also increases the intracellular accumulation of some amino acids including non-utilizable amino acids, such as γ-aminoisobutyric acid or cycloleucine, in the cell pool, by effects on amino acid transport. However, transport of all the naturally-occurring amino acids was not affected in this way so that the overall effect of insulin on protein synthesis was achieved by other means.

4.2   Insulin and the ribosome

It was first shown by Korner[27] that ribosomes from the liver of diabetic animals were much less effective than those from normal animals in synthesizing protein. The defect could easily be reversed by previous insulin injection. At present, however, there is considerable uncertainty as to the means by which this effect is achieved.

One view is that rates of RNA synthesis may be depressed in tissues from diabetic animals. Alternatively, there may be abnormalities in translation associated with the ribosomes in diabetes. Thus, the work of Wool[28] and his colleagues has suggested that poly U is translated much less efficiently by ribosomes from diabetic muscle tissue.

Much more recently there has been a detailed investigation into the way in which insulin influences protein synthesis at the ribosomal level in various tissues.[29] In skeletal muscle (consisting of mixed muscle fibres), insulin increases protein synthesis predominantly by an effect on translation, and here it apparently increases the rate of initiation. The number of ribosomes is also increased. In diabetes, lack of insulin results in markedly diminished protein synthesis. In heart muscle, or in slow twitch, red fibre type skeletal muscle (e.g. the soleus muscle), insulin deficiency does not result in any immediate change in rates of initiation. In the heart over longer periods of time diabetes does result, however, in a significant impairment of initiation. The lack of effect of diabetes in such tissues has been attributed to an increased concentration of glucose 6-phosphate within muscle cells. This in turn could be produced by the operation of the glucose–fatty acid cycle (see below). It has been suggested in this instance that glucose 6-phosphate may directly potentiate initiation. This effect could be quite independent of any effects of insulin.

A quite different situation obtains in liver. Here the major change produced by insulin appears to be a transcriptional one. Thus, direct measurements of messenger RNA by hybridization techniques suggest that the albumin messenger is considerably reduced in insulin-deficient diabetes.

A further possibility is that the degradation of intracellular protein is also controlled by insulin. Insulin reduces the rates of protein degradation in heart and skeletal muscle. This may be due to an effect of insulin on the lysosomal system.

5. THE ACTION OF INSULIN AT THE CELLULAR LEVEL

There have been many theories which have sought to explain the action of insulin. It was at first thought that insulin exercised its effects mainly by increasing the flow of carbohydrate metabolites through oxidative pathways. Later, with the development of new knowledge concerning the nature and functions of enzymes, it was logical to suppose that insulin, as well as other hormones, was concerned with a direct effect on one or more enzyme systems. These ideas culminated in the view that the predominant effects of insulin were on the enzyme hexokinase.[30] It was then thought that there was some degree of direct antagonism between insulin and hormones of the anterior pituitary and that the effect could be relieved by insulin. Although this is no longer believed to be the correct explanation, one of the effects of insulin could be to increase the activity of hexokinase by an *indirect* mechanism. Nevertheless, the failure to demonstrate direct effects of insulin led investigators to other theories concerning its mode of action.

5.1 The permeability theory of insulin action

A number of investigators in the 1930s had suggested that insulin might increase the penetration of sugars into tissue such as muscle. Thus, Pollack in 1936[31] showed that insulin increased the levels of the non-utilizable sugar galactose in the heart and diaphragm muscle in the rat. Since galactose is not metabolized by muscle tissue, this effect indicated that insulin was affecting cellular permeability.

Later, other investigators carried out a series of experiments in which they conclusively showed that insulin increased the penetration of a number of sugars into muscle cells.[32] Thus, sugars such as galactose and arabinose penetrate muscle tissues much more readily in the presence of insulin, although they are not metabolized. In the case of glucose a difficulty arose, since glucose is easily metabolized by most tissues and its intracellular levels cannot easily be measured. However, in a series of further experiments, Morgan and his co-workers,[33] using perfused rat hearts, were able to show that while hearts not treated with insulin had virtually no intracellular-free glucose, the glucose concentration rose transiently under the influence of insulin. These experiments suggested that insulin was influencing transport processes across the cell membrane, which were normally rate limiting. In the presence of insulin, transport could exceed the rate of sugar phosphorylation by hexokinase. The permeability theory for the action of insulin is now very generally accepted and appears to apply to other insulin-sensitive tissues such as adipose tissue.

When simple kinetics are applied to this process, it appears that the V_{max} of glucose transport is affected without changing the K_m. The explanation for this change is uncertain. There may be covalent changes in the structure of a glucose carrier involving perhaps phosphorylation and dephosphorylation reactions. It now seems likely that glucose transport systems may be moved to the plasma membranes from intracellular storage sites. It was quite recently shown that insulin increases the translocation of the glucose transport mechanism from the Golgi area to the plasma membrane in fat cells.[34]

5.2   Insulin and cyclic AMP

The possibility that many of the effects of insulin could be ascribed to a common mechanism, such as a lowering of cyclic AMP levels, is an appealing one but is not supported by much experimental data. Thus, insulin can affect glycogen synthesis in muscle without any concomitant change in muscle cyclic AMP levels (see above). The activation of pyruvate dehydrogenase is also apparently independent of cyclic AMP.

Levels of cyclic AMP, however, appear to fall under the influence of insulin in either perfused liver preparations or in adipose tissue. In adipose tissue, however, more marked effects on cyclic AMP levels are seen only if the tissue has been previously stimulated with adrenaline or other lipolytic hormones.

There is considerable controversy as to how insulin might achieve a lowering of tissue cyclic AMP. One possibility is that it might directly affect either the enzyme adenylate cyclase (which generates cyclic AMP) or the diesterase which breaks it down. Some evidence has now accumulated suggesting that in adipose tissue 'ghosts' (that is, adipose tissue membranes devoid of intracellular contents) there is an inhibition of the cyclase, whereas in homogenates of fat cells insulin causes an increase in the activity of one of the diesterase enzymes. It now seems likely that the increased activity of the liver low K_m diesterase at least, is due once again to a phosphorylation reaction,[35] which is stimulated by insulin.

5.3.   Insulin and other possible second messengers

Another possibility is that the interaction of insulin with its receptor may produce low molecular weight peptides which act as second messengers. This idea has been actively canvassed by Larner.[36] It has been claimed that one low molecular weight peptide is elaborated in the plasma membrane of insulin-sensitive cells, which can cause the specific dephosphorylation of pyruvate dehydrogenase and glycogen synthetase, thus activating them. Another peptide which is smaller and less basic can increase phosphorylation by inhibiting a phosphoprotein phosphatase. It has also recently been suggested that such peptides are produced by means of a specific protease, which is activated when insulin interacts with its receptor. Mild proteolytic digestion of insulin-sensitive cells, such as adipocytes, may release the peptide.

Indeed, under some conditions exposure of cells to trypsin can simulate insulin action. The peptide material does not seem to be induced by a product of insulin digestion, since both concanavalin A and antibody to the insulin receptor are able to mimic the effects of insulin.

The idea that insulin may exert many of its effects through dephosphorylation is not new. As early as 1958, Randle and Smith[37] suggested that insulin might stimulate glucose transport by stimulating the dephosphorylation of a carrier.

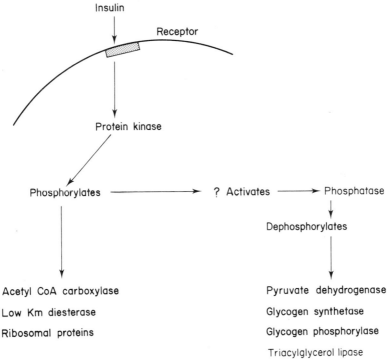

Figure 10.7 Hypothetical scheme to explain the mechanism of action of insulin by phosphorylation and dephosphorylation

As a result of the work of many others, it is now clear that insulin stimulates the activities of several enzymes such as glycogen synthetase, glycogen phosphorylase and pyruvate dehydrogenase by dephosphorylation. Other enzymes such as the low K_m diesterase of liver are phosphorylated and activated by insulin.

The effects of insulin on enzymes could therefore be explained on the basis of phosphorylation or dephosphorylation, as shown diagrammatically in Figure 10.7. There are a number of variations of this scheme. In one proposal by Denton and his colleagues,[23] it is a protein kinase which is released from the plasma membrane. This either phosphorylates directly those enzymes which are activated by phosphorylation mechanisms, or alternatively phosphorylates another peptide which activates a phosphatase—when dephosphorylation takes place.

5.4   Insulin and potassium

Insulin increases the efflux of sodium and influx of potassium across plasma membranes in tissues such as muscle.[38] It is believed that this effect may be secondary to a primary stimulation of Na^+/K^+ ATPase by insulin. Insulin appears to stimulate ATPase directly in isolated lymphocyte membranes as well as in liver membranes. In addition, the effects of insulin on K^+ accumulation are blocked by ouabain. In fact, some control of metabolism could be exerted by changes in intracellular K^+, which is known to affect cyclic AMP-dependent protein kinase.

5.5   Influence of calcium on insulin activity

A number of investigators have considered that Ca^{2+} ions are necessary for the effects of insulin on the stimulation of sugar permeability in tissue. In support of this idea is the work of Clausen[39] which suggested that insulin induces a rise in cytoplasmic calcium in fat cells. There are considerable experimental difficulties in working with calcium on account of the very low levels of free calcium in the cytosol, and lack of precise knowledge of its distribution between cytoplasm and organelles. Moreover, if insulin induced a generalized rise in cytosolic calcium concentrations, a wide variety of disparate metabolic events would take place, e.g. the stimulation of glycogen breakdown as well as the stimulation of pyruvate dehydrogenase. While insulin stimulates pyruvate dehydrogenase, it normally opposes glycogen breakdown. Generalized changes in calcium concentrations (as opposed to local changes) cannot therefore explain the effects of insulin.

6.   INTERACTION OF INSULIN WITH RECEPTORS

Much early work had suggested that insulin was effective at the cell membrane level without any necessity for it to penetrate cells. The binding of insulin specifically to insulin-responsive tissues had been first suggested by Stadie in 1952.[40] By analogy with the posterior pituitary hormones it at first seemed possible that interaction with SH-groups on cell surfaces, by a disulphide interchange mechanism, was one way by which insulin might be effective. However, since insulin in which the intrachain disulphide group was replaced by a methylene bridge was also biologically effective, this explanation seems less likely. Insulin bound to Sepharose is biologically active. Derivatives bound in this way stimulate glucose oxidation and lipogenesis and inhibit lipolysis in isolated adipose tissue cells. This work has led to attempts to solubilize the insulin receptor itself.

It is now clear that the receptor is a glycoprotein of high molecular weight. It is thought to consist of four subunits—2 α units of molecular weight 125 000, and two β units of molecular weight 90 000.[41] These are believed to be linked by disulphide bridges in a manner which resembles the subunit structure of immunoglobulins. The structure is divalent for insulin. It is illustrated in Figure 10.8 (see also Chapter 18). Very recently, it has been proposed that the β subunit of the insulin receptor has intrinsic protein kinase activity, and that this activity is stimulated by insulin. In fact, the β subunit itself is phosphorylated both at tyrosine and serine residues by such enzymes.[41a] The primary effect of insulin may therefore be a kinase stimulation leading to receptor phosphorylation as well as to the phosphorylation of other substrates. Such a proposal gives some support to the ideas of Denton, already discussed.

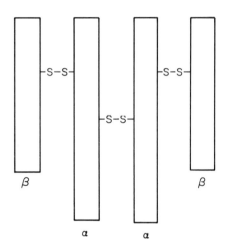

Figure 10.8 Diagrammatic structure of insulin receptor

Some of the changes that take place once insulin has combined with this receptor have already been discussed in Section 5.3. Thus, on one theory it now seems possible that the receptor activates a proteolytic enzyme which in turn produces a small polypeptide second messenger substance. Once insulin has combined in this way with the receptor, both insulin and the receptor proteins may together be internalized. It is now not thought that the insulin receptor complex is active within the cell, but it seems more likely that such a complex is a vehicle for translocating insulin to the lysosomes where it may be broken down. Thus, insulin binding to organelles such as the Golgi complex has been reported. Once the insulin has been released into lysosomal structures, then the receptor protein itself may be recycled to the surface of the cell.[42]

7. MAJOR METABOLIC EFFECTS ENCOUNTERED IN INSULIN-DEFICIENT DIABETES

In animals deprived of insulin by previous treatment with alloxan or other diabetic drugs, such as streptozotocin, a distinct pattern of metabolic changes is readily apparent. A similar pattern also obtains in those human diabetics (juvenile diabetics) who are insulin-deficient through disease.

7.1   Effects on carbohydrate metabolism

In the absence of insulin, tissue uptake of glucose is impaired probably as a result both of defective glucose transport and, in some tissues, defective glucose phosphorylation. Levels of cyclic AMP are increased in tissues such as liver. This in turn results in some diminution in glycogen synthesis and an increase in its breakdown. Levels of glucose rise and in turn produce osmotic changes resulting in loss of water from cells. In addition, the enzymes concerned with gluconeogenesis show an increased activity resulting in an increased flow of carbon through the reversal of glycolysis to yield free glucose. An important secondary change which takes place is that citrate levels which are increased as a result of fatty acid breakdown (see below) directly inhibit enzymes such as phosphofructokinase which in turn impedes glycolysis. This will in turn cause an increase in intracellular levels of fructose 6-phosphate and glucose 6-phosphate. The latter is able to inhibit hexokinase in muscle and therefore impede tissue phosphorylation. This inhibitory mechanism has been termed the 'glucose fatty acid cycle' by Randle and his associates.[43] It is an important factor restraining carbohydrate metabolism in starvation, and diabetes. Pyruvic dehydrogenase activity is also lessened and pyruvate is therefore oxidized less rapidly to carbon dioxide and water.

In man, such changes in carbohydrate metabolism produce a considerable impairment of glucose tolerance. A full discussion of these changes is outside the scope of this book. Further details are given in textbooks of diabetes.[44] However, it should be noted that in man and in animals made diabetic by agents such as alloxan and streptozotocin, blood glucose can easily rise from a normal level of 5 mM to 50 mM.

7.2   Changes in fat metabolism

Owing to insulin lack and elevated levels of cyclic AMP, there is a rise in the activity of the hormone-sensitive lipase which increases breakdown of neutral fat to free fatty acid and glycerol. Fatty acids which are bound to albumin from adipose tissue are then taken to the liver where they are normally oxidized to give carbon dioxide and water. It fatty acid breakdown is excessive, then acetyl-CoA formed by fatty acid breakdown will accumulate and give rise to acetoacetyl-CoA.

7.3 Formation of ketone bodies

Most of the acetoacetate formed is derived by the condensation of acetyl-CoA molecules through the activity of the enzyme β-ketothiolase:

(i) Acetyl-CoA + acetyl-CoA ⇌ Acetoacetyl-CoA + CoA.

The loss of CoA (from acetoacetyl–CoA) is a complicated process involving the intermediate formation of β-hydroxy-β-methylglutaryl-CoA:

(ii) Acetoacetyl-CoA + acetyl-CoA →
   β-hydroxy-β-methylglutaryl-CoA + CoA.

This substance is also an intermediate in sterol formation, and in fact in diabetes excessive quantities of cholesterol appear to be formed from this intermediate. β hydroxy-β-methylglutaryl-CoA then undergoes cleavage to give free aceto-acetate and acetyl-CoA.

(iii) β-hydroxy-β-methylglutaryl-CoA → Acetoacetate + acetyl-CoA.

Acetoacetate itself is easily reduced to β-hydroxybutyrate through an NAD specific β-hydroxybutyrate dehydrogenase. Both acetoacetate and β-hydroxy-butyrate may rise to very high levels in diabetic patients, under conditions of severe lack of insulin. The result of this excessive breakdown of fat is therefore to increase levels of ketone bodies. There is also some evidence to suggest that in diabetes there is, in addition to an increased production of acetoacetate, an impaired removal of it by oxidation. In man, as well as in animals, this may result in a fatal ketoacidosis.

The glycerol also produced by excessive fat breakdown is taken to the liver in the blood where it is reconverted into glucose through gluconeogenesis.

7.4 Lipoprotein lipase

Insulin also increases the activity of lipoprotein lipase. The presence of this enzyme in the endothelial cells of blood vessels results in an increased removal of fatty acids from triglyceride and their incorporation into triglyceride in adi-pose tissue.

The diminution of activity in this enzyme in diabetes and the consequent lipaemia is one of the factors which result in the characteristic milky appearance of the blood in insulin-deficient diabetics.

7.5 Protein metabolism

In the absence of insulin the activity of some of the enzymes concerned with breakdown of protein to amino acids is increased. This particularly effects enzymes such as transaminases resulting in an increased conversion of amino acids to alanine. In the first instance protein breakdown leads to increased intra-cellular levels of amino acids which in turn are converted readily to alanine and glutamine. There are thus increased blood levels of amino acids in diabetes and these are transported from peripheral tissues such as muscle to the liver. Alanine and glutamine seem to predominate.[45] In the liver, due to an increased activity of the enzymes of gluconeogenesis, there is an increased flow back to glucose.

Changes of the type described may, however, readily be reversed by the administration of insulin, although it may take hours, and in some instances, as in man, days for complete reversal to be obtained.

8. SUMMARY OF MAJOR METABOLIC EFFECTS OF INSULIN

Effects of insulin have been summarized in Table 10.1. It remains to be seen whether any single molecular mechanism can explain all these diverse effects.

Table 10.1 Principal metabolic effects of insulin

		Tissue
Short-term effects	1. Increases sugar transport	Muscle, adipose tissue
	2. Increases protein synthesis	Muscle, liver
	3. Inhibits lypolysis	Adipose tissue
	4. Activation and inactivation of enzymes *Activates* (a) glycogen synthetase (b) lipoprotein-lipase (c) pyruvate dehydrogenase (d) acetyl-CoA carboxylase (e) low K_m diesterase *Inactivates* (a) phosphorylase (b) hormone sensitive lipase	Liver, muscle Blood vessel wall Adipose tissue Liver Adipose tissue
Long-term effects	*Activates* Glucokinase *Depresses* Controlling enzymes of gluconeogenesis	Liver Liver

REFERENCES

1. Banting, F. G. and Best, C. H. (1922) *J. Lab. Clin. Med.,* **7**, 251–266.
2. Knowlton, F. P. and Starling, E. H. (1912) *J. Physiol.,* **45**, 146–163.
3. Best, C. H., Dale, H. H., Hoet, J. P. and Marks, H. P. (1926) *Proc. Roy. Soc. B.,* **100**, 55.
4. Lowey, S., Slayter, H. S., Weeds, A. G. and Baker, H. (1969) *J. Mol. Biol.,* **42**, 1–29.

5. Heilmeyer, L. M. G., Ruegg, J. C. and Weiland, T. (Eds.) (1976) *Molecular Basis of Motility* Springer-Verlag, Berlin.
6. Gemmill, U. C. L. and Hamman, L. Jr. (1941) *Bull. Johns Hopkins Hosp.*, **68**, 50.
7. Villee, C. A. and Hastings, A. B. (1949) *J. Biol. Chem.*, **181**, 131–139.
8. Cohen, P. (1982) *Nature*, **296**, 613–620.
9. Hers, H. G. (1976) *Ann. Rev. Biochem.*, **45**, 167–189.
10. Villar-Palasi, C. and Larner, J. (1960) *Biochim. Biophys. Acta*, **39**, 171–173.
11. Berthet, J., Jacques, P., Henneman, G. and deDuve, C. (1954) *Arch. Int. Physiol.*, **62**, 282.
12. Jefferson, L. S., Exton, J. H., Butcher, R. W., Sutherland, E. W. and Park, C. R. (1968) *J. Biol. Chem.*, **243**, 1031–1038.
13. Exton, J. H., Jefferson, L. S., Butcher, R. W. and Park, C. R. (1966) *Amer. J. Med.*, **40**, 709–715.
14. Hers, H.-G. and van Schaftingen, E. (1982) *Biochem. J.*, **206**, 1–12.
15. MacDonald, M. J. and Lardy, H. A. (1978) *J. Biol. Chem.*, **253**, 2300–2307.
16. Walker, D. G. and Rao, S. (1964) *Biochem. J.*, **90**, 360–368.
17. Ashmore, J. and Weber, G. (1968) In: *Carbohydrate Metabolism and its Disorders*, Vol. 1, (Eds. Dickens, F., Randle, P. J. and Whelan, W. J.). Academic Press, London and New York, pp. 335–374.
18. Weingrad, A. I. and Renold, A. E. (1958) *J. Biol. Chem.*, **233**, 267–272.
19. Jungas, R. L. and Ball, E. G. (1962) *Fed. Proc.*, **21**, 202.
20. Coore, H. G., Denton, R. M., Martin, B. R. and Randle, P. J. (1971) *Biochem. J.*, **125**, 115–127.
21. Hughes, W. A., Brownsey, R. W. and Denton, R. M. (1980) *Biochem. J.*, **192**, 469–481.
22. Randle, P. J., Sale, G. J., Kerbey, A. L. and Krebs, A. (1981) *Cold Spring Harbor Conf. Cell Proliferation*, **8**, 687–699.
23. Denton, R. M., Brownsey, R. W. and Belsham, G. J. (1981) *Diabetologia*, **21**, 347–362.
24. Luck, J. M., Morrison, G. and Wilbur, L. F. (1928) *J. Biol. Chem.*, **77**, 151.
25. Sinex, F. M., MacMullen, J. and Hastings, A. B. (1952) *J. Biol. Chem.*, **198**, 615–619.
26. Manchester, K. L. and Young, F. G. (1958) *Biochem. J.*, **70**, 353–358.
27. Korner, A. (1960) *J. Endocrin.*, **20**, 256–265.
28. Wool, I. G., Stirewalt, W. S., Kurichara, K., Low, R. B., Bailey, P. and Oyer, D. (1968) *Recent Progr. Hormone Res.*, **24**, 139–213.
29. Jefferson, L. S. (1980) *Diabetes*, **29**, 487–496.
30. Colowick, S. P., Cori, G. T. and Slein, M. W. (1947) *J. Biol. Chem.*, **168**, 588–596.
31. Pollack, L. and Feher, G. (1936) *Klin. Wsch.*, **15**, 282.
32. Levine, R. and Goldstein, M. S. (1955) *Recent Progr. Hormone Res.*, **11**, 343–380.
33. Morgan, H. E., Henderson, M. J., Regen, D. M. and Park, C. R. (1961) *J. Biol. Chem.*, **236**, 253–261.
34. Kono, T., Robinson, F. W., Blevins, T. L. and Ezaki, O. (1982) *J. Biol. Chem.*, **257** 10942–10947.
35. Marchmont, R. J. and Houslay, M. D. (1980) *Nature*, **286**, 904–906.
36. Larner, J., Cheng, K., Schwartz, C., Kikuchi, K., Tamura, S., Creacy, S., Dubler, R., Galasko, G., Pullin, C. and Katz, M. (1982) *Fed. Proc.*, **41**, 2724–2729.
37. Randle, P. J. and Smith, G. H. (1958) *Biochem. J.*, **70**, 490–500.
38. Zierler, K. L. (1966) *Amer. J. Med.*, **40**, 735–739.
39. Clausen, T., Elbrink, J. and Martin, B. R. (1974) *Acta Endocr. Suppl.*, **191**, 137–143.
40. Stadie, W. C. (1952) *Physiol. Rev.*, **34**, 52–100.
41. Czech, M. P. and Massague, J. (1982) *Fed. Proc.*, **41**, 2719–2723.

41a Gazzano, H., Kowalski, A., Fehlmann, M. and van Obberghen, E. (1983) *Biochem. J.*, **216**, 575–582.
42. Krupp, M. N. and Lane M. D. (1982) *J. Biol. Chem.*, **257**, 1372–1377.
43. Randle, P. J., Garland, P., Newsholme, E. A. and Hales, C. N. (1965) *Ann. N.Y. Acad. Sci.*, **131**, 324.
44. Oakley, W. G., Pyke, D. A. and Taylor, K. W. (1978) *Diabetes and its Management*, 3rd edn. Blackwell, Oxford.
45. Felig, P. (1975) *Ann. Rev. Biochem.*, **44**, 933–955.

Chapter 11

Glucagon

1.   THE DISCOVERY OF GLUCAGON

Shortly after the discovery of insulin, it was shown that the intravenous administration of crude extracts of pancreas caused a rapid rise in blood glucose which preceded the characteristic fall due to insulin.[1] The responsible factor appeared to be different from insulin in properties and was, for example, stable under alkaline conditions. This material was termed glucagon or 'sugar mobilizer'. At a later date the term 'hyperglycaemic-glycogenolytic factor' (or HGF) was used to describe it, and although this term is still sometimes found in the older literature, the earlier name is now the preferred one. The rise in blood sugar produced by the intravenous injection of partially purified insulin is rapid, and is particularly evident when injections are made into the portal vein (Figure 11.1).

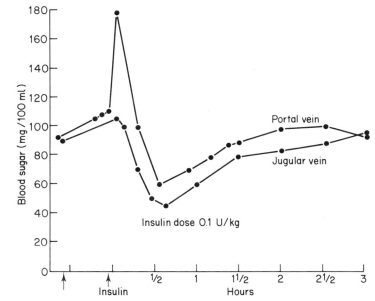

Figure 11.1   Effects of crude insulin injected into the portal or jugular vein of dogs on blood sugar concentration (after Murlin)

2.   CHARACTERISTICS OF GLUCAGON

2.1   Extraction of glucagon

Glucagon was originally prepared as a by-product in the manufacture of insulin. Behrens and his collaborators[2] in 1953 were the first to prepare crystalline glucagon from an amorphous fraction obtained during the preparation of commercial insulin. The material was precipitated by acetone and further purified by fractional precipitation methods.

2.2 The amino acid sequence of glucagon

Using such preparations, the complete sequence of glucagon from pig[3] and beef pancreas was soon established. These were found to be identical. The sequence of glucagons from several other species has now been determined. Species differences in the sequences of glucagon are in general much less than for the corresponding insulins.[4] Guinea-pig glucagon seems, however, to be an exception. In the duck, there is a replacement of residues at positions 16 and 28 of pig glucagon. More profound differences involving other residues have been found in two fish glucagons so far examined. All the glucagons with sequences so far determined, however, have an N-terminal histidine and they uniformly contain tryptophan. The amino acid sequences of glucagons from various species are shown in Figure 11.2.

Species

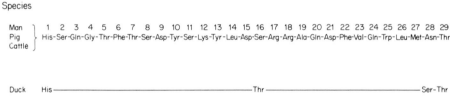

Figure 11.2 Species differences in the sequence of glucagon

The three-dimensional structure of glucagon has been the subject of X-ray crystallography studies by Blundell and his associates.[5] It seems probable that glucagon is stored in granules in the A_2 cells in the trimer form, which is thought to be a stable structure and relatively resistant to enzyme attack. When released into the blood the trimer readily dissociates to give a monomer, which has little secondary structure. This would be expected to be the stable form at the low concentration at which it circulates. It is thought that the monomer adopts a helical structure when binding to the receptor.

2.3 The association of glucagon with A_2 cells in the islets of Langerhans

That the islets of Langerhans are a principal source of glucagons is shown by their ready extraction from such tissues in many species. Thus, they are present only in the endocrine pancreas (Brockmann Bodies) of certain fish where endocrine pancreas and acinar tissues are anatomically distinct. In addition, their association with A_2 cells is shown by specific histochemical techniques in which tryptophan is specifically stained for and more recently similar conclusions have been reached by immunofluorescent studies. Moreover, if the B cells of islets are selectively destroyed by agents such as streptozotocin, then the residue of the islets can still make and secrete glucagon.[6] In some animals, too, such as the duck, the islets consist predominantly of A_2 cells and in these animals the pancreas contains much larger quantities of glucagon than in other species.

2.4   Gut glucagons

The presence of glucagon-like material in parts of the gastrointestinal tract other than in the islets of Langerhans was first suggested by Sutherland and de Duve in 1948.[7] Since then, the nature and properties of this and a number of biochemically related materials have been the subject of considerable debate. Careful work with dogs suggests that glucagon, which is not readily distinguishable from the pancreatic hormone, is present in a number of regions of the gastrointestinal tract but especially in the fundus of the stomach. Here the cells that produce it are similar to the A_2 cells of the endocrine pancreas. Quite recently, it has been suggested that material which appears to be identical to pancreatic glucagon is present in the submaxillary glands. Work of this kind may offer an explanation as to how levels of glucagon may remain quite high in animals, such as the dog, following total pancreatectomy.

In addition to polypeptide material of the same molecular size as pancreatic glucagon, there now appears to be a large number of substances extractable from various regions of the gastrointestinal tract which react with glucagon antibodies.[8] These substances are collectively designated GLI (or glucagon-like immuno-reactivity). The largest of these has a molecular weight of 12 000 daltons and appears to have little biological activity. A number of other peptides which are smaller in molecular size appear to stimulate glycogenolysis and to react with glucagon receptors. Certain of these substances now appear to be identical with biosynthetic precursors of pancreatic glucagon[9] (see Section 4). Such substances include the 100 amino acid peptide glicentin (see Figure 11.3), and a pancreatic glucagon structure lengthened at its C-terminus by a short amino acid extension. There is species variation in the production of these substances. Moreover, they do not appear to be released, as is glucagon, in response to a fall in blood glucose concentration. Indeed, the larger molecular weight material in GLI appears to be released instead as a result of the presence of sugars, such as glucose, in the gastrointestinal tract.

Figure 11.3 Diagrammatic structure of glicentin

It has been presumed that the gene for a glucagon precursor is expressed in many intestinal cells, although processing to yield molecules of pancreatic glucagon is often incomplete.

2.5 Glucagon in blood

As with the gastrointestinal tract, there appear to be several molecular species of glucagon-like material which can cross-react with glucagon antibodies and which circulate in blood. In man, four types have been identified with approximate molecular weights of 2000, 3500 and 9000 daltons, together with a species of high molecular weight (160 000). The origin and functions of such substances are still disputed. Nevertheless, the 3500 molecular weight moiety seems to be identical with pancreatic glucagon. It may also be produced by other cells in the gastrointestinal tract (e.g. gastric A cells). The material with a molecular weight of 9000 daltons may well represent a glucagon precursor. The very high molecular weight material has been termed 'big plasma glucagon'. This has a molecular weight of approximately 160 000 and seems not to be a product of the A cells. It has been suggested that it has biological activity. The 2000 molecular weight material is of unknown function and origin, and it could represent a degradation product of glucagon. After pancreatectomy, all of these four types of circulating glucagon have been detected, again showing their origin from other cells in the gastrointestinal tract.

3.   ASSAY OF GLUCAGON

Older methods relied entirely on bioassay techniques. These utilized the rise in blood glucose produced when glucagon was injected into animals, or later the release of glucose from glycogen in liver slices. In this last process the enzyme phosphorylase is rapidly activated by glucagon and the glucose 1-phosphate so produced is isomerized to glucose 6-phosphate, which itself is converted to glucose through glucose 6-phosphatase. A more recent bioassay method involves the effects of glucagon in liberating glycerol from chick fat cells.[10] This last method appears to be very sensitive and much more so than earlier bioassay techniques.

For routine measurement of glucagon levels radioimmunoassay techniques are almost always used since they combine relative specificity with sensitivity. However, the immunoassay of glucagon has been generally much more difficult than the immunological determination of insulin because of the difficulty of producing an adequate antibody to it in animals. The methods most commonly used rely on the separation of antibody-bound glucagon from free glucagon by various protein precipitation techniques.[11] A difficulty with the immunoassay for glucagon lies in the production of different types of antibody, specific for different sequences in the glucagon molecule. Two major types have been recognized, one which binds to the C-terminal portion of the molecule and one binding between the centre and N-terminus.

Another problem which is frequently encountered during the assay of glucagon is its very ready destruction by enzymes in a variety of tissues or tissue fluids. For this reason the assay is often carried out in the presence of inhibitors of protein breakdown, such as trasylol.

Because of the multiple forms of blood glucagon already referred to, it is difficult to measure pancreatic blood glucagon with accuracy by immunoassay techniques. Fasting blood glucagon concentrations are thought to range from 50 to 80 pg/ml. The turnover of glucagon, like that of insulin, is rapid. It is degraded both in the kidney and the liver.

4. BIOSYNTHESIS AND RELEASE OF GLUCAGON

4.1 Biosynthesis

Like insulin, glucagon is thought to be made from higher molecular weight polypeptide precursors. These are converted within the A_2 cells to the 3500 molecular weight form of the hormone. There is, nevertheless, much less certainty about the specific details of this process than there is for the formation of insulin from its precursors.

The presence in pancreas of higher molecular weight forms of glucagon which are able to react with glucagon antibodies has already been mentioned. Such a peptide was first characterized by Steiner and his colleagues[12] and shown to have a molecular weight of 4500. This material possesses an additional eight amino acids at the carboxy terminus. A similar substance of even higher molecular weight, glicentin, has also been isolated from pig intestine.[13] Its structure is shown diagrammatically in Figure 11.3. It consists of a central glucagon molecule which is preceded at the N-terminus by a much larger peptide. At the C-terminus, the structure closely resembles the octapeptide sequence already described by Steiner and his colleagues. Glicentin has also been demonstrated to be present in the A_2 cells of the pancreas by immunofluorescent techniques.

Glicentin is present along with glucagon in the A_2 granules and it may be secreted with it when glucagon release is stimulated. Another peptide—intermediate in size between glicentin and glucagon—is said to be released in equimolar amounts with glucagon on A_2 cell stimulation.

A second approach to the problem of characterizing glucagon precursors has been to incorporate labelled amino acids into proteins in the A_2 cells and to examine the radioactive products formed. Thus, it was shown that the labelled amino acids were incorporated into peptides of 12 000 molecular weight in fish islets or into material of 9000 molecular weight in guinea pig islets.[14] However, conversion of these substances to the normal form of extractable glucagon of 3500 molecular weight was difficult to carry out in the laboratory.

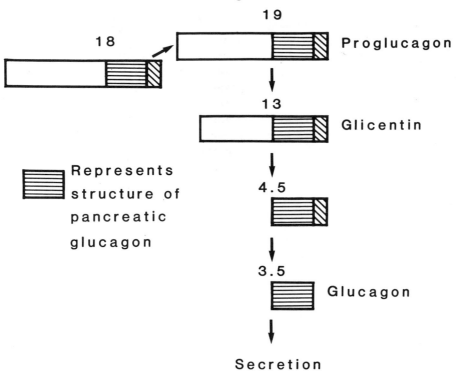

Figure 11.4 Suggested pathway for biosynthesis of glucagon in the rat. Numbers refer to kilodaltons

Recently, detailed studies with rat islets carried out by Steiner's group[15] have suggested a rather complex series of steps for the processing of glucagon precursors. In these experiments, various labelled amino acids were incorporated into rat islets and the products very carefully characterized by gel electrophoresis. The labelled peptides were further examined by tryptic digestion. The data suggest the scheme shown in Figure 11.4. The parent substance has a molecular weight of 18 000. This is increased to an even larger substance of 19 000 molecular weight before there is a progressive degradation of this material to yield glucagon of molecular weight 3500. Most of the stages in this process have still not been fully worked out. Similar large molecular weight precursors of glucagon have also been found in fish islets.

In summary, therefore, the processing of glucagon precursors is a highly complex multi-stage process. This seems to take place both in cells in the intestinal mucosa as well as in the A_2 cells of the pancreas. In the pancreas itself degradation proceeds predominantly to yield glucagon, whereas the process falls short of this in the intestinal cells where higher molecular weight forms containing a glucagon sequence are released. The processing of glucagon precursors is slow, and this may well explain the difficulty in incorporating labelled amino acids into glucagon *in vitro*. Glucagon biosynthesis has recently been reviewed elsewhere.[16]

4.2 Effects of glucose concentration on glucagon release

Glucagon is stored as granules in the A_2 cells of the islets and released by a process which is similar to that seen for insulin.

In early experiments in man it was shown that a lowering of blood glucose resulted in an increased level of glucagon as measured by immunoassay.[17] In animals a rise in blood glucose could depress glucagon levels in pancreatic venous blood. Similar results were shown in a number of species. It was suggested that a fall in blood glucose was the normal physiological stimulus to glucagon release. Thus, release of this hormone was triggered off by a reversal of the physiological stimulus for insulin secretion.

The effects of glucose seem to be directly on the islets of Langerhans since in many experiments on isolated islets, when glucose concentrations were lowered, an elevation in glucagon release into the incubation medium was seen.

4.3 Other metabolites and glucagon release

As with a number of other hormones, amino acids can also induce the release of glucagon. Among the most powerful is arginine (which can also induce release of insulin and growth hormone). A protein meal will therefore result in glucagon release.

Long chain fatty acids, by contrast, suppress the release of glucagon. Both palmitate (bound to albumin), and β-hydroxybutyrate may reduce glucagon secretion.[18] This effect is seen when fatty acids are infused into various animal preparations as well as in isolated pancreas or islet preparations of various kinds. Fatty acids are in any case quite readily oxidized to carbon dioxide by A_2 cells.

4.4 Other hormones and glucagon release

Adrenaline is known to increase cyclic AMP in the A_2 cells of the islets of Langerhans and promote an increased release of glucagon[19]—acting predominantly through β-receptors. This is in contrast to its effects in inhibiting insulin release. Pancreozymin can also increase the release of glucagon from perfused pancreas. This effect may be of some physiological importance in that nutrients in the gastrointestinal tract may increase pancreozymin release and reinforce the effects of amino acids on the secretory mechanism. Somatostatin impairs glucagon release, although its physiological role is open to question.

There is now evidence that insulin is necessary for regulating the release of glucagon, both in various animal preparations of pancreas, as well as in man. Thus, the effects of glucose in suppressing glucagon release are not seen unless insulin in relatively high concentration reaches the A_2 cells of the islets. For this reason it has been supposed that lack of insulin in human diabetes may lead to the very high levels of circulating glucagon often encountered in the untreated juvenile form of the disease. Somatostatin has been used clinically to correct the excessive glucagon levels in this condition.

4.5 Biochemical processes leading to glucagon release

Several workers have suggested that the secretion of glucagon is accompanied by a *fall* in energy production in the A_2 cells, a situation which seems to contrast with that in the B cells during insulin secretion. Thus, substances which disrupt metabolism (e.g. malonate, 2,4-dinitrophenol and iodoacetate) lower islet ATP levels, and all of these promote glucagon release. As explained above, insulin is needed for high levels of glucose (e.g. 16 mM) to suppress glucagon release *in vitro*. In fact, recent findings strongly suggest that insulin increases glucose metabolism by A_2 cells in isolated islets. Direct measurements suggest that insulin actually increases ATP in such preparations.[20]

There is at present uncertainty as to the exact role of calcium ions in glucagon release, although the inhibiting effect of glucose on glucagon release is dependent on extracellular calcium.[21]

5.   BIOCHEMICAL EFFECTS OF GLUCAGON

Glucagon is a hormone which unequivocally functions by increasing the level of cyclic AMP within cells. In fact, the effects of glucagon may generally be closely simulated by artificially raising cyclic AMP in tissues, for example by the use of phosphodiesterase inhibitors such as theophylline.

5.1 Effects on blood glucose levels

The most widely recognized response to glucagon is the rise in blood glucose which rapidly follows its administration to animals. Although mammals generally respond in this way, there is wide variation in the extent of the species response. Thus, cats show a much greater response than do other mammals and there is frequently a prolonged response in reptiles and birds. In man, the response to intravenous glucagon is smaller than in many other species. Factors such as dietary habits (which may influence glycogen levels in the liver) could be partly responsible for some of these differences.

The major source of the increased glucose in the blood is through glycogen breakdown in the liver (see Chapter 10). The liver is, in fact, the primary target organ for glucagon. As might be expected, there is a depletion of liver glycogen following glucagon administration, although after 24 hours glycogen may well be restored to normal owing to a secondary release of insulin. Glucagon may also produce a depletion in glycogen in cardiac muscles, although without concomitant changes in blood glucose in this instance.

There is also a concomitant depression of glycolysis as well as an increase in glycogen breakdown. This will result in glucose 6-phosphate residues being hydrolysed (through glucose 6-phosphatase), so as to raise blood glucose, rather than undergoing further breakdown through the glycolytic pathway. There are two enzymes which are inhibited by glucagon in the glycolytic pathway. These are phosphofructokinase and pyruvate kinase.

Glucagon, phosphofructokinase and pyruvate kinase

It was originally supposed that glucagon inhibited the activity of phosphofructokinase by a direct phosphorylation of the enzyme through a cyclic AMP-dependent protein kinase. It was later shown by Hers and his collaborators[22] that this effect was mediated by an alteration in the levels of a new modulator of phosphofructokinase, fructose 2,6-bisphosphate. This substance in micromolar quantities is a powerful stimulator of phosphofructokinase. Its concentration is lowered by glucagon. The structure is shown in Figure 11.5. This material reverses the inhibition of phosphofructokinase by ATP and enhances the affinity of the enzyme for the substrate fructose 6-phosphate. According to Hers, glucagon prevents the formation of fructose 2,6-bisphosphate as well as increasing its breakdown. It achieves these effects by increasing liver cyclic AMP concentration, which in turn diminishes the activity of the enzyme responsible for phosphorylating fructose 6-phosphate to give fructose 2,6-bisphosphate. Similarly, the activity of the specific bisphosphatase enzyme (which hydrolyses fructose 2,6-bisphosphate) is increased due to cyclic AMP (Figure 11.6). In both cases the enzyme is thought to be phosphorylated. In many respects the control of phosphofructokinase by the bisphosphate mechanism resembles what takes place in the presence of AMP. This last substance reciprocally increases phosphofructokinase activity while at the same time diminishing fructose 1,6-bisphosphatase action. Glucagon also directly inhibits pyruvate kinase by promoting phosphorylation of the enzyme, opposing the effects of insulin.

5.2   Effects on fat metabolism

Glucagon's effects on fat metabolism are much less clear cut than its effects on blood sugar, or its effects on the metabolism of carbohydrates generally. In addition, there may be quite marked species differences in the responses of tissues to glucagon. Thus, while glucagon is lipolytic in rat liver and rat adipose tissue, it is much less effective as a lipolytic agent in man. The mechanism of the lipolytic effect of glucagon is believed to lie in the activation of adipose tissue

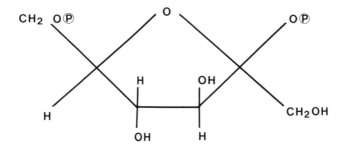

Figure 11.5 Structure of fructose 2,6-bisphosphate

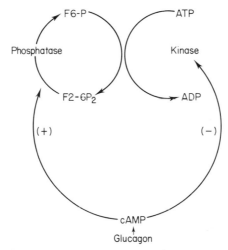

Figure 11.6 Formation and breakdown of
fructose 2,6-bisphosphate

lipase by phosphorylation under the influence of locally increased cyclic AMP levels.

Glucagon also lowers the activity of the enzyme acetyl-CoA carboxylase, which controls the rate limiting step of lipogenesis. Since malonyl-CoA inhibits the activity of acyl-carnitine transferase I (the enzyme which regulates fatty acid entry into mitochondria), there will be an increase in fatty acid uptake and oxidation in the mitochondrial compartment.[23] By these mechanisms glucagon will therefore inhibit fatty acid synthesis. In addition, the synthesis of fat will also be restricted by the lowering of glycolytic activity already referred to. For instance, the diminished flux of metabolites through glycolysis will lower glycerol 3-phosphate. These effects of glucagon are illustrated in Figure 11.7.

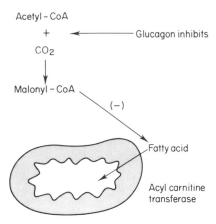

Figure 11.7 Effect of glucagon on lipid
metabolism

5.3   Effects on protein and amino acid metabolism

A direct effect of glucagon in increasing gluconeogenesis has been consistently
reported. In whole animals, glucagon may produce a marked increase in nitrogen excretion leading to a negative nitrogen balance. These changes may be
accompanied by a fall in blood amino acid levels as amino acids are taken up by
the liver and other tissues. There is also a concomitant increase in the excretion
of urea and in the activity of several enzymes, such as transaminases, which are
concerned with protein breakdown.

At the tissue level, effects of glucagon in increasing gluconeogenesis can
readily be demonstrated. For example, in perfused liver preparations [14]C-
lactate is readily converted into ^{14}C-glucose. Such a change can be seen within 2
minutes of adding glucagon to the preparation. Exton and Park[24] first suggested that glucagon might directly stimulate certain rate limiting steps in gluconeogenesis. Initially it was thought that one or other of the reactions between
pyruvate and phosphoenolpyruvate was involved (Figure 11.8). Measurements
of metabolic intermediates in perfused liver treated with glucagon preparations
support the idea that both pyruvate carboxylase as well as phosphoenolpyruvate carboxykinase may be stimulated.

In addition, as explained above, the inactivation of pyruvate kinase and phosphofructokinase will necessarily inhibit glycolysis and encourage gluconeogenesis. Since these changes appear to take place rapidly and at physiological
levels of circulating glucagon, they represent an important normal control
mechanism in switching carbohydrate metabolism from glycolysis to gluconeogenesis.

The responses described are short-term effects which are seen after an

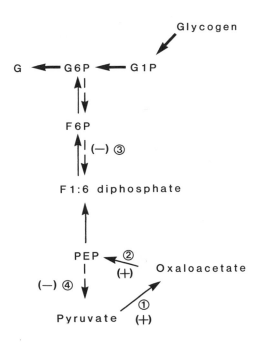

Figure 11.8 Glucagon and gluconeogenesis. Glucagon stimulates pyruvate carboxylase (1), phosphoenolpyruvate carboxykinase (2); it inhibits pyruvate kinase (4) and phosphofructokinase (3)

extremely short interval of time once the tissue has been exposed to the hormone. In addition, there are a number of longer-term effects due to glucagon. Thus, there is an increase in the absolute levels of phosphoenolpyruvate carboxykinase.[25] Such effects are seen both with cultures of neonatal liver as well as in whole animal preparations. In these instances glucagon is seen to be increasing enzyme synthesis and the effect can be inhibited by inhibitors of polypeptide synthesis such as cycloheximide.

6. ACTION OF GLUCAGON AT THE CELLULAR LEVEL

6.1 Glucagon receptors and adenylate cyclase

It is generally supposed that glucagon is effective by raising tissue levels of cyclic AMP. This is achieved by direct stimulation of the enzyme, adenylate cyclase.

Possible mechanisms for the activation of the cyclase have recently been the subject of intensive study. It was first shown by Rodbell[26] that GTP was necessary for the activation of adenylate cyclase by glucagon. Subsequent workers showed the presence of a protein which specifically interacted with guanine-containing nucleotides termed the 'guanine nucleotide regulatory protein' or 'G protein'. Various models have been proposed to explain how this protein may be involved in stimulating the cyclase. One such model due to Limbird[27] is illustrated in Figure 11.9.

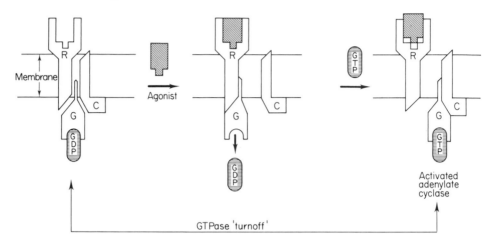

Figure 11.9 Interaction of glucagon with receptor leading to the activation of adenylate cyclase. R refers to receptor, G to G-protein, and C to adenylate cyclase

In any of these models the fluidity of the plasma membrane is assumed. This allows lateral movement of proteins. The initiating event is the interaction between a hormone such as glucagon and the receptor (Stage 2). As a consequence, GDP is dissociated from the G protein. GTP then becomes attached to the G protein in its place. The G protein and GTP then activate the catalytic component of the cyclase, hence producing cyclic AMP (Stage 3). Cyclic AMP will continue to be synthesized until GTP is hydrolysed. When this takes place the G protein is dissociated from the catalytic subunit and the system returns to the resting state (Stage 1).

The termination of the reaction depends therefore on the hydrolysis of the GTP. If this is prevented, as by cholera toxin or by certain synthetic analogues of guanine nucleotides, adenylate cyclase activity is greatly potentiated. Cholera toxin is thought to alter the G protein by ADP ribosylation of one of its components.

6.2 Glucagon and mitochondrial function

There is a considerable body of evidence to suggest that glucagon rapidly affects mitochondrial functions. The effect leads to increases in succinate dehydrogenase activity, adenosine nucleotide formation and to the stimulation of pyruvate carboxylase. It may also stimulate hepatocyte oxygen consumption, an effect reversible by insulin. Oxaloacetate rises in mitochondria prepared from liver previously perfused with glucagon. One way in which glucagon could influence mitochondrial metabolism is by its effects on cation fluxes. Thus, under appropriate circumstances glucagon can be shown to produce a net Ca^{2+} efflux from liver mitochondria.[28] To what extent such calcium fluxes are important in regulating liver metabolism is uncertain at present.

7.   GENERAL METABOLIC SIGNIFICANCE OF GLUCAGON

7.1   Glucagon in normal animals

The exact role of glucagon in metabolic control is still debatable. While insulin seems to be quite essential for metabolic homeostasis in mammals, the same does not necessarily seem the case for glucagon. In higher animals, glucagon perhaps plays the role of a fine regulator of metabolic processes. This is especially seen with regard to its short-term effects in regulating blood glucose. In reptiles and birds, as has been suggested above, however, it may be particularly important in maintaining blood glucose during fasting.

An early suggestion was that the role of glucagon was primarily to counteract excessive rebound hypoglycaemia which followed ingestion of a carbohydrate meal.[29] Thus, when glucose is taken by mouth, the rise in blood sugar may frequently overstimulate the islets of Langerhans so that a short rebound-hypoglycaemic phase is produced. If glucagon is simultaneously measured in the blood, a rise in this hormone accompanies the later fall in blood sugar. Glucagon in this sense will prevent an excessive fall in blood glucose following carbohydrate ingestion.

Amino acids may also produce a rise in blood glucagon, and this is seen following a protein meal. It has also been suggested that once again the extra production of glucagon following protein feeding will counter-balance any excessive insulin produced by amino acids stimulating the B cells of the islets. The physiological importance of the effects of fatty acids in inducing a fall in blood glucagon are much more controversial. It may be, however, that in some animals a suppression of glucagon release is a device for controlling excessive fat breakdown. This will prevent an undue rise in fatty acids and in ketone bodies in situations such as starvation.

7.2   Role in starvation

Glucagon levels rise in the early phase of fasting. This is due both to the fall in blood glucose as well as to the rise in blood amino acids. Such a response may be especially important in increasing gluconeogenesis so that blood glucose may be adequately maintained. In fact, in some animals, such as reptiles or birds, glucagon is the predominant pancreatic hormone and pancreatectomy leads to hypoglycaemia due to loss of A_2 cells, rather than to diabetes, in these species

7.3   Metabolic role of glucagon in diabetes in man

In insulin-deficient human diabetes as well as in experimental diabetes in animals, blood glucagon levels are often very high. As explained in an earlier section this may well be an effect of insulin lack on A_2 cell function. There has been controversy over whether such high levels of glucagon are necessary for the development of ketosis in diabetes. Lack of insulin itself can produce ketosis, but both this and hyperglycaemia are certainly exaggerated by the presence of excessive amounts of circulating glucagon.

Under more normal circumstances, glucagon is probably less essential than insulin in the overall regulation of metabolism. Depancreatized patients need only the administration of insulin for the maintenance of life. The diabetes so produced is, nevertheless, rather unstable.

REFERENCES

1. Collens, W. S. and Murlin, J. R. (1929) *Proc. Soc. Exp. Biol. Med.*, **26**, 485.
2. Staub, A., Sinn, L. and Behrens, O. K. (1955) *J. Biol. Chem.*, **214**, 619–632.
3. Bromer, W. W., Sinn, L. G. and Behrens, O.K. (1957) *J. Amer. Chem. Soc.*, **79**, 2807–2810.
4. Sundby, F. (1976) *Metabolism*, **25**, 1319–1321.
5. Blundell, T. and Wood, S. (1982) *Ann. Rev. Biochem.*, **51**, 123–154.
6. Howell, S. L., Edwards, J. C. and Whitfield, M. (1971) *Horm. Metab. Res.*, **3**, 37–43.
7. Sutherland, E. W. and de Duve, C. (1948) *J. Biol. Chem.*, **175**, 663–674.
8. Murphy, R. F., Buchanan, K. D. and Flanagan, R. W. J. (1981) In: *Hormones in Normal and Abnormal Human Tissues*, Vol. 2 (Eds. Fotherby, K. and Pal, S. B.). Walter de Gruyter, Berlin, New York, pp. 187–221.
9. Tager, H. S. and Markese, J. (1979) *J. Biol. Chem.*, **254**, 2229–2233.
10. Langslow, D. R. and Hales, C. N. (1970) *Lancet*, **i**, 1151–1152.
11. Henquin, J. C., Malvaux, P. and Lambert, A. E. (1974) *Diabetologia*, **10**, 61–68.
12. Tager, H. S. and Steiner, D. F. (1973) *Proc. Nat. Acad. Sci. USA*, **70**, 2321–2325.
13. Moody, A. J., Holst, J. J., Thim, L. and Lindkaer-Jensen, S. (1981) *Nature*, **289**, 514–516.
14. Hellerström, C., Howell, S. L., Edwards, J. C., Andersson, A. and Östenson, C-G. (1974) *Biochem. J.*, **140**, 13–21.
15. Patzelt, C., Tager, H. S., Carroll, R. J. and Steiner, D. F. (1979) *Nature*, **282**, 260–266.
16. Hellerström, C. (1984) In: *Handbook of Experimental Pharmacology*, Vol. 1 *Glucagon*, (Ed. Lefèbre, P.). Springer-Verlag, Berlin, Heidelberg, New York, pp. 121–138.
17. Unger, R. H., Eisentraut, A. M., McCall, M. S. and Madison, L. L. (1962) *J. Clin. Invest.*, **41**, 682–689.
18. Edwards, J. C. and Taylor, K. W. (1970) *Biochim. Biophys. Acta*, **215**, 310–315.
19. Leclercq-Meyer, V., Brisson, G. R. and Malaisse, W. J. (1971) *Nature, New Biology*, **231**, 248–249.

20. Östenson, C-G., Andersson, A., Brolin, S. E., Petersson, B. and Hellerström, C. (1976) In: *Glucagon: Its Role in Physiology and Clinical Medicine* (Eds. Foa, P., Bajaj, J. S. and Foa, N.). Springer-Verlag, New York, Heidelberg, Berlin, pp. 243–245.
21. Leclercq-Meyer, V., Marchand, J. and Malaisse, W. J. (1978) *Diabetes,* **27**, 996–1004.
22. Hers, H. G. and van Schaftingen, E. (1982) *Biochem. J.,* **206**, 1–12.
23. McGarry, J. D. and Foster, D. W. (1980) *Ann. Rev. Biochem.,* **49**, 395–420.
24. Exton, J. H. and Park, C. R. (1969) *J. Biol. Chem.,* **244**, 1424–1433.
25. Yeung, D. and Oliver, I. T. (1968) *Biochemistry,* **7**, 3231/3239.
26. Rodbell, M., Krans, H. M. J., Pohl, S. L. and Birnbaumer, L. (1971) *J. Biol. Chem.,* **246**, 1872–1876.
27. Limbird, L. E. (1981) *Biochem. J.,* **195**, 1–13.
28. Baddams, H. M., Chang, L. B. F. and Barritt, G. J. (1983) *Biochem. J.,* **210**, 73–77.
29. Bürger, M. (1947) *Z. ges. inn. Med.,* **2**, 311.

Chapter 12

Hormones of the gastrointestinal tract: A survey

<table>
</table>

1. INTRODUCTION

The polypeptide hormones of the gastrointestinal tract now form a formidable array. Thus, in addition to the 'classical' gastrointestinal hormones secretin, pancreozymin-cholecystokinin, and gastrin, it is now necessary to add at a minimum gastric inhibitory peptide (GIP), motilin, neurotensin, somatostatin, bombesin, vasoactive intestinal peptide (VIP) and glicentin.[1] The discovery of these new hormones has in general not followed the classical pattern of endocrinology: that of extraction of a gland to find an 'active principle' whose general biological functions are already clearly identified—an example of this is the extraction of insulin from the pancreas. The recently discovered hormones have frequently been first identified by the sophisticated techniques of preparative separation and identification which are now available and only later has it been possible to try to identify a biological function for them.[2] A good example is the 'pancreatic polypeptide' which is chemically well characterized, is released into the bloodstream and yet has only ill-defined biological functions. The situation is further complicated by problems of nomenclature, which have given some hormones (gastric inhibitory peptide and somatostatin are good examples) names which relate to their first identified localization and function. Yet it has subsequently emerged that the name may not be wholly appropriate to describe the most important functions of the hormone as they have been later discovered.

1.1 Peptidergic nervous system

To complicate the situation still further, many of the homones of the gastrointestinal tract, in addition to localization in secretory cells of various areas of the tract, may also be present in the parasympathetic nervous system in the gut, and in the brain—see Table 12.1. For instance, the vagus nerve contains VIP, substance P, enkephalin, somatostatin, gastrin and serotonin in addition to acetylcholine and a small amount of noradrenaline. Many of these peptides have been shown to be released on vagal stimulation. The significance of this release of

Table 12.1 Some gastrointestinal tract hormones and their presence in nerves in the gut and central nervous system

Hormone	Endocrine cell type in the gut	Nerve fibres in gut	Nerve fibres in CNS
Bombesin	P cells	+	+
Enteroglucagon	L cells	−	+
Gastrin	G cells	+	+
GIP	K cells	−	−
Motilin	Enterochromaffin-2 cells (EC$_2$)	−	+
Neurotensin	N cells	−	+
Pancreatic polypeptide	PP cells (pancreas)	+	−
Pancreozymin–CCK	I cells	+	+
Secretin	S cells	−	−
Substance P	Enterochromaffin-1 cells (EC$_1$)	+	+
Somatostatin	D cells (pancreas)	+	+
Vasoactive intestinal peptide	H cells	+	+

gastrointestinal hormones from peptidergic nerves has yet to be worked out in detail. Somatostatin is discussed in detail in Chapter 3. Enkephalins and their role are described in Chapter 5.

There follows below a brief survey of the functions of the hormones and postulated hormones of the gastrointestinal tract, concentrating in particular on their proposed actions as hormones. In many cases the list of actions is long and diverse, and problems remain in determining which are physiologically relevant, and which merely reflect pharmacolgical effects of the peptides which are demonstrable only at high concentrations.[1]

2. GASTRIN

Gastrin was isolated from the antral mucosa by Gregory and Tracy[3] and was quickly shown to have powerful effects on gastric acid secretion. Subsequently it has been shown to have effects on almost all the secretory and smooth muscle cells of the gut.

2.1 Chemistry

Gastrin, as originally isolated, was in two similar forms consisting of 17 amino acids, and differing only in the presence or absence of a sulphate group in the tyrosine residue—see Figure 12.1. This has been termed G17. A substantial proportion of gastrin in plasma shows a distinctly higher molecular weight than this

pyroGLU-GLY-PRO-TRP-LEU-GLU-GLU-GLU-GLU-GLU-ALA-
TYR-GLY-TRP-MET-ASP-PHE-NH$_2$

Figure 12.1 Structure of human gastrin I. Gastrin II has
a sulphate ester on Tyr-12

form, and is termed 'big gastrin'.[4] It has been isolated in small quantities from gastric mucosa. It contains 34 amino acids and has been termed G34. Tryptic digestion of G34 produces a fragment which closely resembles G17 and for this reason it has been suggested that G34 may be a precursor form. Immunocytochemistry has shown that both G17 and G34 may originate in the G cells of the antrum. Their distribution along with those of the other hormones discussed in this chapter is shown in Figure 12.2.

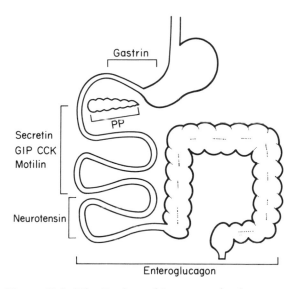

Figure 12.2 Distribution of hormones in the gastro-
intestinal tract. Adapted from Bloom, S. R. and
Polak, J. M. (Eds.) (1981) *Gut Hormones*, 2nd edn,
Churchill-Livingstone, Edinburgh

Gastrin shares with pancreozymin–CCK (see p. 325) and with the amphibian skin peptide caerulein, a common C-terminal pentapeptide. The three molecules differ in their sulphation—thus sulphation of the tyrosine at position 7 from the C-terminus confers a CCK-like pattern of activity while sulphation at position 6 from the C-terminus or absence of sulphation confers gastrin-like properties. The C-terminal pentapeptide (pentagastrin) is readily synthesized and has many of the biological activities of the whole G17 molecule.

2.2   Regulation of secretion

As already stated, G17 and G34 both circulate in the blood, the concentration of G34 being some fourfold higher, partly because its half-life of 34 minutes is far longer than that of G17 (6 minutes). However, in terms of its potency (that is, the concentration required to stimulate half maximal gastric acid secretion) G17 is five times more potent than G34. Another higher molecular weight form of gastrin (big big gastrin) has also been identified, but it is not clear whether it has any physiological significance; it does not bind to affinity columns in the way that other gastrin molecules do.

Feeding provides the physiological stimulus to gastrin secretion and it seems that both G17 and G34 are secreted, although G17 may appear more rapidly than G34. The presence of peptides and amino acids in the gastric mucosa may be important to provide adequate stimulation to the secretion process. There may also be some degree of stimulation by cholinergic innervation—by the vagus and at least in some species by sympathetic innervation. The presence of an acid pH in the stomach may provide a powerful inhibition of further gastrin release. In addition to these influences which may be physiologically important, a number of other factors have been shown to stimulate gastrin secretion. These include bombesin, prostaglandin E, Ca^{2+} and hypoglycaemia. Secretion is inhibited by somatostatin, secretin, vasoactive inhibitory peptide, glucagon and gastric inhibitory peptide.

2.3   Actions

The most important actions of gastrin are to stimulate secretion of gastric acid and of pepsin and growth of acid-secreting gastric mucosa. Several other actions of gastrin are known, and are listed in Table 12.2. Some of these resemble those of pancreozymin–CCK, e.g. contraction of the gall bladder, probably because the pentapeptide comprising the carboxyterminal end of the molecule is identical in these two hormones.

The effects of gastrin on gastric acid secretion are well defined and of the various forms of gastrin, G17 is the most potent in exerting this effect. While many other hormones may exert effects on gastric acid secretion in some

Table 12.2   Actions of gastrin

Increases
Secretion of pepsin and gastric acid
Pancreatic enzyme secretion
Water and electrolyte secretion by stomach and pancreas
Water and electrolyte absorption from small intestine
Muscle contraction of stomach, intestine, gall bladder
Secretion of insulin, glucagon, somatostatin

circumstances, gastrin, along with histamine and acetyl choline, is probably the most important. The secretion of acid on feeding may occur in at least two phases: cephalic, associated with the sight, smell or taste of food, and gastric, when the food enters the stomach. Passage of food to the small intestine may inhibit gastric acid secretion in many circumstances, and despite the presence of high concentrations of gastrin in the duodenum, its release from this source after feeding has not been conclusively demonstrated.

The exact mechanism by which gastrin stimulates the secretion of acid is not completely clear. Two possible methods have been considered:

(a) that gastrin has a direct interaction with parietal cell receptors, and

(b) that histamine is involved in modulating the gastrin response. The evidence for this derives from the fact that histamine receptor antagonists (e.g. cimetidine) are powerful inhibitors of gastrin- and food-stimulated gastric acid secretion.

It has been possible to reconcile these two theories by suggesting that the parietal cells have receptors for both gastrin and histamine but that histamine is required for the full expression of the direct effects of gastrin interaction with its receptors; acetyl choline, on the other hand, acts on a third distinct class of receptors which can, as expected, be blocked by atropine. It has recently been shown that injection of gastrin into the hypothalamus causes increased acid secretion in rats while injection in the cerebrum was without effect. It is not clear how these effects are mediated.

3.   SECRETIN

Secretin was the first hormone to be identified and named.[5] However, its purification was first achieved by Jorpes and Mutt in 1961. It was sequenced and synthesized in 1966 and is a 27-amino acid polypeptide (Figure 12.3); it has structural similarities with glucagon, a total of 14 amino acids occupying the same positions in the two molecules. Secretin does not appear to exist in multiple forms, nor to have active fragments.

	1	2	3	4	5	6	7	8	9	10	11	12	13	14	15	16
Glucagon:	HIS	SER	GLN	GLY	THR	PHE	THR	SER	ASP	TYR	SER	LYS	TYR	LEU	ASP	SER
GIP:	TYR	ALA	GLU	GLY	THR	PHE	ILE	SER	ASP	TYR	SER	ILE	ALA	MET	ASP	LYS-
Secretin:	HIS	SER	ASP	GLY	THR	PHE	THR	SER	GLU	LEU	SER	ARG	LEU	ARG	ASP	SER
VIP:	HIS	SER	ASP	ALA	VAL	PHE	THR	ASP	ASN	TYR	THR	ARG	LEU	ARG	LYS	GLN-

	17	18	19	20	21	22	23	24	25	26	27	28	29 - - - 43
Glucagon:	ARG	ARG	ALA	GLN	ASP	PHE	VAL	GLN	TRP	LEU	MET	ASN	THR
GIP:	ILE	ARG	GLN	GLN	ASP	PHE	VAL	ASN	TRP	LEU	LEU	ALA	GLN- - - GLN
Secretin:	ALA	ARG	LEU	GLN	ARG	LEU	LEU	GLN	GLY	LEU	VAL	-(NH₂)	
VIP:	MET	ALA	VAL	LYS	LYS	TYR	LEU	ASN	SER	ILE	LEU	ASN	-(NH₂)

Figure 12.3 Structure of porcine secretin, vasoactive intestinal peptide, gastric inhibitory peptide and glucagon. (Reproduced from *Textbook of Endocrinology* (Ed. Williams, R. H.) 6th edition, W. B. Saunders Inc., 1981)

The distribution of secretin is shown in Figure 12.2; it originates in the cells of the mucosa of the upper portion of the small intestine.

3.1   Regulation of secretion

The most important physiological stimulus to secretion is the presence of acid in the small intestine. The secretin then in turn increases the secretion of bicarbonate from the pancreas, eliciting production of a watery alkaline pancreatic juice into the small intestine which serves to neutralize the acid in the intestine. To a lesser extent it also promotes the secretion of pancreatic enzymes, whose secretion is predominantly controlled by the actions of PZ-CCK. Secretin also serves to decrease gastric acid secretion and reduce contraction of the pyloric sphincter, thereby reducing the rate of emptying of the stomach. Feeding does not itself appear to be an adequate stimulus for secretion.

3.2   Actions

As indicated in Table 12.3, secretin has been shown to have multiple effects, but probably the most important are those on the pancreatic acinar cells which regulate the secretion of water and bicarbonate, with relatively little effect on the secretion of protein. Conversely, pancreozymin-CCK acts purely to alter rates of secretion of digestive enzymes.[6]

After ingestion of a meal it is now clear that there is a significant increase in plasma secretin levels and that this increase can be prevented by pre-treatment with cimetidine which will inhibit gastric acid secretion or by neutralization of gastric acid by ant-acids. This circulating secretin will in turn influence secretion of bicarbonate and of water since reduction of circulating secretin levels by the use of specific antisera to the hormone markedly diminishes the bicarbonate and water response to the meal. These effects may be exerted on the pancreatic cells via binding to specific receptors in the membrane which are coupled to adenylate cyclase, the binding leading to stimulation of this enzyme and to production of cyclic AMP.

Secretin will inhibit the release of gastrin, and of gastric acid. It will also lower oesophageal sphincter pressure and stimulate insulin secretion, but there are considerable doubts as to whether these represent physiological actions as distinct from pharmacological effects seen only at high concentrations.

Table 12.3   Actions of secretin

Increases	Decreases
Pancreatic secretion of H_2O, bicarbonate, electrolytes and protein	Gastric HCl and gastrin secretion and duodenal motility
Gall bladder secretions	
Brunner gland secretion in intestine	

4. PANCREOZYMIN–CHOLECYSTOKININ

It was originally thought that cholecystokinin (CCK) produced gall bladder contraction,[7] while a separate hormone, pancreozymin, induced the secretion of enzymes in the pancreatic juice.[8] However, it has subsequently emerged that these represent different facets of the actions of a single molecule which is now referred to as pancreozymin–CCK (PZ–CCK).[9]

4.1 Chemistry

Pancreozymin–CCK exists in at least two chemical forms, one containing 39 and one 33 amino acids (Figure 12.4). The latter differs from the former by the deletion of 6 amino acids at the N-terminus. The relative amounts of pancreozymin-CCK 33 and pancreozymin-CCK 39 in intestine are not known, nor is it clear whether the 39 amino acid-form is a precursor molecule. The 5 C-terminal amino acid residues of pancreozymin-CCK are identical to those of pentagastrin. The C-terminal octapeptide contains the complete range of biological activities of pancreozymin-CCK and is 2–10 times more active on a molar basis than the complete molecule.[9] It is also clear that pancreozymin-CCK 8 exists in tissues in its own right at high concentration in the gut, and also in the brain. It is likely that the octapeptide is derived from the 39 or 33 amino acid forms by enzymatic cleavage, rather than synthesized as a separate molecule. The distribution of pancreozymin-CCK in the gut is shown diagrammatically in Figure 12.2.

LYS-ALA-PRO-SER-GLY-ARG-VAL-SER-MET-ILE-LYS-ASN-LEU-GLN-SER-LEU-ASP-PRO-
10

SER-HIS-ARG-ILE-SER-ASP-ARG-ASP-TYR-MET-GLY-TRP-MET-ASP-PHE-NH$_2$
20   30

Figure 12.4 Structure of porcine pancreozymin-cholecystokinin 1–33

4.2 Regulation of secretion

Immunoassay of pancreozymin-CCK has proved difficult and most early work relied on bioassays involving gall bladder contraction or pancreatic enzyme secretion. These showed that amino acids and fatty acids, particularly long-chain fatty acids, were potent stimuli of PZ-CCK secretion. There is still little information available from immunoassay to quantitate the effects of various stimuli on immunoreactive PZ-CCK in the blood where its half-life is very short (3 minutes), nor is it clear which molecular species predominates in serum, PZ-CCK 39, PZ-CCK 33 and PZ-CCK 8 all having been detected.

4.3   Actions

Probably the most important actions of pancreozymin-CCK physiologically are to increase the secretion of enzymes in the pancreatic juice, to stimulate gall bladder contraction and to increase gastrointestinal motility. It has also been shown to have effects in inhibiting absorption of fluid and Na^+, K^+ and Cl^- ions from the jejunum and ileum. A number of other effects have been observed, but in many cases these may be pharmacological rather than physiological. These are listed in Table 12.4.

At the molecular level, pancreozymin-CCK is known to bind to specific cell-surface receptors on its target cells, with a consequent increase of intracellular cyclic GMP levels and mobilization of Ca^{2+}. It also has a strong trophic action on the acinar pancreas and can increase both weight and functional capacity of the gland.

Table 12.4   Actions of pancreozymin–CCK

Increases	Decreases
Secretion of enzymes and electrolytes from pancreas	Secretion of acid and pepsin, and electrolyte absorption from small intestine
Gall bladder contraction	
	Appetite and food intake
Secretion of insulin, glucagon and pancreatic polypeptide	
Secretion from Brunner glands	
Contraction of gastric antrum	

4.4   Pancreozymin-CCK in the brain

Recent evidence based on immunofluorescence studies and on immunoassay of brain extracts has shown conclusively that CCK is present in the brain in quite high concentrations, specifically in cortex, amygdala and corpus striatum. Three lines of evidence have been proposed to suggest that pancreozymin-CCK might act as a neurotransmitter in the brain. These are (a) that it is produced in nerve cells and stored in nerve terminals, (b) that the mechanisms of its release resemble those of neurotransmitters, and (c) that degradation mechanisms for the hormone exist at the terminals. However, little is known of the physiological importance of the peptide in these brain tissues.

5.   VASOACTIVE INTESTINAL PEPTIDE (VIP)

5.1   Occurrence

Aqueous extracts of mammalian lungs are highly vasoactive and this activity cannot be explained by the histamine and prostaglandin content of lung tissue.

Said and Mutt[10] extracted and partially purified a vasodilator peptide from porcine lung and then, using the same bioassay as for the lung peptide (measurement of femoral blood flow and arterial blood pressure), they discovered that peptide fractions from porcine duodenum also contained a vasodilator which they soon purified to homogeneity and named vasoactive intestinal peptide (VIP).[11] Its structure is shown in Figure 12.3.

Use of the techniques that enabled Mutt and Jorpes to extract and purify secretin yielded a VIP-rich peptide as a side fraction. Three additional separation procedures produced a highly purified preparation of VIP, from which traces of impurities could be virtually eliminated by a final gel filtration step.

VIP is now believed to occur throughout the gastrointestinal tract. Higher concentrations are present in the colon, ileum and jejunum than in the upper portions of the gut, and relatively high levels are also found in the pancreas. VIP in both gut and pancreas is contained mainly within nerve elements. In the intestinal wall it is located predominantly in the nerve-rich muscular layer. VIP has also been found in nervous tissue. The highest levels of immunoreactivity are found in the cerebral cortex, hippocampus, amygdala and corpus striatum. In man the highest content is in the median eminence: little VIP is found in the cerebellum or in the brainstem. VIP also occurs widely in the peripheral nervous system where it is a major component of the 'peptidergic' system of nerves. Subcellular fractionation of VIP-rich parts of the brain showed VIP to be concentrated in the synaptosomal fraction along with two established neurotransmitters, dopamine and noradrenaline. This localization is in keeping with a role for the peptide in synaptic function.

5.2   Actions

The full range of actions of VIP is only partially known; actions that have been described are summarized in Table 12.5.

Peptides showing the most extensive structural similarities to VIP are secretin and glucagon. These three peptides also share important activities. Thus, secretin and VIP relax gastrointestinal and other smooth muscle and stimulate

Table 12.5   Actions of vasoactive intestinal polypeptide

Increases	Decreases
Vasodilation	Water absorption (intestine)
Bronchodilation	Gastric acid and pepsin secretion
Water and bicarbonate secretion	
Glucagon and insulin secretion	
Adrenal steroidogenesis	
Glycogenolysis (liver)	
Lipolysis (adipocytes)	
Prolactin secretion	

bicarbonate secretion; glucagon and VIP stimulate glycogenolysis and myo-cardial contractility; GIP and VIP inhibit gastric acid secretion and stimulate insulin secretion (Table 12.5).

The entire sequence of VIP is not required for significant hormonal activity. VIP-like biological activity is present in the C-terminal peptide, VIP_{18-28}, and this activity increases with increasing chain length.

5.3 Physiological role

The wide distribution of VIP, its presence in neurones and nerve terminals and its ability to influence numerous body functions make it unlikely that the peptide functions as a circulating hormone. It seems probable that it may serve as a paracrine secretion, i.e. as a local hormone. Possible functions of VIP include acting as a neurotransmitter or, more likely, a neuromodulator in the central nervous system. The localization of VIP in hypothalamic nuclei, its apparent secretion into portal hypophysial blood and its ability to stimulate pituitary adenylate cyclase and to promote the secretion of prolactin, luteinizing hormone and growth hormone after injection into the third ventricle of the brain, point to a possible role for VIP in modulating hypothalamic–pituitary function. Its vasodilator activity in many vascular beds raises the possibility that it may mediate increases in blood flow to these organs. VIP may mediate certain responses that have been attributed to 'non-adrenergic inhibitory' nerves. Peptides, rather than purines, might be the mediators of this system of nerves. As discussed above, peptides which might fulfil such a role include, in addition to VIP, somatostatin, substance P, enkephalins, bombesin and gastrin.

6. GASTRIC INHIBITORY PEPTIDE

GIP was first isolated as a side fraction in the purification of PZ-CCK from the duodenum. Its existence had been postulated by Brown *et. al.*[11] in a fraction which was able to inhibit the secretion of gastric acid: hence the term gastric inhibitory peptide. Subsequently it has become clear that GIP has at least 3 other functions: stimulation of intestinal secretion, stimulation of insulin secretion and stimulation of glucagon secretion.[12]

Purification and amino acid sequencing of this material has been accomplished and revealed a polypeptide of 43 amino acid residues. The amino acid sequence of GIP together with the sequences of glucagon, secretin and VIP are illustrated in Figure 12.3.

Immunocytochemical studies have been used to identify GIP-containing cells in the gastrointestinal tract of man, pig and dog. GIP cells were predominantly localized in the lower and middle zone of the mucosa of the duodenum and the jejunum. The cell of origin has been identified as the K cell characterized by granules with an average diameter of 300 nm.

6.1   Regulation of secretion

Radioimmunoassay for GIP using a guinea-pig antibody raised against GIP conjugated to albumin by the carbodiimide technique has been described. All available antisera appear to be directed toward the C-terminal portion of the GIP molecule. Fasting serum levels of GIP range from non-detectable to approximately 500 pg/ml. After either a solid or a liquid meal, circulating levels increase at least fivefold. Triglycerides produced a sustained rise in GIP. Ingestion of glucose elevated GIP levels within 45–60 minutes, whereas intravenous glucose was without effect. Attempts were made to correlate this with the increase in insulin secretion. After administration of intravenous glucose there was no change in GIP from basal levels. After oral glucose, GIP levels rise within 10 minutes. Time courses for release of GIP and insulin are almost identical.

Duodenal infusion of an amino acid mixture has also been shown to stimulate GIP. Peak GIP levels (15 minutes) preceded the insulin peak (30 minutes). Both somatostatin and glucagon inhibit GIP release and a possible role as local (paracrine) and feedback modulators could be considered. A feedback regulation of GIP secretion by insulin has also been suggested.

6.2   Actions

GIP is insulinotropic in man, dog and rat when administered in doses similar to those found postprandially.[13] The prevailing glucose concentration is important for the action of GIP. A threshold glucose concentration is necessary for GIP to stimulate insulin release from isolated islets. GIP is therefore a potent insulinotropic polypeptide and a possible candidate for the gastrointestinal insulinotropic hormone. In the isolated pancreas, GIP produces biphasic glucagon release. GIP stimulates somatostatin release from the perfused canine pancreas.

In dogs, GIP infusion produces inhibition of acid secretion from denervated stomach pouches. Maximum inhibition occurs at an infusion rate which produces circulating GIP levels observed after feeding. Pepsin secretion and motor activity in the stomach are both inhibited by GIP. Since GIP does not stimulate adenylate cyclase in the small intestine or in the colon, its effect is unlikely to be mediated by cyclic AMP.

7.   MOTILIN

Instillation of acid into the duodenum causes inhibition of gastric motor activity and delayed gastric emptying. Evidence for the involvement of a hormonal fac-

PHE-VAL-PRO-ILE-PHE-THR-TYR-GLY-GLU-LEU-GLN-

ARG-MET-GLN-GLU-LYS-GLU-ARG-ASN-LYS-GLY-GLN

Figure 12.5 The structure of porcine motilin

tor in such a reflex resulted from dog experiments, in which increased motor activity of denervated and transplanted pouches of stomach occurred when the duodenal pH was raised by one unit. The search for a possible stimulatory factor began with screening of various preparations of secretin and cholecystokinin. A polypeptide was finally purified from a side fraction produced during the isolation of secretin and the name motilin was chosen because of its ability to stimulate motor activity in fundic pouches.[14] Amino acid analysis of motilin demonstrated the presence of a high content of glutamate or glutamine. The complete amino acid sequence has been determined (Figure 12.5).

Motilin was initially localized to 5-hydroxytryptophan (5-HT)-containing enterochromaffin (EC) cells. Most cells were detected in the duodenum and upper jejunum. Radioimmunoassays for motilin have been developed with antibodies raised against natural porcine motilin and synthetic (14-Met) motilin. Species differences in relative gastrointestinal distribution may occur, since the jejunal : duodenal : ileal ratios have been reported as 100 : 7 : 0.04 (pig) 100 : 442 : 1.3 (man) and 100 : 156 : 2.8 (dog). Only one molecular form of motilin has been detected in plasma and duodenal extracts.

7.1   Regulation of secretion

In dogs with denervated stomach pouches, the infusion of 50 ml of Tris buffer into the duodenum resulted in a rapid increase in motilin levels at a time when duodenal pH increased by 0.7 units; this was accompanied by increased motor activity of the body of the stomach. Maximum levels were reached at 5 minutes, when motor activity was also maximal, but duodenal pH had returned to the pre-infusion level. Conversely, duodenal alkalinization with Tris decreases motilin levels in man. Possibly, species differences may account for the apparently conflicting results. Increases in circulating motilin occur in response to intraduodenal acid in dog and in man. The motilin response is not accompanied by changes in motility.

7.2   Actions[15]

Studies on strips of duodenal and colonic circular muscle and fundic muscle from stomach have shown that porcine motilin increased the contractile response. The contractile response to motilin is blocked by the calcium transport antagonist verapamil, although motilin has no effect on calcium uptake. Motilin-induced contractions are associated with an increase in intracellular cyclic GMP, whereas an increase in endogenous cyclic AMP is accompanied by a reduction in tone.

In vivo studies with motilin showed it to be a powerful stimulant of motor activity in pouches of the stomach of conscious dogs. Cholinergic blockade with atropine caused a strong reduction in fundic motor activity in response to motilin. Gastric emptying of liquid meals was stimulated by motilin in dogs. In man, motilin inhibited gastric emptying of a liquid meal, whereas porcine motilin stimulated solid emptying. These different results may reflect species differences. In the dog, the effect of motilin is on the body of the stomach and is modulated by vagal tone.

During investigations on the effect of motilin on gastric motility, it was observed that pepsin secretion was also stimulated. An increased gastric mucosal blood flow correlated with the pepsin secretion, and it was speculated that the motilin effect may be mediated by this increase. A direct effect on peptic cells is equally possible. A stimulatory effect on basal volume, bicarbonate, and protein output from the pancreas and inhibition of secretin-stimulated bicarbonate output in conscious dogs have also been reported.

8.   PANCREATIC POLYPEPTIDE

During the isolation of chicken insulin and of bovine and porcine glucagon and insulin, gel filtration revealed a new polypeptide fraction eluting from the column between glucagon and insulin. The function of these peptides was unknown so they were referred to as avian (APP) and bovine (BPP) pancreatic polypeptide.

8.1   Isolation

Pancreatic polypeptides isolated from the pancreas of pig, sheep or human are very similar to APP (Figure 12.6) each containing 36 amino acids in a single chain. Amino acid sequences of pancreatic polypeptide and glucagon appear to be unrelated; the structures of the mammalian PPs isolated so far are unrelated to other gastrointestinal hormones.

GLY-PRO-SER-GLN-PRO-THR-TYR-PRO-GLY-ASP-ASP-ALA-
10

PRO-VAL-GLU-ASP-LEU-ILE-ARG-PHE-TYR-ASP-ASN-LEU-
20

GLN-GLN-TYR-LEU-ASN-VAL-VAL-THR-ARG-HIS-ARG-TYR-NH$_2$
30

Figure 12.6   The structure of avian pancreatic polypeptide

PP-producing cells in mammalian tissues were identified by immunohistochemical procedures. In the human pancreas the pancreatic polypeptide cells were located mainly at the periphery of the islets; some were also found scattered in the exocrine parenchyma and even in the epithelium of the ducts.

Pancreatic polypeptide cells were also found in the mucosa of the stomach, duodenum, ileum, colon and rectum. In most species the distribution of pancreatic polypeptide cells in the pancreas was similar to that in man. The cytoplasm of the pancreatic polypeptide cells contains electron-dense granules with closely applied limiting membranes.

8.2.   Release of pancreatic polypeptide

Food is one of the most effective stimulants of pancreatic polypeptide release. Protein appeared to be more effective than fats and carbohydrates for the release of pancreatic polypeptide. Only ingested protein, fat and glucose had the ability to release pancreatic polypeptide; given intravenously they were ineffective. Intraduodenal infusion of dilute HCl was reported to have no effect on pancreatic polypeptide release in some species, but it increased secretion of the polypeptide in dog and man. An amino acid mixture stimulated the isolated slices of the uncinate process of dog pancreas to secrete pancreatic polypeptide, but individual amino acids were ineffective.

Hormones such as gastric inhibitory peptide and vasoactive intestinal peptide were reported to cause secretion of pancreatic polypeptide from the isolated pancreas, whereas somatostatin, which is known to suppress the release of gastrin, secretin, glucagon and pancreozymin-CCK, definitely decreased secretion of pancreatic polypeptide *in vitro* and *in vivo*. Vagotomy does not significantly alter basal concentration of pancreatic polypeptide; vagal mediation of its release can be mimicked by administration of insulin, cholinergic agents or electrical stimulation of the vagus. Secretion of pancreatic polypeptide from the isolated pancreas is enhanced by acetylcholine.

8.3   Metabolic effects

Avian pancreatic polypeptide and bovine pancreatic polypeptide were originally isolated as contaminants of insulin or glucagon so the biological effect of pancreatic polypeptide was first examined on carbohydrate and lipid metabolism. In contrast to the actions of glucagon and insulin, BPP had no effect on blood sugar at doses that caused marked gastrointestinal actions. Bovine pancreatic polypeptide failed to release insulin from isolated pancreas *in vitro* and no effect on lipid metabolism has been reported. Bovine pancreatic polypeptide had no diuretic action and no effect on Na^+, K^+, Mg^{2+} or Ca^{2+} excretion.

Bovine pancreatic polypeptide slowly stimulated gastric volume and acid secretion. When gastric acid secretion was induced by pentagastrin, secretion of HCl was inhibited. Both inhibitory and stimulatory effects of bovine pancreatic polypeptide occurred only at high doses.

In fasted dogs pancreatic secretion, bicarbonate and enzyme outputs were all decreased by bovine pancreatic polypeptide, while the same functions induced by intraduodenal infusion of amino acids (via the release of endogenous pancreomyzin-CCK) were inhibited by pancreatic polypeptide in a dose-related manner. Bovine pancreatic polypeptide has various actions on gut motility. Thus large doses cause increased gut motility in the antrum, duodenum and colon of the dog. At low doses it relaxed the intraluminal pressure in the dog antrum, pylorus, duodenum, ileocaecal sphincter and descending colon. Pancreatic polypeptide may play a role in the physiological regulation of food intake, depending on its concentration.

9. NEUROTENSIN

Neurotensin was first isolated from bovine hypothalamus but was subsequently identified in bovine and human intestine (ileum).[16] The structure of the hormone is the same from all these sources and consists of a 13-amino-acid peptide. The structure is shown in Figure 12.7.

pyroGLU-LEU-TYR-GLU-ASN-LYS-PRO-ARG-ARG-PRO-TYR-ILE-LEU

Figure 12.7 The structure of bovine neurotensin

The peptide is synthesized in the so-called N cells of the mucosa of the ileum which contains 90% of the total neurotensin of the body. Little is known of the mechanism and regulation of its biosynthesis, but it seems clear that in the blood there may be biologically active forms circulating which have molecular weights less than that of the 13-amino-acid peptide in the cells of origin. The postulated roles of neurotensin are several, and as with some of the other hormones discussed in this chapter it is difficult to separate the physiological actions from the purely pharmacological effects which have also been described (see Table 12.6). In addition to the roles of neurotensin in the peripheral circulation which are listed here, it may also act as releasing factor for the anterior pituitary hormones, ACTH, luteinizing hormone, thyroid-stimulating hormone and growth hormone.

Table 12.6 Some actions of neurotensin

Increases	Decreases
Vasodilation	Plasma volume
Vascular permeability	Gastric acid and secretin secretion
Gastrin secretion	

10. BOMBESIN

Bombesin is a 14-amino-acid peptide first isolated from frog skin. Its structure is shown in Figure 12.8. Subsequently it has been shown to have a wide distribution in the mammalian gut, both in peptidergic nerves and in apparently endocrine cells, especially in the antrum and duodenum.[17] Nerve fibres which contain bombesin can be demonstrated biochemically in the submucosal area of the gut wall throughout the whole length of the intestine. It is not clear whether bombesin circulates in significant quantities in peripheral blood in normal circumstances, although its release can be elicited in calves in some conditions.

pyroGLU-GLN-ARG-LEU-GLY-ASN-GLN-TRP-ALA-VAL-GLY-HIS-LEU-MET-NH$_2$

Figure 12.8 The structure of bombesin

Bombesin has been suggested to have a wide variety of actions in various systems, e.g. the central nervous system, gut smooth muscle, pancreas, pituitary, kidney and heart. It is far from certain from the information so far available whether all or indeed any of these functions have any physiological as distinct from pharmacological relevance.

11. GLICENTIN

There was considerable interest following the introduction of glucagon immunoassays in the finding that glucagon appeared to be present not only in the A cells of the pancreatic islets (Chapter 11) but also in a relatively large quantities in the gut. Subsequent detailed studies have demonstrated that this material is not in fact identical to pancreatic glucagon but rather glucagon-like immunoreactivity (GLI) or a whole series of such peptides, the best characterized of which is glicentin[18] which contains 100 amino acids. The highest concentration of glicentin is found in 'L' cells of the jejunum and duodenum, and it may also be present in the pancreatic A cells along with glucagon itself. Glicentin has no receptors on liver cells and yet increases hepatic glycogenolysis, possibly after degradation to glucagon or a similar molecule. Use of specific antisera has shown that glicentin circulates in the blood, but a specific physiological function for it has yet to be identified.

REFERENCES

1. Bloom, S. R. and Polak, J. M. (Eds.) (1981) *Gut Hormones,* 2nd edn Churchill-Livingstone, Edinburgh.
2. Pearse, A. G. E., Polak, J. M. and Bloom, S. R. (1977) *Gastroenterology,* **72**, 746–761.

3. Gregory, R. A. and Tracy, J. H. (1964) *Gut,* **5**, 103–117.
4. Walsh, J. H. and Lam, S. K. (1980) In: *Clinics in Gastroenterology* Vol. 9 (Ed. Creutzfeldt, W.), W. B. Saunders, Philadelphia, pp. 567–591.
5. Bayliss, W. and Starling, E. H. (1902) *J. Physiol.,* **28**, 325–353.
6. Rayford, P. L., Miller, J. A. and Thompson, J. C. (1976) *New Eng. J. Med.,* **294**, 1093–1101 and 1157–1163.
7. Ivy, A. C. and Goldberg, E. (1928) *Amer. J. Physiol.,* **86**, 599–613.
8. Harper, A. A. and Raper, H. S. (1943) *Amer. J. Physiol.,* **102**, 115–125.
9. Jorpes, J. G. and Mutt, V. (1973) In: *Handbook of Experimental Pharmacology,* Vol. XXXIV, Springer-Verlag, Berlin, pp. 1–179.
10. Said, S. I. (1980) In: *Gastrointestinal Hormones* (Ed. Glass, G. D. J.). Raven Press, New York, pp. 245–273.
11. Brown, J. C., Mutt, V. and Pedersen, R. A. (1970) *J. Physiol.,* **209**, 57–64.
12. Brown, J. C., Dryburgh, J. R., Ross, S. A. and Dupré, J. (1975) *Recent Progr. Hormone Res.,* **31**, 487–532.
13. Brown, J. C., Frost, J. C., Kwank, S., Otte, S. and McIntosh, C. H. S. (1980) In: *Gastrointestinal Hormones* (Ed. Glass, G. D. J.). Raven Press, New York, pp. 223–232.
14. Brown, J. C., Mutt, V. and Dryburgh, J. R. (1971) *Canad. J. Physiol. Pharmacol.,* **49**, 399–405.
15. Christophides, N. D. and Bloom, S. R. (1981) In: *Gut Hormones,* 2nd edn (Eds. Bloom, S. R. and Polak, J. M.). Churchill-Livingstone, Edinburgh, pp. 273–279.
16. Floyd, J. C., Fajans, S. S., Pek, S. and Chance, R. E. (1977) *Recent Progr. Hormone Res.,* **33**, 519–570.
17. Walsh, J. H., Reeve, J. R. and Vigna, S. R. (1981) In *Gut Hormones,* 2nd edn (Eds. Bloom, S. R. and Polak, J. M.). Churchill-Livingstone, Edinburgh, pp. 413–418.
18. Sundby, F., Jacobsen, H. and Moody, A. J. (1976) *Horm. Metab. Res.,* **8**, 366–371

Chapter 13

Parathyroid hormone and calcitonin

1. PARATHYROID HORMONE

1.1 Introduction

Parathyroid hormone has at least two alternative names—PTH, which is simply an abbreviation for parathyroid hormone, and parathormone, this last mostly in the American literature. We use here parathyroid hormone.

The thyroid originates from an invagination of the floor of the pharynx, and in the adult exists as two lobes on either side of the trachea. It has a very rich vascular supply and has one of the highest rates of blood flow of any organ. The gland is composed of numerous follicles, each consisting of a spherical area of colloid, and a single layer of cells surrounding it which consists of follicular cells

producing thyroxine and triiodothyronine, and parafollicular cells which synthesize and secrete calcitonin. The parathyroid glands lie on the surface of the
lobes of the thyroid in the neck. In humans there are usually 2–4 such glands
each with an abundant blood supply and composed of two major cell types, the
chief cells which are responsible for the production of parathyroid hormone,
and the oxyphil cells whose function is not clear.

1.2 Chemistry

The role of the parathyroid glands in the prevention of tetany and the maintenance of normal concentrations of plasma calcium has been appreciated since the
late nineteenth century, but a reliable method for the extraction of an active
principle from the parathyroid was not found till many years later. In 1925, Collip[1] showed that hot hydrochloric acid would extract a potent substance from
bovine parathyroid glands which corrected hypocalcaemia and tetany in
parathyroidectomized dogs. Subsequently 8 M urea and cysteine in cold
hydrochloric acid has also been used to extract active hormone; treatment of
these extracts by solvent and salt fractionation, followed by countercurrent distribution, yielded a highly purified preparation. Isolation of a highly purified
polypeptide was also accomplished by gel filtration and ion-exchange
chromatography on carboxymethylcellulose.
Two closely-related forms of the bovine hormone have been isolated, termed
bovine parathyroid hormones (BPTH) I and II. Form I is devoid of threonine,
whereas form II contains one residue of threonine and one less valine residue.
Porcine and human parathyroid hormones exist in only one form. The structures of the bovine, porcine and human hormones are compared in Figure 13.1.
Earlier work indicated that the full 84-amino acid sequence of bovine
parathyroid hormone is not required for biological activity, the biological
activity being associated with a 30–40 amino acid fragment at the amino-terminal region of the molecule. The successful synthesis of a biologically active
amino-terminal fragment (residues 1–34) of the bovine parathyroid hormone
sequence confirmed that amino-terminal fragments of the molecule were the
ones of most importance for biological activity. It is now clear that in order to
be active, a fragment of parathyroid hormone must contain a sequence no
shorter at the carboxyl end than residue 27 and no shorter at the amino end than
residue 2.[2]

1.3 Assay

1.3.1 Bioassay

Three main types of bioassay have been used. In the first, *in vivo*, method
elevation of serum calcium is measured after injection of extracts into

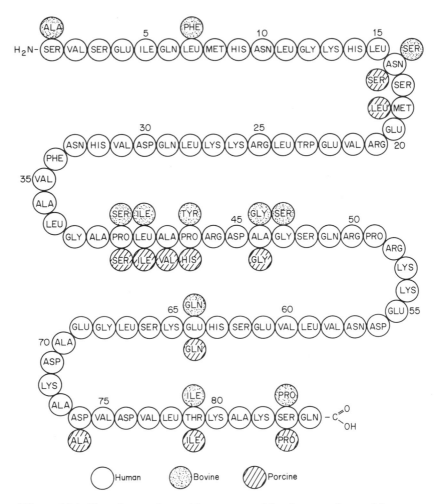

Figure 13.1 Complete amino acid sequence of bovine, porcine and human parathyroid hormones. Reproduced with permission from *Textbook of Endocrinology*, (Ed. Williams, R. H.) 6th edition, 1981, W. B. Saunders Inc.

parathyroidectomized rats. In the second, *in vitro*, method, activation of kidney adenylate cyclase activity by parathyroid hormone is measured. The first has the usual problems of handling large numbers of samples and of accuracy; the second is more easily quantitated but is rather non-specific.

A third type of bioassay, which is much more sensitive than either of the other two, is a cytochemical assay based on the activation by parathyroid hormone of glucose 6-phosphate dehydrogenase activity in guinea-pig kidney tubules. This assay is extremely sensitive, but not readily amenable to the handling of large numbers of samples.

1.3.2 Immunoassay

The estimation of levels of parathyroid hormone in blood by immunoassay has been confused in part by the apparent heterogeneity of the circulating forms of the hormone.[4] Any fragment of the hormone which does not contain amino acids 2–27 is likely to be biologically inert but may still cross-react with many parathyroid hormone antisera. As a result of this problem, attempts have been made to establish sequence-specific radioimmunoassays based on antisera with varying specificities, and this sort of analysis has suggested that much of the hormone measured in traditional immunoassays of parathyroid hormone is indeed biologically inactive. Nevertheless it seems likely that the major circulating form of parathyroid hormone is the one which is extracted from the gland (1–84), which has both biological and immunological activity.

1.4 Biosynthesis

Studies of biosynthesis were greatly facilitated by the successful development of an *in vitro* system using slices of bovine parathyroid glands.[3] Incorporation of radioactive amino acids into parathyroid hormone was detected in extracts from the gland by using gel filtration and chromatography on carboxymethylcellulose to analyse radioactive proteins. One peak of radioactive peptide was synthesized within the gland, larger in size than PTH. It was biologically active in that it produced hypercalcaemia in rats and bone resorption *in vitro* and was chemically similar to PTH in that it reacted with antisera to parathyroid hormone. Subsequently it was demonstrated that this larger peptide, proparathyroid hormone, was indeed a precursor of parathyroid hormone, and that it was composed of 96 residues, 6 more than the 84 residues of parathyroid hormone.

It was shown that proparathyroid hormone is labelled more rapidly than parathyroid hormone during short incubations of parathyroid tissue with ^{14}C-labelled amino acids and that the pro-hormone was progressively converted to the hormone. If the labelled amino acids were removed after short initial incubations and replaced with unlabelled amino acids ('pulse-chase' experiments), the radioactivity in the proparathyroid hormone decreased, whereas that in parathyroid hormone continued to increase. Similar results were obtained when protein synthesis was inhibited during the chase period with puromycin, indicating that the continued synthesis of protein was not necessary for the transformation of proparathyroid hormone to parathyroid hormone. Sequence studies have shown that proparathyroid hormone differs from parathyroid hormone in that it has an additional hexapeptide sequence at the amino terminus of the hormone (Figure 13.2). This additional prohormone-specific peptide is very basic.

The results of extensive assays of the biological activity of native proparathyroid hormone show that the prohormone is biologically active when

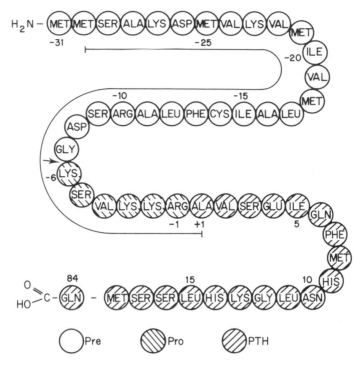

Figure 13.2 The structures of the biosynthetic intermediates in the production of parathyroid hormone. The sequence of pre-pro-parathyroid hormone (residues −31 to 84) and pro-parathyroid hormone (−6 to 84) are indicated, along with the first 18 residues of parathyroid hormone itself (+1 to 18.) Reproduced with permission from Habener, J. F. *et al.* (1977) *Recent Progr. Hormone Res.* **33**, 249–299. Copyright (1977) Academic Press

tested by hypercalcaemia assays in rat, and chick *in vivo* or in *in vitro* assays of either bone resorption in tissue culture or activation of renal adenylate cyclase (see below); however, the activity of the prohormone in all systems is less than that of parathyroid hormone.

As has also been found with other polypeptide hormones, it has become clear that there exists an additional precursor, termed pre-proparathyroid hormone. This has an additional extension of 25 residues at the amino terminus and a very short half-life (∼ 30 seconds). It probably facilitates the transfer of the nascent polypeptide chain through the membrane of the endoplasmic reticulum to the cisternal space, where it is rapidly removed.

Newly-synthesized proparathyroid hormone is released from the polyribosome to the cisternal space of the endoplasmic reticulum. Conversion of pro-parathyroid hormone begins within 10 minutes, the time shown to be required for newly synthesized proteins to reach the Golgi complex. A longer time, up to an hour, is required for transfer of newly formed hormone to the secretory granule (Figure 13.3).

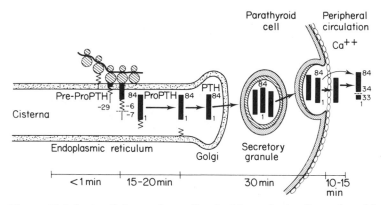

Figure 13.3 Intracellular pathway for the biosynthesis of parathyroid hormone, showing the initial biosynthesis, cleavage of pre- and pro-hormones and storage of the hormone in secretory granules before secretion. Adapted and reproduced with permission from Habener, J. F. *et al.* (1977). *Recent Progr. Hormone Res.* **33**, 249–299. Copyright (1977) Academic Press

The kinetics of the biosynthesis and conversion of proparathyroid hormone to parathyroid hormone in bovine gland slices are illustrated in Figure 13.3. After 15–20 minutes of incubation, small amounts of labelled parathyroid hormone first begin to accumulate. Between 15 and 30 minutes, the formation of labelled parathyroid hormone from proparathyroid hormone proceeds at a linear rate. The half-time for the conversion of proparathyroid hormone to parathyroid hormone is approximately 10 minutes, as determined either from steady-state data or by direct measurements in pulse-chase experiments. These observed rates of conversion of proparathyroid hormone to parathyroid hormone are considerably faster than those found for conversion of proinsulin to insulin where the half-time of conversion to insulin was found to be about 1 hour

Little is presently known about the physical properties of the enzyme that transforms proparathyroid hormone to parathyroid hormone because it has not yet been completely characterized. It is clear, however, that it must have trypsin-like specificity.

1.5 Secretion

As already discussed, there are forms of immunoreactive hormone in the circulation that differ in size and relative immunological activity from hormone in the glands. These multiple forms of circulating immunoreactive hormone have made it difficult to apply the immunoassay as a general assay for hormone in plasma. Nevertheless, the results obtained by immunoassay agree with classical experiments showing that blood calcium concentration is the primary determinant of parathyroid hormone secretion.[5,6] Infusions of calcium

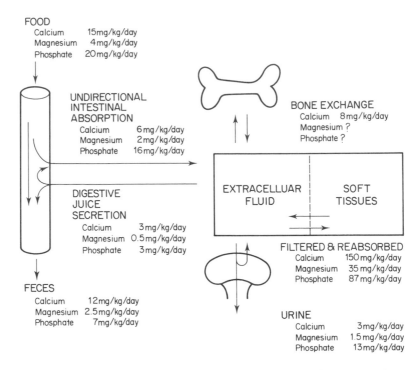

FOOD
Calcium	15mg/kg/day
Magnesium	4mg/kg/day
Phosphate	20mg/kg/day

UNDIRECTIONAL
INTESTINAL
ABSORPTION
Calcium	6mg/kg/day
Magnesium	2mg/kg/day
Phosphate	16mg/kg/day

BONE EXCHANGE
Calcium	8mg/kg/day
Magnesium	?
Phosphate	?

DIGESTIVE
JUICE
SECRETION
Calcium	3mg/kg/day
Magnesium	0.5mg/kg/day
Phosphate	3mg/kg/day

EXTRACELLUAR
FLUID

SOFT
TISSUES

FECES
Calcium	12mg/kg/day
Magnesium	2.5mg/kg/day
Phosphate	7mg/kg/day

FILTERED & REABSORBED
Calcium	150mg/kg/day
Magnesium	35mg/kg/day
Phosphate	87mg/kg/day

URINE
Calcium	3mg/kg/day
Magnesium	1.5mg/kg/day
Phosphate	13mg/kg/day

Figure 13.4 Schematic diagram of the exchange of calcium, magnesium and phosphate between bone, gut and kidney in adult man. Reproduced with permission from *Textbook of Endocrinology* (Ed. Williams, R. H.) 6th Edition, W. B. Saunders Inc., 1981

or ethylenediamine tetra-acetic acid (EDTA) raise or lower blood calcium concentrations, respectively, and induce a corresponding fall or rise in concentration of hormone in plasma. Blood phosphate itself has no direct effect on hormone secretion. Phosphate acts on parathyroid hormone secretion only indirectly by lowering blood calcium. High concentrations of magnesium, on the other hand, can suppress parathyroid hormone secretion.

Studies *in vitro* confirm that calcium is the principal inhibitor of parathyroid hormone secretion.[7] The parathyroid glands are unique among the endocrine systems in their secretory responses to calcium. Parathyroid cells respond to changes in calcium concentration without apparent interaction with any other controlling agent such as a trophic hormone; in other endocrine glands, lack of calcium seems to exert an inhibitory effect on the primary secretagogues. Excess calcium is usually without effect.

The parathyroid gland contains sufficient stored hormone to provide basal secretory needs of approximately 7 hours. When hormone secretion is stimulated, the rate of secretion of hormone increases approximately five fold.

The stores of hormone in the parathyroid glands are relatively low when compared to other endocrine glands, such as some pituitary or pancreatic islet cells, where stores are sufficient to meet secretory demands for at least 24 hours.

Factors other than calcium are known to be able to regulate rates of parathyroid hormone secretion. Thus, adrenaline and isoproteronol (β-adrenergic receptor agonists) increase release of parathyroid hormone *in vitro* and *in vivo*, an effect which is prevented by the β-adrenergic blocking agent propranolol.[8] It seems likely that these agents exert their effects via stimulation of adenylate cyclase and elevation of cyclic-AMP levels, and it has also been shown that cyclic-AMP or theophylline (a phosphodiesterase inhibitor) can lead to stimulation of secretion of the hormone. It is not clear whether the effect of calcium on secretion rates is mediated via a cyclic AMP-dependent mechanism, presumably by a direct effect on adenylate cyclase. It has also been suggested that calcitonin might increase parathyroid hormone secretion by a direct action on the parathyroid cells.

After secretion, parathyroid hormone is short-lived; the hormone is rapidly cleared from the circulation with a half-life of less than 20 minutes. Since radio-immunoassay may overestimate some fragments and underestimate others, the actual rate of disappearance of intact hormone from the circulation may be even more rapid.[9]

1.6   Calcium handling by the body

Plasma calcium accounts for about 2% of the total body calcium in normal conditions and of this fraction only the unbound portion, which accounts for 50% of the total, participates in the majority of the biological functions of extracellular calcium. These include secretion, maintenance of membrane integrity and potentials, blood clotting and a variety of calcium-dependent enzyme processes. The remainder of the calcium in the blood is protein-bound and is thought not to be subject directly to endocrine regulation.

The major organs which are involved in calcium homeostasis are intestine, kidney and bone (Figure 13.4). The intestinal cells of the duodenum are the major sites of reabsorption of calcium, magnesium and phosphate, and the rate of absorption of calcium is under endocrine regulation by parathyroid hormone, and also by vitamin D (see below). The kidney will filter all free calcium, magnesium and phosphate, and each of these is reabsorbed in the proximal and distal tubules so that urinary phosphate is 5–20% of the quantity filtered, while the corresponding figures for calcium and magnesium are 0.5–5% and 2–10%, respectively. Parathyroid hormone exerts one of its actions by altering the rate of reabsorption of calcium (and possibly magnesium) by the distal tubule.

Bone represents a major storage site of calcium in an array of salts resembling hydroxyapatite ($Ca_{10}(PO_4)_6(OH)_2$). The cells which are responsible for laying down these stores of mineral within the collagenous bone matrix are called osteoblasts. Once they have become trapped within the matrix of the bone (with lacunae around them and with a good blood supply) they become relatively inactive and are termed osteocytes. At the inner face of the bone are cells responsible for reabsorption (resorption) of bone mineral and these are termed osteoclasts. The activities of both osteoblasts and osteoclasts are under hormonal control.

1.7 Actions of parathyroid hormone

In mammals with intact parathyroid glands the free calcium level of plasma and extracellular fluid is maintained at a concentration of approximately 1.5×10^{-3} M. This control is the result of the integrated action of parathyroid hormone on renal function and on calcium transport into and out of bone, assisted by stimulation of calcium absorption through the gastrointestinal tract.

1.7.1 Effects on kidney

The hormone causes an enhanced reabsorption of calcium from the glomerular-filtrate; conversely, parathyroidectomy results in an immediate increase in calcium excretion.[11] The exact mechanism of the effect is unclear: it may be linked with increased sodium clearance by a common mechanism in the renal tubule. Thus, the replacement of sodium by choline or ouabain will inhibit the calcium reabsorption by the proximal renal tubule. It seems likely that cyclic AMP may be the mediator of the effects of parathyroid hormone on renal tubule function.

At the same time, parathyroid hormone increases renal excretion of phosphate, although it is still not clear whether this is a direct effect or if it might be secondary to changes in sodium and/or bicarbonate reabsorption. Dibutyryl cyclic AMP is known to exert similar influences on sodium and phosphate in the proximal tubule so it seems possible that this nucleotide might again represent a common mechanism of action.

1.7.2 Effects on bone

Parathyroid hormone exerts direct effects on bone handling of calcium within one hour of its addition *in vitro*, initially to mobilize calcium directly from bone stores[12] and subsequently to induce synthesis of lysosomal enzymes which in turn facilitate bone resorption.[13] This latter effect is prevented by inhibitors of protein synthesis. These two actions are both mediated by effects on osteoclast function, and there is an increased number of osteoclasts in the bone after prolonged exposure to parathyroid hormone, while the number of osteoblasts tends to be diminished. Receptors for parathyroid hormone are found on both osteoblasts and osteoclasts, and binding of the hormone has been shown to lead to activation of adenylate cyclase, suggesting that in both cases the effects may be mediated through cyclic-AMP.

It has been shown that theophylline or dibutyryl cyclic-AMP injected *in vivo* can mimic the hypercalcaemic action of parathyroid hormone.[14] This effect was observed in nephrectomized animals but not in animals with intact parathyroid glands. The anomalous results which were obtained in animals with their parathyroid and thyroid glands intact may reflect a potentiating effect of theophylline on calcitonin release.

1.7.3   Effects on intestine

PTH itself has no direct effect on calcium absorption by the small intestine. However, it does have indirect effects via its regulation of 1,25-dihydroxy-cholecalciferol production in the kidney. This vitamin D derivative then increases Ca^{2+} absorption from the intestine.

2.   CALCITONIN

2.1   Introduction

Calcitonin, sometimes termed thyrocalcitonin, is produced in mammals principally by the parafollicular or 'c' cells of the thyroid, which comprise about 0.1% of the volume of the gland.[15] They may occur singly or in clusters and are larger than the follicular cells. They contain the normal complement of organelles of a polypeptide-secreting cell, including specific hormone-storage granules. The scattered distribution of these cells throughout the thyroid follicles has made it rather difficult to perform meaningful biochemical studies. Calcitonin has also been identified in the hypothalamus of several vertebrates, including man.[16] In more primitive species it may be present mainly in the ultimobranchial body and lung.

2.2   Chemistry[17] and synthesis

The calcitonins show certain common structural features (Figure 13.5) including a 1,7-amino-terminal disulphide bridge, constant chain length of 32 amino acids, and a carboxyl-terminal prolinamide. The locations of the nine residues common to all known forms are also illustrated in Figure 13.5. The entire 32 amino acid chain appears to be required for biological activity. Fragments of the molecule are totally inactive. There is markedly enhanced biological potency of the fish hormones (salmon and eel calcitonins) compared to the calcitonins from other species. The biological activity of salmon I or II or eel calcitonins *in vivo* in the rat varies from 2500 to 3500 U/mg, compared to 50 to 200 U/mg for porcine, bovine, ovine or human calcitonins. The high potency of the piscine molecules cannot yet be attributed to a specific part of the molecule. The high

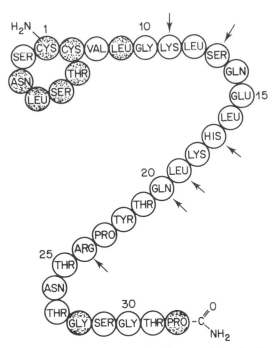

Figure 13.5 Structure of salmon calcitonin 1. The shaded residues are common to eight molecules whose structure is known. The arrows indicate residues common to fish calcitonins which are the biologically most active forms.
Based on ref. 15

sensitivity of biological response to salmon calcitonin make this molecule particularly suitable for use in studies on radioligand–receptor interactions.

Recent use of recombinant DNA techniques (see Chapter 18) has allowed sequencing of the calcitonin precursor RNA. It seems likely that, as with pro-opiomelanocortin (Chapter 5), this precursor may be cleaved to yield more than one hormone. In fact, two additional potential hormone products have been identified, one of which may predominate in the hypothalamus (calcitonin gene related peptide),[18] and one in the thyroid. This latter may have calcium lowering activity[19] and has been termed katacalcin.

2.3 Secretion and actions

The rate of secretion of calcitonin can be regulated in a feedback manner by the concentration of free calcium in plasma, but it now seems likely that gastrointestinal hormones (gastrin and/or pancreozymin-cholecystokinin) may also influence secretion under normal physiological conditions.

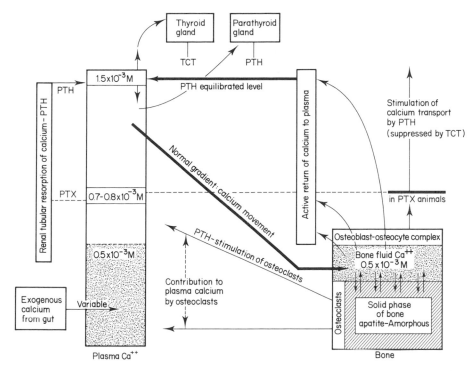

Figure 13.6 Illustration of the sites of action of parathyroid hormone and calcitonin. **PTH**—parathyroid hormone; **TCT**—calcitonin; **PTX**—after parathyroidectomy. Redrawn from ref. 12

Calcitonin produces hypocalcaemia by a direct effect on bone. A reduction in serum calcium could be produced in rats by calcitonin after removal of kidney, of the gastrointestinal tract, or of the parathyroid glands. Kinetic studies after administration of calcitonin to rats preinjected with ^{45}Ca showed[28] that the hypocalcaemic effect could be explained most simply as a result of bone resorption. Experiments *in vitro* using foetal rat long bones, also demonstrated that calcitonin can act directly on bone. The effect was seen as an inhibition of parathyroid hormone-induced resorption, but the action of calcitonin on bone resorption was also observed in the absence of parathyroid hormone. Talmage and co-workers suggest that early effects of calcitonin are most logically explained by reduced calcium efflux from bone cells and bone fluid (secondary to increased phosphate influx), the effect on bone resorption requiring too extended a period of time to account for the rapid hypocalcaemic effect.

The production of hypocalcaemia by calcitonin in nephrectomized rats demonstrated that the urine is not a major route of loss of the calcium leaving the blood under the influence of calcitonin, and that the kidney is not in any other way an obligatory mediator of the hypocalcaemic effect of calcitonin in rats. Nevertheless it has been clearly demonstrated that calcitonin can produce increased excretion of electrolytes in humans and in experimental animals. Possible effects of parathyroid hormone and calcitonin on the renal tubule are shown in Figure 13.6. It can also increase the urinary excretion of sodium in rats, sheep, rabbits and man. In human beings, the effect is transient, urinary sodium returning to normal concomitantly with compensatory increases in plasma renin and aldosterone.

There is little evidence that thyroidectomy significantly affects urinary electrolyte excretion in a way that is specifically corrected by administration of physiological doses of calcitonin. In the absence of such evidence it is unlikely that the documented effects of administered calcitonin on the urinary excretion of phosphate, sulphate, sodium, potassium, calcium and magnesium are significant physiologically. The effect of calcitonin on the renal conversion of 25-hydroxycholecalciferol to 1,25-dihroxycholecalciferol may be a physiologically important one.

Calciferol (vitamin D) is the most important factor responsible for increasing the efficiency of calcium absorption in the gut; the significance of parathyroid hormone in this process is still uncertain. Gray and Munson showed that the thyroid gland, through calcitonin, protects against hypercalcaemia of dietary origin in rats.[20] It had been shown previously, in experiments with eviscerated rats, that the gastrointestinal tract was not essential for the hypocalcaemic effect of calcitonin. However, gastrin, pancreozymin-cholecystokinin, and certain synthetic analogues are powerful stimulants of calcitonin secretion and as such may be important components of the hormonal system which serves to protect against hypercalcaemia of dietary origin.

The observed simultaneous and parallel reduced levels of calcium and phosphate in the blood are consistent with the presumed basic action of calcitonin on bone to inhibit resorption, an action for which there is evidence in both *in vivo* and *in vitro* systems. However, inhibition of bone resorption may not adequately explain the effects on serum calcium and phosphate produced rapidly by an injection of the hormone or by a rapid increase in the endogenous secretion rate. Perhaps a more likely explanation is that calcitonin can cause a rapid transfer of inorganic phosphate out of blood and at the same time the major effect of calcitonin on calcium is to inhibit its entry into the plasma. The two effects may be interrelated, the excess phosphate tending to complex with calcium in the bone fluid and reduce calcium efflux. Calcitonin, depending on the dose level, can reduce the plasma concentrations of magnesium, inorganic sulphate, hydroxyproline and citrate.

The reduction of blood (and urine) hydroxyproline by calcitonin is consistent with the inhibitory effect of the hormone on bone metabolism. The effect on hydroxyproline has been demonstrated most convincingly in experimental animals in which the rate of bone metabolism has been increased by hyperparathyroidism.

2.4 Metabolism

Calcitonin disappears rapidly from plasma and it is metabolized quickly, in apparent contrast with the rather long duration of its hypocalcaemic effect. In all species studied, salmon calcitonin has a longer duration of action than porcine or other mammalian forms of calcitonin and it is this property of the salmon hormone as well as its high potency that has led to its choice for therapeutic use.

There is a correlation between the relative durations of action of salmon and porcine calcitonin, the rates of disappearance from plasma and the metabolic clearance rates, but it would not appear to be sufficient to explain the higher potency of salmon calcitonin, which has been observed with *in vitro* as well as *in vivo* systems.

Assessment of the rate of destruction of calcitonin by various organs by measuring arteriovenous differences in the dog showed that salmon calcitonin lost little activity by passage through liver or skeletal muscle and bone, whereas porcine calcitonin was rapidly metabolized. These observations in the dog *in vivo* are in good agreement with *in vitro* findings in the rat.

2.5 Interactions with parathyroid hormone

Calcitonin and parathyroid hormone are antagonistic in their effects on the concentration of serum calcium, the rate of transfer of calcium and phosphate from bone to blood and the rate of bone resorption.

The effectiveness of calcitonin in parathyroidectomized animals and hypoparathyroid patients indicates that the actions of calcitonin are not dependent on the presence of the parathyroid gland and are not necessarily due to an inhibitory effect on the secretion of parathyroid hormone. Parathyroid hormone itself has no direct effect on calcitonin-producing cells, their rate of secretion or on calcitonin itself. After thyroidectomy, because of the absence of calcitonin, parathyroid hormone has a greater effect on the serum calcium, the rate of transfer of calcium and phosphate from bone to blood and the rate of bone resorption. The physiological antagonisms of calcitonin and parathyroid hormone are the result of opposite effects on bone metabolism rather than on effects on their respective rates of secretion.

An antagonism between the effects of the two hormones on bone *in vitro* has also been demonstrated. The antagonism does not appear to be direct or competitive. Parathyroid hormone promotes bone resorption in the absence of calcitonin and calcitonin inhibits bone resorption in the absence of parathyroid hormone.

The effects of the two hormones on the renal handling of inorganic phosphate are not in opposition. Both hormones promote urinary phosphate excretion, the effect of parathyroid hormone is within the physiological range, whereas that of calcitonin may be confined to pharmacological dose levels. In general it may be concluded that parathormone is physiologically a far more important regulator of serum calcium than is calcitonin. Another instance of an antagonistic pair of hormones of which one is of major and one only minor importance may be seen in the interrelationship between insulin and glucagon (Chapters 9 and 11) in the regulation of blood glucose concentrations.

REFERENCES

1. Collip, B. J. (1925) *J. Biol. Chem.*, **63**, 395–438.
2. Habener, J. F., Kemper, B.W., Rich, A. and Potts, J. T. (1977) *Recent Progr. Hormone Res.*, **33**, 249–299.
3. Hamilton, J. W. and Cohn, D. V. (1969) *J. Biol. Chem.*, **244**, 5421–5429.
4. Berson, S. A. and Yalow, R. S. (1968) *J. Clin. Endocrinol.*, **28**, 1037–1049.
5. Copp, D. H. and Davidson, A. G. F. (1961) *Proc. Soc. Exp. Biol.*, **107**, 322–344.
6. Care, A. D., Sherwood, L. M., Potts, J. T. and Aurbach, G. D. (1966) *Nature*, **209**, 55–57.
7. Sherwood, L. M., Mayer, G. P., Ramberg, C. F., Kronfeld, D. S., Aurbach, G. D. and Potts, J. T. (1968) *Endocrinology*, **83**, 1043–1051.
8. Fischer, J. A., Blum, J. W. and Binswanger, U. (1973) *J. Clin Invest.*, **52**, 2434–2440.
9. Fischer, J. A., Binswanger, U. and Dietrich, F. M. (1974) *J. Clin. Invest.*, **54**, 1382–1394.
10. Copp, D. H. (1969) *J. Endocrinol.*, **43**, 137–161.
11. Pullman, T. N., Lavender, A. R., Aho, I. and Rasmussen, H. (1960) *Endocrinology*, **67**, 570–582.
12. Talmage, R. V. and Meyer, R. A. (1976). In: *Handbook of Physiology*, Section 7, Vol. 7 (Ed. Aurbach, G. D.). American Physiological Society, Washington, D.C., pp. 343–351.
13. Vaes, G. (1968) *Nature*, **219**, 939–940.
14. Wells, H. and Lloyd, W. (1969) *Endocrinology*, **81**, 139–144.
15. Potts, J. T. and Aurbach, G. D. (1976) In: *Handbook of Physiology*, Section 7, Vol. 7 (Ed. Aurbach, G. D.). American Physiological Society, Washington, D.C., pp. 423–430.
16. Fischer, J. A. *et al.* (1981) *Proc. Nat. Acad. Sci. USA*, **78**, 7801–7805.
17. Rodan, S. B. and Rodan, G. A. (1974) *J. Biol. Chem.*, **249**, 3068–3074.
18. Amara, S. G., Jonas, V., Rosenfeld, M. G., Ong, E. S. and Evans, R. M. (1982) *Nature*, **298**, 240–244.
19. MacIntyre, I., Hillyard, C. J., Murphy, P. K., Reynolds, J. J., Gaines Das, R. E. and Craig, R. K. (1982) *Nature*, **300**, 460–462.
20. Gray, T. K. and Munson, P. L. (1969) *Science*, **166**, 512–513.

Chapter 14

Erythropoietin, angiotensin, plasma kinins and related substances

1. GENERAL

The polypeptides described in this chapter are in general derived from larger proteins present in the circulation and in turn synthesized in organs such as the kidney or liver. They are cleaved from their parent proteins by enzymatic mechanisms, some of which, in the case of the angiotensins and the plasma kinins, are relatively complicated. Many of these small polypeptides are still not well characterized, although a number have been synthesized chemically. Some, like the kinins, may be predominantly active as local hormones. However, both erythropoietin and the angiotensins may be regarded as typical polypeptide hormones which produce effects at multiple sites.

2. ERYTHROPOIETIN

2.1 Historical

That tissue hypoxia might induce increased red cell production was first suggested at the end of the nineteenth century.[1] Much later it was shown that blood from an hypoxic rat could cause marrow hyperplasia when injected into a normal rat—implying that some circulatory factor was involved in the response to lack of oxygen.[2] Later still it was shown that plasma from anaemic animals could induce extra red cell production when administered to normal animals. The substance responsible for these effects was termed erythropoietin.

2.2 Isolation and properties

There have been several attempts to purify erythropoietin from plasma. These have often involved using the blood of animals made anaemic in which erythropoietin concentrations are greatly increased. A more than one million-fold purification has been achieved using the plasma of sheep made anaemic with phenylhydrazine.[3] The material so obtained has a molecular weight of 45 000 with a high (30%) carbohydrate content. The carbohydrate includes sialic acid. Erythropoietin is therefore a glycoprotein.

2.3 Site of origin of erythropoietin

It is now generally thought that erythropoietin is mainly produced in the kidney. Nevertheless, as with many other hormones, while one tissue may be the principal one associated with its production, it may also be made in other tissues, such as the liver. In the kidney, it now seems that erythropoietin release is associated with the glomerulus, although the exact site of its production is still in dispute. In support of this view, it has been shown that there is an association between fluorescent labelled antibodies to erythropoietin and the kidney glomerulus. It is thought that an enzyme called erythrogenin is produced in kidney tissue which converts a protein precursor in the blood to erythropoietin.[4] The system has obvious analogies with the angiotensin system with which the kidney is also associated.

2.4 Assay

A bioassay technique has customarily been used. This depended originally on red cell production in anaemic animals. More recently, the utilization of labelled iron (^{59}Fe) in starved rats has been used. Adequate immunoassays are now available.

2.5 Mode of action

Erythropoietin induces the proliferation and further differentiations of certain stem cells on the route to the fully formed red cell. Erythropoietin is known to stimulate iron uptake, i.e. to increase the formation of haemoglobin in red cells, and also to promote the release of reticulocytes from the marrow. The synthesis of rate-limiting enzymes (such as aminolaevulinic acid synthetase) may also be affected.

In addition, erythropoietin may directly affect the synthesis of red cell structural proteins. Quite possibly the primary effect of erythropoietin is on transcriptional processes in a primitive stem cell. Such a transcriptional effect has been seen with highly purified erythropoietin preparations.[5] The earliest changes that can be detected are an increased synthesis of messenger RNA in bone marrow cells. It is noteworthy that erythropoietin appears to interact with a tissue receptor in such cells in order to produce its regulatory activity.

2.6 Action at the cellular level

As with many other polypeptide hormones, it has frequently been suggested that erythropoietin is effective through second messengers.[6] Thus, the injection of cyclic AMP (or its dibutyryl derivative) into animals results in an increased red cell production. However, this effect appears to be blocked by an antiserum made against erythropoietin. Experiments designed to test directly the hypothesis that cyclic nucleotides mediate the effects of erythropoietin have frequently given contradictory results. It has generally been difficult to show that when either bone marrow cells or foetal liver cells are incubated with erythropoietin there is always a rise in tissue cyclic AMP. Suggestions have also been made that cyclic GMP is the responsible messenger, although this view is by no means generally accepted. More recently, it has been suggested that Ca^{2+} ions are necessary for target tissue to respond to erythropoietin.[7]

3. ANGIOTENSINS

3.1 Historical

A substance which increased blood pressure in animals was extracted from kidney homogenates by Tigerstedt and Bergman in 1898[8] and named renin by them. Later investigators, and especially Goldblatt[9] in the 1930s, showed that pressor material could be released into the circulation by restricting renal blood flow. The material produced by kidneys in this way was not hypertensive itself, but appeared to be an enzyme which acted upon a circulating factor called angiotensinogen to produce a rise in blood pressure. Several enzymatic products of angiotensinogen are now known (angiotensins I, II and III). Such materials have a far wider spectrum of biological activity than their effects as pressor agents. Thus, angiotensin II controls aldosterone release from the zona glomerulosa of the adrenal cortex and hence modulates salt and water metabolism. Angiotensin I may also increase catecholamine release from the adrenal medulla, as well as facilitating sympathetic nervous transmission. The renin-angiotensin system has been reviewed in detail elsewhere.[9a]

3.2   Production of angiotensins

The manner in which angiotensins are produced from angiotensinogen is shown in Figure 14.1. Angiotensinogen is a protein which is produced in the liver, from which it passes into the circulation. In the blood, angiotensinogen is acted upon by the enzyme renin, which specifically hydrolyses a Leu–Leu bond. The product termed angiotensin I is then further acted upon by a circulating carboxy-dipeptidase (converting enzyme) causing the loss of the two carboxyl terminal amino acids, yielding angiotensin II. An amino peptidase may then cleave the N-terminal aspartic acid so as to yield an active heptapeptide, angiotensin III.

3.3   Renin

The enzyme renin is produced in the juxta-glomerular apparatus of the kidneys, in which it is present in granular form and from which it is secreted into the plasma. The enzyme is a glycoprotein of molecular weight 37 000–43 000. There is some evidence to suggest that it is present as a biosynthetic precursor of molecular weight approximately 60 000, which possesses only slight biological activity. Such material, which has been termed 'big renin', appears to circulate in plasma.[10] The activation of the precursor depends on enzymatic proteolysis although there is some uncertainty about the nature of the enzymes involved.

The release of renin depends upon changes in arterial wall tension in the blood vessels supplying the glomerulus. In addition, the sodium level reaching the macula densa of the kidneys greatly influences renin discharge. A low plasma sodium promotes release. The renal nerve supply may influence renin discharge, particularly through the sympathetic system. The release of renin is also increased by posterior pituitary hormones. There is some evidence that renin release from the kidney is controlled by changes in cyclic AMP levels at its

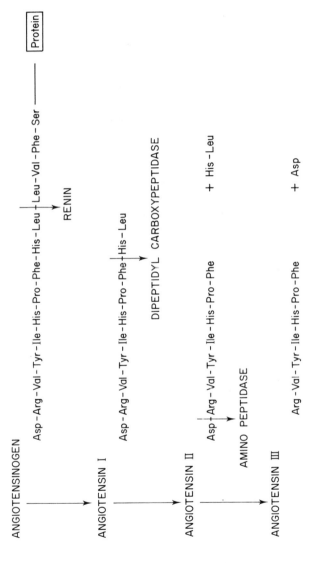

Figure 14.1 Production of angiotensins

site of production. Thus, derivatives of cyclic AMP (such as the dibutyryl derivative), glucagon and theophylline (a phosphodiesterase inhibitor), all increase renin output.

3.4   Angiotensinogen (renin substrate)

This glycoprotein—the substrate for renin—is produced by the liver, from which it is released into the blood. It is thought to have a molecular weight of approximately 50 000 (in pig liver), although a higher molecular weight material has been found in human liver. Its hepatic production is under the control of several other endocrine factors. These include glucocorticoids, oestrogens, ACTH and angiotensin II. The exact mechanism of angiotensinogen production is not yet well understood.

3.5   Mode of action of angiotensins

The broadest spectrum of biological activity is exhibited by angiotensin II, although both angiotensin I and angiotensin III show effects in their own right. Some of the most dramatic effects of the octapeptide are concerned with the cardiovascular system. Thus, it induces contraction of smooth muscle in vessel walls—perhaps through a translocation of Ca^{2+} in muscle cells. This effect produces a sustained rise in blood pressure, It also affects cardiac musculature directly through specific receptors so as to produce a positive inotropic effect.

3.5.1   *Effects on the adrenal cortex*

It was first shown in 1960 that angiotensin II stimulates aldosterone release.[12] This appears to be a direct effect of the octapeptide on the zona glomerulosa of the adrenal cortex. The effect has been shown in perfused adrenals as well as in adrenal cell suspensions. It is thought that renin increases the rate of conversion of cholesterol to pregnenolone, perhaps through increasing local cyclic AMP concentration—although this is far from certain. Protein synthesis may also be involved, since aldosterone synthesis due to angiotensin II is blocked by various protein synthesis inhibitors. The zona glomerulosa is also sensitive to angiotensin III.

The general effects of lowering plasma sodium will therefore be to increase aldosterone secretion indirectly as a result of an increase in renin release.

3.5.2   *Other effects of angiotensin II*

Angiotensin II has also been shown to possess direct effects on the central nervous system so as to increase blood pressure. It may also increase the release of the hormone vasopressin.

3.5.3 Structure-function relationships

Structure–function relationships have been investigated in considerable detail for the angiotensins, and many analogues have now been prepared. For the effects of angiotensin II to be exerted on blood pressure, the minimal chain length is the carboxy-terminal hexapeptide—although this structure has only 1% of the activity of the full octapeptide. The aromatic amino acids in the structure appear to be completely essential. Modification of the C-terminal phenylalanine appears to block the angiotensin receptor. Such modified angiotensins have been used extensively in pharmacology as angiotensin antagonists.

3.6 Assay

Angiotensin I, angiotensin II, renin and the converting enzyme may now all be measured by immunoassays which have superseded the older bioassay techniques.[13] Using such methods, these substances have recently been detected in cells from the central nervous system.

4. SUBSTANCE P AND THE TACHYKININS

Von Euler and Gaddum[14] first described material present in extracts of intestinal mucosa which produced contraction of jejunal musculature and reduced the blood pressure in rabbits. This substance was termed substance P. Substance P is found in peripheral nerves as well as in nerves supplying endocrine glands. In its biological activity it may closely resemble the effects of cholinergic drugs.[15] Substance P is a neurotransmitter. A number of closely related substances with similar properties have now been described. These are found both in nervous tissue in invertebrates (e.g. eledoisin, in octopus salivary glands and physalaemin in the skin of toads). All these substances are undecapeptides and contain a C-terminal, Gly–Leu–Met, with phenylalanine in position 5 of the chain.

5. THE PLASMA KININS[16]

The term kinin refers to short peptides derived enzymatically from circulating globulins. Such peptides have powerful effects in causing dilatation of small blood vessels and in increasing capillary permeability. The parent substances which are thought to be globulins are termed kininogens. Two substances have so far been well characterized among this group of peptides. These are:

(1) bradykinin—a nonapeptide, and

(2) kallidin—a decapeptide (which possesses an extra lysine on the N-terminus of the bradykinin molecule).

The structure of these is shown in Figure 14.2.

Arg-Pro-Pro-Gly-Phe-Ser-Pro-Phe-Arg

Bradykinin

Lys-Arg-Pro-Pro-Gly-Phe-Ser-Pro-Phe-Arg

Kallidin

Figure 14.2 Structure of bradykinin and kallidin

The formation of these substances in plasma is very rapid and may take place when blood is in contact with foreign surfaces such as particulate matter or with damaged tissues. It may involve an enzymatic cascade process similar to those observed in blood-clotting. Such substances may be considered as local hormones, perhaps of particular importance in the response to tissue injury. The enzymes involved in converting kininogens to kinins include the kallikreins, a complex family of 25–30 homologous serine proteases.[17]

REFERENCES

1. Miescher, F. (1893) *Cor. Bl. Schewerz-Artze,* **23**, 809.
2. Reissmann, K. P. (1950) *Blood,* **5**, 372.
3. Goldwasser, E. and Kung, C. K. H. (1971) *Proc. Nat. Acad. Sci. USA,* **68**, 697–698.
4. Fisher, J. W. (1972) *Pharmacol. Rev.,* **24**, 459–508.
5. Weiss, T. L. and Goldwasser, E. (1981) *Biochem. J.,* **198**, 17–21.
6. Graber, S. E., Bomboy, J. D. Jr., Salman, W. D. Jr and Krantz, S. V. (1972) *J. Lab. Clin. Med.,* **90**, 162–170.
7. Misiti, J. and Spivak, J. L. (1979) *J. Clin. Invest.,* **64**, 1573–1579.
8. Tigerstedt, R. and Bergman, P. G. (1898) *Skand.-Archiv. Physiol.* **8**, 223.
9. Goldblatt (1938) *Harvey Lectures,* **33**, 237.
9a. Gibbons, G. H., Dzau, V. J., Farhi, E. R. and Barger, A. C. (1984) *Ann. Rev. Physiol.* **46**, 291–308.
10. Day, R. P., Luetscher, J. A. and Zager, P. G. (1976) *Amer. J. Cardiol.,* **37**, 607.
11. Allisson, D. J., Tanigawa, M. and Assaykeen, T. A. (1972) In: *Control of Renin Secretion* (Ed. Assaykeen, T. A.). Plenum, New York, p. 33–47.
12. Laragh, J. M., Angers, M., Kelly, W. G. *et al.* (1960) *J. Amer. Med. Assoc.,* **174**, 234.
13. Fishman, M. C., Zimmerman, E. A. and Slater, E. E. (1981) *Science,* **214**, 921–923.
14. Von Euler, U. S. and Gaddum, J. H. (1931) *J. Physiol.,* **72**, 74.
15. Erspama, V., Falconier-Erspamer, G. and Linari, G. (1977) In: *Substance P* (Eds. von Euler, U. S. and Pernow, B.). New York, Raven Press, pp. 67–74.
16. Roch e Silva (1977) In: *Chemistry and Biology of the Kallikrein–Kinin System in Health and Disease,* Dept. of Health, Education and Welfare. Publication No. (N.I.H.) 76–291. U.S. Govt. Printing Office, Washington D.C. 1977, pp. 7–15.
17. Mason, A. J., Evans, B. A., Cox, D. R., Shine, J. and Richards, R. I. (1983) *Nature,* **303**, 300–307.

Chapter 15

Structure–function relationships among the polypeptide hormones

1. INTRODUCTION

To the organic chemist the structure and chemistry of a naturally occurring protein or polypeptide is of limited interest. The interest of such compounds lies mainly in their biological activities. A major goal of the biochemist studying a protein or polypeptide hormone must therefore be to explain how a molecule of rather unremarkable chemical properties is able to bring about specific and often very remarkable biological effects, and to explain what features of the structure of the hormone are responsible for these actions. Some general features of the study of such structure–function relationships will be discussed in this chapter, and their application in the case of a few families of polypeptide hormones will be presented.

The crucial event in the action of a hormone on its target tissue is the interaction with its cellular receptor. Receptors were postulated as mediators of drug action by early pharmacologists—they were thought necessary to explain the specificity and the amplification seen in drug action. The concept was adopted by endocrinologists, and recent studies on the molecular nature of receptors have confirmed, amplified and refined the early postulate (see Chapter 17).

Binding of hormone to receptor in the target tissue is now recognized as the first event in the action of the hormone. Subsequent events are probably largely a consequence of the activity of the 'activated' hormone–receptor complex, and once this has been formed the structure of the hormone itself may be relatively less important. Study of the relationship between the structure of a hormone and its function thus becomes primarily study of the features of the structure of the hormone which determine binding to (and activation of) the receptor. Nevertheless, it must be remembered that many features other than binding to the receptor are of importance in determining the overall function of a hormone, including storage and secretion, transport and stability in the circulation, eventual removal from the circulation and possibly, in some cases, entry into the target cell. All of these factors must be considered when structure–function relationships are being considered.

2. THE INVESTIGATION OF STRUCTURE–FUNCTION RELATIONSHIPS

Ideally the investigation of structure–function relationships in this field would be centred on a detailed study of the three-dimensional structure of hormone–receptor complexes. In practice, however, detailed knowledge of the structure of polypeptide–hormone receptors is very limited and cannot yet form the basis for structure–function studies. Alternative approaches have to be used, and these are mainly based on studies with hormone analogues—involving use of either naturally-occurring variants or artificial analogues made in the laboratory.

2.1 Naturally-occurring analogues

Polypeptide hormones show a great deal of species variation.[1,2] In some cases variation of structure is accompanied by variation of biological activity, while in others a good deal of structural variation can be accommodated without marked variation in biological actions. Comparison of species variants of a particular hormone using several different assays can provide information about the features of that hormone which are (or are not) necessary for biological (or other) activity.

In addition to variants of hormones arising from species differences, structurally related hormones may exist in a single organism (or species). This is a consequence of the occurrence of polypeptide hormones in families—a theme which has recurred frequently in this book. Comparison of the properties of such structurally related hormones in a range of assays can, again, provide information about structure–function relationships. Thus, comparison of the properties of the closely related peptides oxytocin and vasopressin provides valuable information about the relation of structure to function in the neurohypophysial hormones (see Section 4).

A third type of naturally-occurring variant which may throw light on structure–function relationships is that arising as a consequence of genetic mutation. Analysis of the effects of specific, small mutations on biological and other properties has been extremely valuable in delineating the relation of structure to function in the case of some proteins and polypeptides (analysis of mutations in haemoglobin perhaps provides the most elegant example[3]) but has proved relatively less important as a way of studying polypeptide hormones. Nevertheless, the analysis of some rare mutations, as well as some which occur with quite high frequency, has provided valuable information, as will be exemplified later in this chapter.

2.2 Chemical and enzymic modification

Alteration of the structure of a polypeptide by chemical or enzymic means can provide analogues whose biological and other properties may be altered. This type of approach has yielded valuable information about the relation of structure to function in many instances. A wide range of methods is available for specifically modifying proteins,[4,5] although the method is limited, of course, to the more reactive side-chains. Amino acids with unreactive side chains (such as valine, leucine, phenylalanine, etc.) cannot normally be modified without drastically altering other features of the structure of the polypeptide. Chemical modification can therefore provide little information about the (often considerable) importance of such amino acids in the functioning of a polypeptide.

Limited cleavage by proteolytic enzymes can provide fragments of a polypeptide for structure–function studies and may provide information about the particular region(s) of a polypeptide in which activity lies.

2.3 Synthetic analogues

Peptide chemists are able to synthesize polypeptide hormones of moderate size and have had some success in the chemical synthesis of small proteins. Complete synthesis of analogues for structure–function studies has been mainly confined to fairly small peptides, but here an enormous amount of work has been carried out, with considerable success. The method allows, in principle, the construction of *any* analogue of a peptide, so here alteration of unreactive amino acid side-chains is as easy as alteration of reactive ones. Even for a small peptide the number of possible analogues which can be constructed is extremely large, and prediction of which ones are likely to be most interesting is difficult. Thus, in the case of the oxytocin–vasopressin family many hundreds of synthetic analogues have been described, but the possibilities are far from exhausted.

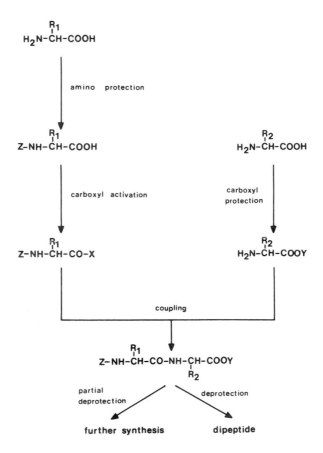

Figure 15.1 Scheme illustrating the chemical synthesis of a dipeptide using the 'classical' approach

The principles of peptide synthesis[6,7]

Peptide synthesis now comprises a substantial branch of organic chemistry, but no attempt will be made to present a detailed account of the methodology here. A few of the main principles will be summarized, however. Two main approaches have been developed for chemical synthesis of peptides: (1) the classical approach which involves carrying out reactions in solution (usually in organic solvents) and isolation and purification of intermediates, and (2) the solid state approach of Merrifield[7] in which amino acid derivatives are attached sequentially to a growing peptide chain attached to a solid resin support. In general the former method is more rigorous but more laborious, and the latter is quicker, readily automated, but liable to produce heterogeneous products.

The principles involved in the 'classical' type of peptide synthesis are illustrated in Figure 15.1 for the synthesis of a dipeptide. In order to link the two amino acids together in a specific fashion, the amino terminus of one and the carboxyl terminus of the other are 'blocked' by specific groups. Reactive side-chains (amino groups, sulphydryl groups, etc.) also have to be blocked to prevent them reacting during coupling. The blocked amino acid derivatives are then coupled together to form a peptide bond. Blocking groups are then removed to give an unmodified dipeptide, or, if the dipeptide is to be used as the basis of further peptide synthesis, the amino or carboxyl terminal blocking

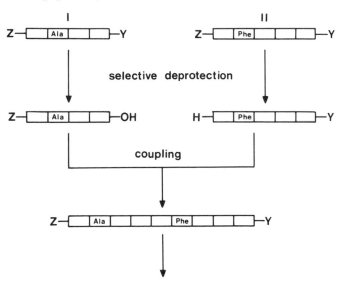

Figure 15.2 Scheme illustrating the chemical synthesis of an octapeptide from two 'protected' tetrapeptides, using the 'classical' approach

group is selectively removed and the dipeptide derivative is coupled specifically to another amino acid or peptide derivative. The intermediates produced at each stage are normally purified and characterized in order to remove unwanted products of side reactions and to ensure that the synthesis continues in the specific direction desired. A large peptide is produced by coupling together smaller ones, as illustrated in Figure 15.2.

It is clear from this brief summary that essential features of an effective peptide synthesis are suitable coupling agents and suitable blocking agents, which can be readily and selectively removed when necessary. Advances in

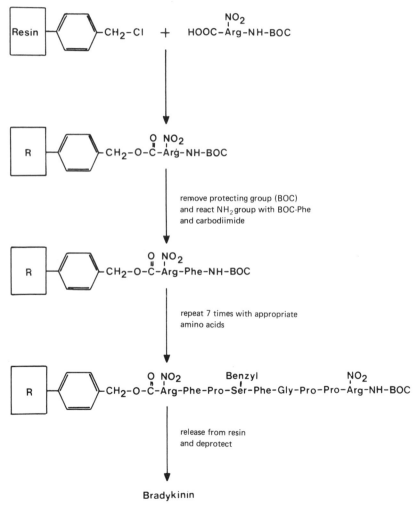

Figure 15.3 Diagram illustrating the chemical synthesis of bradykinin, using the solid phase method

peptide synthetic chemistry over the past 30 years have been largely dependent upon the development of such reagents. An important feature of the successful reagents and reaction schemes is that they avoid racemization; most biologically active peptides contain only L amino acids, and partial conversion to the D isomer leads to loss or modification of activity and must be carefully avoided.

Use of the solid phase method for peptide synthesis is illustrated in Figure 15.3. The blocking groups and coupling procedures are similar to those used for the classical approach, but now the first (usually C-terminal) amino acid of the peptide is attached to a resin support, and amino acid derivatives are attached to it sequentially. After each reaction step, excess reagents are readily removed by simply washing the resin, so extensive purification of the intermediates is obviated (and, indeed, impossible). When synthesis is complete, the peptide is cleaved from the resin and side-chain blocking groups are removed chemically. The method is readily automated, reagents being delivered to a column or chamber containing the resin in a pre-programmed sequence. A disadvantage of the method is that incomplete reaction at any step (or premature termination)

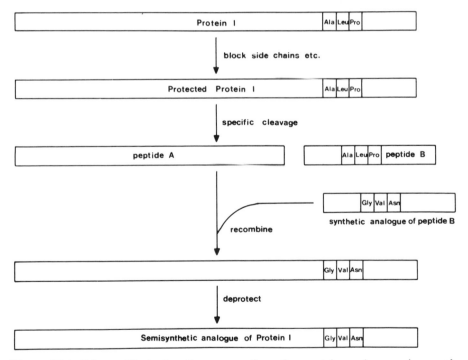

Figure 15.4 Scheme illustrating the preparation of a protein analogue using semi-synthesis. Protein I is modified to block reactive groups and then cleaved specifically to give a large peptide A, and a small peptide, B. A and B are separated, and peptide A is then recombined with an analogue of peptide B prepared by chemical synthesis. After deprotection, a semisynthetic analogue of protein I results

introduces heterogeneity into the structure of the polypeptide which is very difficult to eliminate from the final product. Nevertheless, the method has proved
very valuable for the synthesis of small and medium-size peptides, and more
recent technical developments suggest that it will become increasingly useful.

2.4   Peptide and protein semisynthesis[8]

Despite the progress made in total chemical synthesis of peptides, synthesis of a
large peptide or small protein remains a formidable task, and is impracticable
for the production of large numbers of analogues. An alternative approach has
therefore been developed which involves combining a fragment of a naturally-
occurring peptide with a fragment produced by peptide synthesis. The approach
has been termed 'protein semisynthesis' and is illustrated in Figure 15.4. It has
the advantage that an analogue of a protein can be made which involves
relatively little synthetic work. The disadvantage is, of course, that one is dependent upon the natural protein for major fragments, and the ways in which these
can be obtained are often rather limited. Use of the semisynthetic approach will
be illustrated by reference to insulin (Section 6).

2.5   Use of genetic engineering to prepare hormone analogues

In the past few years it has become clear that the new techniques of genetic
engineering allow transfer of a 'gene' for a polypeptide hormone to a bacterium.
Expression of the gene in its new host allows the bacterium to produce the corresponding hormone (see Chapter 18). The approach has now been successfully
applied to several naturally-occurring hormones, and there seems no reason to
suppose that it cannot be applied also to the production of hormone analogues—both fragments and analogues with altered amino acid sequence. For
example, the peptide hormone somatostatin has been produced in bacteria,
using a completely synthetic 'gene'. Analogues of somatostatin could be produced in the same way using DNA sequences slightly different from that used to
produce the natural hormone (i.e. substituting the codon for one amino acid by
that for another—site-directed mutagenesis). The approach has not yet been
widely applied, but there seems every reason to suppose that it will be successful
within the next few years. It can only be used to produce analogues containing
the 'common' amino acids (i.e. the 20 amino acids found in proteins), but this
still leaves great scope for variation.

2.6   Affinity labelling of receptors; cross-linking studies

The technique of affinity labelling has been extensively used to study binding of
ligands to proteins, particularly substrate analogues to enzyme-active sites, antigens to antibodies and neurotransmitters to receptors. It has recently been successfully applied to covalently bind polypeptide hormones to their receptors.

The principle of the method is that a derivative of the ligand is prepared which contains a reactive group ('warhead'). When the ligand binds to its receptor site, the reactive group reacts with a group on the receptor, thereby covalently linking the ligand to the receptor. Subsequent study of this covalently linked complex allows information to be obtained about the nature of the ligand–receptor binding, etc. In some cases the 'warhead' is only potentially reactive, and is activated (for example by irradiation with UV light) only when it has been allowed to bind to the receptor;[9] this reduces the chances of the warhead reacting with another protein before the ligand has associated with the receptor. The use of this approach to study interaction of insulin with its receptor is discussed in Section 6.

Affinity labelling allows covalent linking of ligand and receptor. Alternative approaches to such cross-linking have also been developed. If a labelled hormone is incubated with a receptor preparation, and the resultant hormone–receptor complex is treated with a bivalent reagent which reacts with side-chains on proteins, such as the amino group of lysine, the hormone is often covalently cross-linked to its receptor. The labelled hormone–receptor complex can then be characterized by physicochemical techniques, especially polyacrylamide gel electrophoresis in the presence of a detergent.

2.7   Use of monoclonal antibodies in structure–function studies[10,11]

Another technique which has been only recently developed, and so far relatively little applied, but which possesses great potential for structure–function studies, uses monoclonal antibodies. These are homogeneous, monospecific antibodies which are synthesized by cloned hybridoma cell lines produced by fusing antibody-producing cells from an immunized animal (usually a mouse or rat) with cultured myeloma cells.[10] Each clone of cells produces a single, homogeneous, monospecific antibody. Such an antibody is quite different from that produced normally by immunizing an animal directly, which is invariably highly heterogeneous (polyclonal), containing many different antibody species, with different affinities and antigen specificities.

Such monoclonal antibodies can be used as highly specific reagents to probe the structure of the original hormone antigen. By investigating the nature of hormone–antibody complexes using several different monoclonal antibodies, the nature of the antigenic determinants on the hormone can be defined. Furthermore, by studying whether the interaction of the antibody with the hormone blocks biological or receptor-binding activity, it is possible to obtain information about the possible involvement of a given region of the hormone molecule (i.e. a region corresponding to a particular antigenic determinant) in biological activity. For example, use of monoclonal antibodies has shown the existence of several distinct antigenic determinants in human growth hormone, only some of which interact with the closely-related human placental lactogen.[12]

Some of these antibodies completely block binding of hormone to receptor, whereas others do not.[13]

Monoclonal antibodies have many other potential uses in hormone studies, which should also be mentioned here. Since each clone produces a single antibody, it is possible to produce a highly homogeneous and specific antibody even if a heterogeneous antigen preparation is used for immunization. Many different clones may have to be screened to find one that produces an antibody of interest. Once this has been found it will provide a specific antibody which can then be used for many purposes, including radioimmunoassay, further purification of the original antigen (by affinity chromatography) and biological studies.

3. THE VALUE OF STRUCTURE–FUNCTION STUDIES AND OTHER APPLICATIONS OF HORMONE ANALOGUES

It will be clear from the foregoing that most of the studies on the relationship of structure to function in polypeptide hormones have been carried out using hormone analogues. Before going on to deal with specific examples it will be worth considering some general aspects of the use of such analogues.

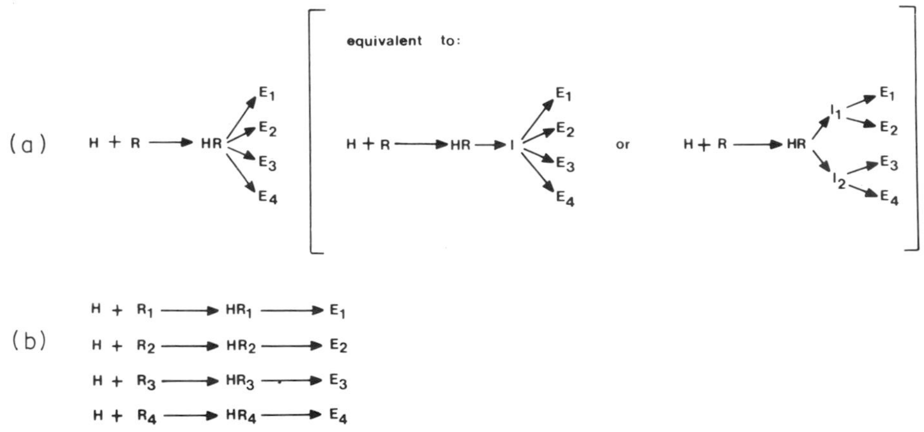

Figure 15.5 Diagram illustrating how multiple actions of a hormone may be mediated by a single receptor or multiple receptors. In (a), interaction of a hormone with a single receptor leads to several different effects (E_1, E_2 etc.), with or without the mediation of intermediates (I, I_1, I_2). In (b), interaction of the hormone with different receptors (R_1, R_2 etc.) brings about different effects (E_1, E_2 etc.)

3.1 Multiple actions of hormones

Most polypeptide hormones have more than one biological action. In considering such multiple actions a fundamental question which can be asked is: Are all the different actions of the hormone mediated by the same receptor, or is more than one type of receptor involved? (Figure 15.5). Use of analogues of the hormone can often resolve this problem. Thus, if only a single receptor is involved, and assuming that the actions of the hormone on the responding cell are mediated entirely by the hormone–receptor complex, then an analogue of the hormone should have similar (altered) actions on each of the hormonal effects—if Effect 1 is reduced by 50%, the other effects should be reduced by the same amount (Figure 15.5). On the other hand, if the various effects of the hormone are mediated by different receptors, an analogue may have quite different actions on these effects, because it may interact quite differently with the different receptors. Thus, Effect 1 may be reduced by 50% while the others are unchanged or even increased.

In practice, of course, many other factors may be involved in hormone action in addition to those shown in the simple models of Figure 15.5. Nevertheless, this simplified approach can be quite a useful one, as will be illustrated with regard to the neurohypophysial hormones in the next section.

3.2 Cinical applications of hormone analogues

Although analogues of polypeptide hormones are valuable for structure–function studies, much of the extensive work carried out on them has been motivated by the search for new drugs, with new medically applicable properties. This stems largely from the fact that many hormones do have multiple actions—often including unwanted side effects. Thus, somatostatin as well as inhibiting release of growth hormone also has many other effects, including inhibition of insulin secretion and effects on other physiological processes (Chapter 3). Although the hormone is potentially of great value as an inhibitor of growth-hormone release in conditions where oversecretion presents a serious medical problem, for example acromegaly, the associated other effects severely limit its usefulness for such purposes. Consequently, the search has been underway for derivatives of somatostatin in which effects other than those on growth-hormone release are reduced. These efforts have been rewarded by the discovery of a derivative (Phe^4–somatostatin) which appears to be an almost completely selective growth-hormone release-inhibiting hormone.[14]

In addition to such analogues in which the balance of the various hormonal actions is altered, some derivatives have been described which inhibit the actions of the natural hormone. These presumably bind to the receptor but fail to induce the conformational or other changes which normally follow from hormone–receptor binding and which serve to bring about the effects of the hormone on the target cell. Binding of the analogue inhibits the action of the natural hormone by preventing the latter from occupying the binding site of the receptor—a situation analogous to inhibition of an enzyme by a competitive inhibitor. Clearly, such inhibitory analogues are of considerable potential use both for experimental studies and clinical work. They can be used to inhibit the unwanted actions of a hormone that is produced in excessive amounts by a tumour or hyperactive endocrine gland. Examples of analogues which act as inhibitors will be given in the next few sections.

4.   NEUROHYPOPHYSIAL HORMONES[15,16]

As has been discussed already in Chapter 4, oxytocin and vasopressin and their relatives in lower vertebrates possess a wide range of biological and pharmacological activities, and many biological assays based on these have been devised. A large number of analogues of these hormones have been prepared, and their activities in these various assays have been studied and provide information about structure–function relationships. Indeed, it is probably in this peptide hormone family that the use of analogues to study such relationships has been most fully investigated.

Table 15.1   Activities of some analogues of oxytocin and vasopressin in various bioassay systems

| | Vasopressin assays (potency relative to arginine vasopressin) | | Oxytocin assays (potency relative to oxytocin) | |
| | Vasopressor | Antidiuresis | Uterus contraction *in vitro* | Milk ejection |
Peptide	(rat)	(rat)	(rat)	(rabbit)
Arginine vasopressin	1.00	1.00	0.033	0.144
Oxytocin	0.013	0.013	1.00	1.00
Lysine vasopressin	0.67	0.63	0.011	0.133
Arginine vasotocin	0.61	0.63	0.26	0.47
Isotocin	0.0002	0.0005	0.33	0.67
Des-amino arginine vasopressin	0.93	3.25	0.060	0.178
Des-amino oxytocin	0.003	0.038	1.67	1.19
Phe^3 arginine vasopressin	0.31	0.88	0.0004	0.007
Phe^2 lysine vasopressin	0.138	0.050	0.0007	0.006
Phe^2 oxytocin	0.001	0.001	0.071	0.31
Phe^2 arginine vasotocin	0.31	0.27	0.004	0.022
Phe^2 lysine vasotocin	0.08	0.003	0.002	0.027

Note: For each assay system, the potency relative to that of the appropriate 'parent' peptide is shown. Note that for any one analogue potencies in the two vasopressin assays may differ markedly, as may potencies in the two oxytocin assays.

4.1 Naturally-occurring analogues of oxytocin and vasopressin

The exsistence of a number of naturally-occurring analogues of oxytocin and vasopressin in lower vertebrates has already been discussed in Chapter 4 (Figure 4.2) and their activity in various assay systems has also been considered (Table 15.1). This data provides some useful information about structure–function relationships. Thus, comparison of the various natural analogues suggests strongly that the presence of arginine or lysine (or presumably other basic amino acid residues) at residue 8 is necessary for a peptide to have high activity in vasopressin assays (assays utilizing antidiuresis or elevation of rat blood pressure—see Chapter 4). (The residue numbers and structure of peptides in this family can be seen in Figure 4.2. or Figure 15.6). Arginine at residue 8 gives a higher potency in such assays (using the rat) than lysine, although in the pig, lysine vasopressin, which occurs there naturally, is more active than arginine vasopressin. The structural requirements for oxytocin-like activity (measured in assays using uterus contraction, milk ejection or *depression* of blood pressure in the fowl) are less critical than those for vasopressin—even vasopressin shows substantial activity in oxytocin assays (Table 15.1).

It is also clear that the various oxytocin-like (and vasopressin-like) actions can vary relative to one another in the natural analogues. Thus, the ratio of uterine contraction : milk ejection activity for oxytocin is 1 : 1 (by definition, because of the way the international unit was first defined), but for lysine vasopressin this ratio is 1 : 12 (Table 15.1). This has considerable significance in relation to the models of hormone–receptor binding considered in Section 3.1. It will be discussed further below, when synthetic analogues of oxytocin and vasopressin have been considered.

A final important conclusion to be drawn from consideration of the natural analogues is that for activity on water or salt transport in amphibia, the structural features seen in vasotocin are particularly crucial; alteration at residue 3 *or* 8 very drastically reduces the activity of this hormone.

4.2 Chemical modification of oxytocin and vasopressin

Relatively little work has been carried out on derivatives produced by direct modification of the natural hormones, probably because chemical synthesis of analogues has proved so convenient and successful. One important early study showed that reduction of the disulphide bridge linking residues 1 and 6 (Figure 15.6) in oxytocin or vasopressin abolishes virtually all biological activity.

4.3 Synthetic analogues of oxytocin and vasopressin[15,16]

As has been mentioned already, an enormous amount of work has been done here—hundreds of analogues have been synthesized in the laboratories of du Vigneaud, Berde, Boissonas, Schwartz, Rudinger, Walter, Sawyer and others. It

is not possible to present this work in detail here, but some of the most important results will be summarized.

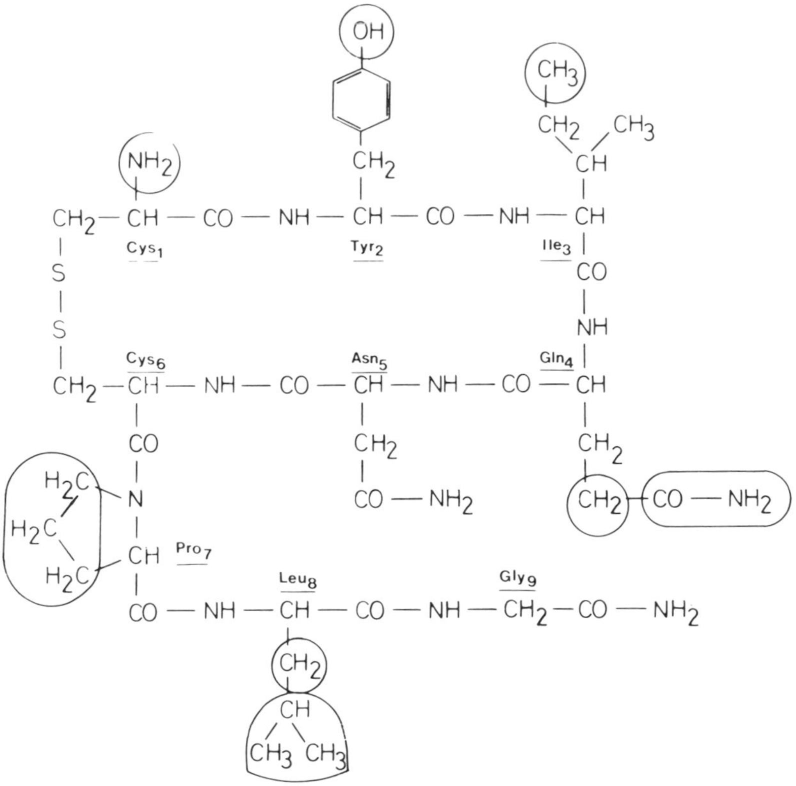

Figure 15.6 The structure of oxytocin. Each of the groups that is encircled
can be removed (by preparation of a synthetic analogue) without complete
loss of activity—at least 10% of the activity of the natural hormone is
retained

4.3.1 The disulphide bridge (ring)

The disulphide bridge of these peptides completes a ring comprising 20 atoms
(Figure 15.6). Increasing the size of this ring by inserting extra amino acids (or
even an extra methylene group) abolishes all biological activity for both oxytocin and vasopressin. Some of these analogues act as inhibitors of the natural
hormones. However, replacement of one sulphur atom by a methylene ($-CH_2-$)
allows retention of some activity for both oxytocin and vasopressin. This is an
important result because it rules out the possibility that a major feature in the
action of the hormones involves disulphide interchange between (and disulphide
bridge formation with) hormone and receptor; such a mechanism for binding of

hormone to receptor was proposed in the early 1960s. An analogue of oxytocin in which both sulphur atoms are replaced by selenium is *more active* than natural oxytocin. These various analogues involving alterations to the nature and size of the 20-atom ring suggest that the size of this ring is of crucial importance, but the nature of the individual members in it is less important.

4.3.2 The amino terminus

Acylation (including acetylation) of the α-amino group of oxytocin or vasopressin reduces the activity to less than 1% of that of the natural hormones. Some acylated analogues of this kind act as inhibitors of the natural hormones. Methylation of the α-amino group also causes marked loss of activity. However, derivatives in which the α-amino group is completely removed retain high activity, and in some assays activity is actually *increased* (Table 15.1). This increase in activity may be because these analogues have an increased stability in the circulation (some enzymes which degrade oxytocin and vasopressin are aminopeptidases, attacking at the α-amino group). It is clear therefore that the free α-amino group is not required for activity. Acylation or methylation presumably lead to loss of activity not because they block a crucial residue, but because the addition of a bulky group to the amino terminus hinders binding for steric reasons.

Although des-amino oxytocin (like des-amino vasopressin) possesses enhanced biological activity, it is completely unable to bind to the binding-protein neurophysin (Chapter 4). This binding apparently involves electrostatic interaction between the protein and the α-amino group of the peptide. The high activity of the des-amino peptides thus also confirms that interaction between neurophysin and the peptides is not necessary for activity at the target cell.

4.3.3 Shortening of the 'tail' (residues 7–9)

Analogues of oxytocin in which any of residues 7–9 is removed show very low activity, although alterations of individual residues may in some cases be accommodated with much less loss of activity.

4.3.4 Changes at residue 2

A few derivatives in which residue 2 is altered from tyrosine to phenylalanine will be considered as representative of the hundreds of analogues which have been studied. In all cases this substitution causes the activity in most assay systems to fall, but substantial activity is retained (Table 15.1). Milk ejection and pressor activity are favoured, in most cases, relative to uterotonic and antidiuretic activity. Lysine vasopressin with phenylalanine substituted for tyrosine at residue 2 is a derivative of some clinical importance, since its relatively high vasopressor activity (compared with antidiuretic activity) is much more marked in man. The analogue can be used to induce local constriction of blood vessels, for example to reduce local bleeding during surgery; the relatively low antidiuretic activity reduces the attendant, unwanted side-effects.

4.3.5 Changes in other residues

The results obtained from large numbers of analogues in which individual residues are substituted for oxytocin are summarized in Figure 15.6. The groups which are circled in this figure represent those which can be removed with retention of substantial biological activity (at least 10% that of natural oxytocin). A similar diagram could be constructed for vasopressin.

4.4 Conclusions regarding structure–function relationships in the neurohypophysial hormones

From the many studies on analogues of these hormones, some general conclusions can be drawn

(a) No individual group (amino acid side-chain) appears to be essential, with the possible exception of asparagine-5. This is particularly apparent from the summary presented in Figure 15.6. Rather, it seems that the overall conformation of the molecule is vital (hence alteration of the size of the 20-membered ring, or the 'tail' comprising residues 7–9, completely abolishes activity) but individual side-chains are less essential (although most make some contribution to the binding). The analogues also suggest that there is no covalent interaction between hormone and receptor.

(b) The relative activities of analogues in different assays vary widely. Some extremes are illustrated in Table 15.2 with regard to vasopressin assays. Although vasopressin is active in both the rat vasopressor assay and the antidiuretic assay, derivatives can be found in which activity is confined almost

Table 15.2 Activities of some analogues of oxytocin and vasopressin in two different vasopressin bioassay systems

Analogue	Assay system (potency in international units/mg)		Ratio of potencies Vasopressor: antidiuretic
	Vasopressor (rat)	Antidiuretic (rat)	
Orn[8] Phe[2] oxytocin	120	0.6	200 : 1
Arginine vasopressin	400	400	1 : 1
Phe[3] oxytocin	3	30	1 : 10
O-methyl-Tyr lysine vasopressin	2	80	1 : 40

entirely to one or the other of these assays. Similar variable ratios can be demonstrated for the oxytocin assays, but here the test animals vary, and the results are therefore, perhaps, less convincing. The fact that the ratio for the two vasopressin assays can vary so widely suggests that the receptors in the kidney (antidiuretic actions) and blood vessels (pressor actions) are different. It seems likely that a similar conclusion can be drawn for the different actions of oxytocin. This provides a nice example of the use of analogues to discriminate between the multiple actions of a hormone (see Section 3.1).

(c) Several of the synthetic analogues with low activity inhibit the actions of the natural hormones. Presumably they compete for receptor sites, but the receptor–analogue complex is not active. Such inhibitory analogues are potentially useful as drugs for treatment of patients with overproduction of particular hormones (for example, overproduction of vasopressin frequently occurs because of its production by ectopic tumours).

(d) Several of the synthetic analogues have properties which make them useful for clinical purposes. The example of Phe^2 lysine vasopressin has already been mentioned. Another analogue used clinically is 1-des-amino-8-D-arginine vasopressin, which has a high ratio of antidiuretic : vasopressor activity and also a prolonged activity (lack of the α-amino group and presence of D-arginine leads to resistance to enzymic degradation in the circulation). This derivative has been used for the treatment of diabetes insipidus.

5.   THE CORTICOTROPIN–MELANOTROPIN FAMILY[17–19]

As explained in Chapter 5, corticotropin, melanotropin and the endorphins and enkephalins form a family of peptides, several of which derive from a common precursor. A considerable amount is known about structure–function relationships in this group of peptides. Attention here will be concentrated on the melanotropic and corticotropic activities.

5.1   Melanotropic activity

Although the α and β melanotropins are the most active peptides in assays involving stimulation of melanophore darkening in amphibian skin (see Chapter 5), corticotropin itself contains significant melanotropic activity (about 1% that of α-MSH). This presumably reflects the fact that corticotropin contains the complete sequence of α-MSH (see Figure 5.7), but without an acetylated amino terminus and with a long polypeptide chain instead of the C-terminal amide of α-MSH. Clearly, these last two features account for its much lower activity than α-MSH.

Chemical modification and chemical synthetic studies have defined in some detail the features of these peptides that are necessary for melanotropic activity.

Table 15.3 Melanotropic activities of some analogues of α-MSH

Peptide	Melanotropic activity (units/mg)
α-MSH	10^7
α-MSH$_{6-9}$(His-Phe-Arg-Trp)	40
α-MSH$_{4-10}$	10^4
α-MSH$_{1-8}$	0
α-MSH$_{1-10}$	10^4
α-MSH$_{1-13-NH_2}$	10^6
ACTH (pig)	10^5
β-MSH (pig)	4×10^6

The smallest peptide which retains some activity is the tetrapeptide comprising residues 6–9 of α-MSH (Table 15.3). Its activity is only 4×10^{-6} that of MSH (but the extreme sensitivity of the MSH assays makes this quite easy to detect). As the chain length is extended, the activity increases (Table 15.3) A peptide comprising residues 4–10 of α-MSH has a thousandth the activity of α-MSH. However, a peptide comprising residues 1–8 of α-MSH has no activity, suggesting that the tryptophan at residue 9 may be essential. Addition of residues 1–3 onto α-MSH 4–10 does not increase activity, but adding the last three residues and an amide group produces a 100-fold increase in activity. Extending the C-terminus further (e.g. by adding residues which occur in corticotropin) causes a drop in activity. Addition of an acetyl group to the amino terminus of α-MSH$_{1-13-NH_2}$ increases activity a further 10-fold. Interestingly, acetylating the amino terminus of corticotropin increases its MSH-like activity substantially.

Some work has been carried out on the alteration of the individual side-chains in α-MSH. Alteration of the methionine at residue 4 (by oxidation with hydrogen peroxide or by synthesizing a derivative in which the sulphur is replaced by a methylene group) leads to substantial loss of activity. Replacement of the arginine at residue 8 by ornithine or citrulline also leads to a drastic loss of activity.

Overall it can be concluded that for melanotropic activity a core of residues; His–Phe–Arg–Trp, appears to be necessary for minimal activity, but most other parts of the structure also contribute to the expression of full activity in MSH. Presumably each makes a contribution in ensuring effective binding to the receptor.

5.2 Steroidogenic activity

Corticotropin-like activity is usually assessed in assays which follow steroid production by adrenocortical tissue (Chapter 5). Degradative studies on corticotropin showed that much of the C-terminal sequence could be removed

Table 15.4 Corticotropic activities of some analogues of corticotropin (ACTH)

Peptide	Corticotropic activity* (units/μmole)
$ACTH_{1-39}$	474
$ACTH_{1-24}$	390
$ACTH_{1-23}$	247
$ACTH_{1-19}$	110
$ACTH_{1-17}$	11
$ACTH_{1-17-NH_2}$	69
$ACTH_{1-16}$	0.1
$ACTH_{1-10}$	0.003
$ACTH_{4-10}$	0.0004

* Determined in an *in vitro* assay.

without substantial loss of activity. Synthetic analogues confirmed this (Table 15.4). Clearly, the carboxyl terminal sequence is not very important for biological activity. Corticotropin$_{1-24}$ has activity almost as great as the natural hormone (1–39) in *in vitro* assays. In *in vivo* assays, however, its relative activity is rather lower, suggesting that it is more susceptible to degradation than the natural hormone and that the C-terminal 15 residues promote the stability of corticotropin in the circulation.

Although $ACTH_{1-24}$ retains almost full activity, shorter derivatives (i.e. with amino acids progressively deleted from the carboxyl terminus) are less active. The hormone contains a 'basic core' (Lys–Lys–Arg–Arg) between residues 15 and 18, and derivatives in which this is partly or completely deleted possess very much lower activity (Table 15.4). However, even residues 1–10 and 4–10 retain slight, but significant, activity, demonstrating that this basic region is not completely essential. Interestingly these minimal peptides with steroidogenic activity include the minimal sequence for MSH-like activity. The importance of the basic region of residues 15–18 is almost certainly due to its positive charge, as is shown by the fact that putting an amide group onto the carboxyl terminus of $ACTH_{1-17}$ increases the steroidogenic activity of the peptide about 6-fold. Presumably, the region binds to a corresponding negatively charged region of the receptor.

The peptide corresponding to residues 11–24 of ACTH acts as a competitive inhibitor of the natural hormone. A possible explanation for this is shown in Figure 15.7. The sequence containing residues 4–10 contains the amino acids necessary for activation of the receptor, while residues 11–24 increase binding only. The peptide containing just 11–24 thus prevents binding of the natural hormone without being able to activate the receptor.[20]

Acetylation of the amino terminus of ACTH causes a big drop in steroidogenic activity, but conversion of the amino terminus from serine to glycine does not. Presumably the α-amino group is required for activity, but the precise

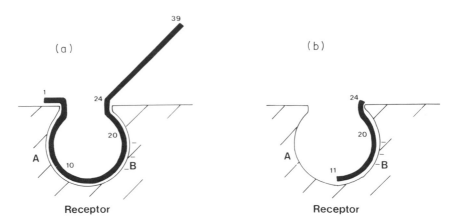

Figure 15.7 Diagram showing how a peptide corresponding to $ACTH_{11-24}$ may inhibit the action of native ACTH. In (a) the binding of $ACTH_{1-39}$ to the receptor is illustrated; binding of the sequence corresponding to residues 4–10 is thought to be essential for activity. In (b) binding of the inhibitory peptide $ACTH_{11-24}$ is illustrated; by occupying part of the receptor this may block binding of the natural ACTH, although unable itself to activate the receptor

nature of the amino-terminal amino acid is not important. Unlike the situation in MSH, modification of methionine at residue 4 can be carried out without a big change in steroidogenic activity.

In conclusion, it is clear that the essential core for melanotropic and steroidogenic activities is similar (and possibly identical), but that features away from this core sequence (residues 5–10) make quite different contributions to the ability of the peptides to bind to their different receptors. Presumably the receptors themselves are similar but have diverged in evolution so that they interact quite differently with MSH and ACTH.

5.3   Receptor binding and activation of adrenal adenylate cyclase[21]

With the development of receptor-binding assays for ACTH, it became clear that analogues of the hormone could also be used to study the relation between receptor binding, stimulation of adenylate cyclase and stimulation of steroidogenesis (see also Chapter 5). In an interesting study,[21] illustrated in Table 15.5, the ability of various synthetic analogues of ACTH to activate adrenal adenylate cyclase, bind to adrenal membrane receptors and inhibit activation of adenylate cyclase was compared. Of the peptides studied only $ACTH_{1-39}$ and $ACTH_{1-24}$ had (equal) activity on adenylate cyclase (although the assay may not have been sufficiently sensitive to detect very slight activity in $ACTH_{6-39}$). However, all peptides which contained residues 9–18 of ACTH could inhibit binding of ACTH to its receptor (presumably by binding themselves) and the

Table 15.5 Effects of synthetic corticotropin (ACTH) peptides on adrenal adenylate cyclase and binding of ^{125}I-ACTH to adrenal membranes

Peptide	Activity in adenylate cyclase assay (U/mg)	Concentration required for 50% inhibition of ACTH-stimulated adenylate cyclase (μM)	Concentration required for 50% inhibition of ^{125}I-ACTH$_{1-24}$ binding (μM)
ACTH$_{1-39}$	144	—	9.3
ACTH$_{1-24}$	148	—	1.3
ACTH$_{6-39}$	0	<1	0.25
ACTH$_{9-24}$	0	18	3.3
ACTH$_{9-19-NH_2}$	0	30	1.0
ACTH$_{9-18}$	0	80	15
ACTH$_{9-20}$	0	80	25
ACTH$_{12-39}$	0	>500	50
ACTH$_{9-16}$	0	N.A*	N.A†
ACTH$_{1-8}$	0	N.A	N.A

Based on the data of Ontjes *et al.* (1977).[21]
*Not active at concentrations up to 500 μM.
†Not active at concentrations up to 100 μM.

ability to inhibit receptor binding correlated closely, although not perfectly, with ability to inhibit the actions of the hormone on activation of adenylate cyclase. This study illustrates very elegantly that binding of hormone to a receptor and activation of events immediately associated with the receptor are by no means identical processes. Use of synthetic analogues thus provides a way of dissecting these early steps of hormone action and of assigning roles to different features of the structure of the hormone.

6. INSULIN[22]

6.1 Three-dimensional structure

Little has been said about the specific three-dimensional structures of either the oxytocin/vasopressin peptide family or the ACTH family in the preceding section. This is because no detailed information is available. These peptides may not have a specific conformation in solution, but exist as a dynamic equilibrium between a number of possible conformations. The peptides presumably assume defined conformations when they bind to the receptor, but information about these is sparse.

The situation is quite different in the case of the small protein-hormone insulin. This has a defined conformation which has been shown to occur in crystals, and which almost certainly occurs also in solution (see Chapter 9), although the hexameric and dimeric organization of molecules seen in the crystal probably is lost (through dissociation) in solution at the concentrations found in blood. It is clear that this defined three-dimensional structure is what interacts with the insulin receptor, and to some extent therefore the interaction between hormone and receptor can be more readily defined than in the case of the smaller hormones discussed so far (see also Chapter 9).

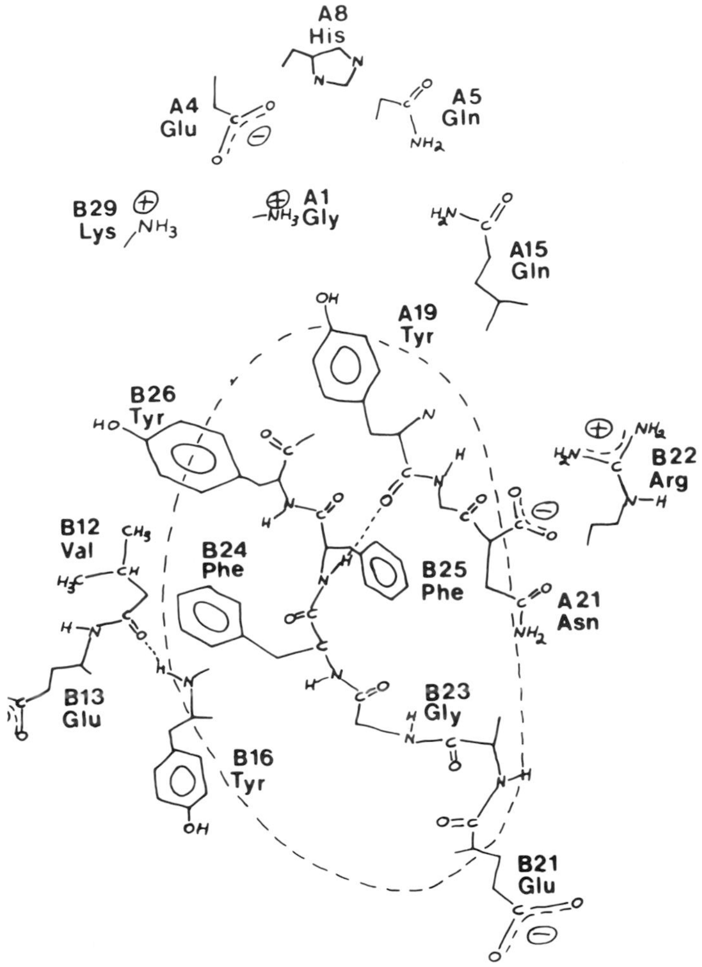

Figure 15.8 Diagram illustrating the region of the insulin molecule (based on the known three-dimensional structure) thought to interact with the receptor. The dotted line encloses those parts of the hormone thought to be most important for receptor binding. (Reprinted with permission from Blundell, T. L., Pitts, J. E. and Wood, S. P. (1982) *Critical Rev. Biochem.* **13**, 141–213. Copyright (1982) CRC Press, Inc., Boca Raton, Florida)

Any change in the tertiary structure of insulin appears to cause drastically reduced activity as well as loss of receptor binding. Furthermore, a hydrophobic face is recognizable on the insulin monomer (which is covered in the dimer), which is highly conserved in evolution, and alteration of which leads to loss of activity. The hydrophobic face is thought to be involved directly in interacting with the receptor.[22]

6.2   Analogues of insulin

Species specificity of insulin (Chapter 9) provides a series of naturally-occurring analogues of the hormone. Chemical modification of the hormone has also been used extensively to provide analogues. Complete chemical synthesis of the hormone has been successfully achieved in several laboratories, but the difficulty of this procedure makes it relatively unsuitable for production of analogues. Semisynthesis of insulin derivatives has proved very successful and popular, however (see Section 6.3).

The results obtained by studying both naturally-occurring analogues and those produced by chemical modification are best interpreted in terms of the known three-dimensional structure of the molecule. Examination of the tertiary structure provided by X-ray crystallographic analysis reveals a hydrophobic face which in the crystal structure (and presumably also in concentrated solution) is involved in dimerization (Figure 15.8). In dilute solution (and certainly at physiological concentrations) dimers dissociate and it has been proposed that the hydrophobic face which is thus exposed is then available for interaction with the receptor, and represents the receptor-binding region of the molecule. In accordance with this hypothesis, species variation in this hydrophobic region is very unusual, and where it does occur (as in, for example, the guinea pig) it leads to marked loss in receptor-binding and biological activity.

The properties of derivatives produced by chemical modification can also be interpreted in terms of this hydrophobic region. A particularly interesting study[23] examined the properties of derivatives produced by acylating the α-amino group on the N-terminal glycine residue of the A chain of beef insulin (the three amino groups of insulin, the α-amino groups of the A and B chains and the ε-amino group of the lysine at residue B-29, have different reactivities and can be modified selectively). Acyl groups introduced were acetyl (CH_3CO-), t-butyloxycarbonyl (t-Boc: $(CH_3)_3\ C-O-CO-)$ and thiazolidine and the conformations of the acylated insulins were studied (by X-ray crystallography and circular dichroism) as well as their activity in receptor-binding and biological (fat-cell lipogenesis) assays. All the acylated insulin derivatives had reduced biological and receptor-binding activity (the two being very closely correlated) which could be interpreted in terms of changes in conformation of particular side-chains in the proposed receptor-binding region. This study, and many others, suggest strongly that, as has been seen for other peptide hormones, several residues are involved in binding of hormone to the receptor, the effects are additive and loss or disruption of any one residue reduces but does not abolish receptor-binding. As in the case of the other hormones considered, receptor-binding seems to be a consequence of non-covalent bonds, which in the case of insulin probably involve mainly hydrophobic interactions and hydrogen bonds.

Studies on naturally-occurring analogues of insulin and derivatives produced by chemical and enzymic modification have been used to show that the metabolic effects of the hormone (glucose oxidation) are dependent upon different structural features from the growth-promoting (somatomedin-like) effects.[24] It was proposed that the two types of effect may involve separate functional domains within the insulin molecule. These results accord well with the idea that the metabolic and growth-promoting effects of insulin are mediated by different receptors, binding to receptors for somatomedin C being primarily responsible for the actions on growth (see Chapter 7).

6.3   Semisynthetic insulin derivatives[8]

Insulin has provided a good model for studies using the semisynthetic approach which has been outlined in Section 2.4, although so far the alterations achieved have been relatively modest ones close to either the amino or carboxyl terminal ends of the A and B chains.

An example of modification at the N-terminus is shown in Figure 15.9.[25] The N-terminal sequence of the B chain can be removed by selective blocking of the other two free amino groups in insulin and then application of the Edman degradation (a chemical method which sequentially removes amino acid residues from the N-terminus of a polypeptide chain). Alternative amino acids can then be reintroduced onto the shortened B chain, so that after removal of the blocking groups a derivative is produced in which B-3 asparagine has been replaced by another amino acid. A range of such derivatives has been studied in almost all of which activity is reduced, but still significant. Again, the reduced activity can be interpreted in terms of changed conformation at the receptor-binding site.

A simple, but important, example of semisynthesis at the C-terminus of insulin involves conversion of pig insulin into human insulin.[26] The two differ only at residue B-30, which is alanine in the pig hormone and threonine in

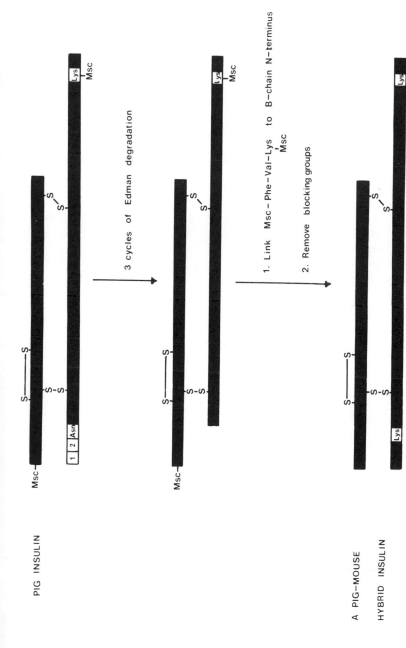

Figure 15.9 Scheme illustrating the production of a pig–mouse hybrid insulin by semisynthesis. The amino groups of the N-terminal glycine (A-chain) and the lysine at residue B29 on pig insulin are blocked selectively (Msc: CH_3-SO_2-CH_2-CH_2-O-CO-). Three amino acid residues are then removed sequentially from the N-terminus of the B chain, using the Edman degradation, and replaced by a synthetic tripeptide. This allows substitution of asparagine at B3 (as found in pig insulin) by lysine (as found in mouse insulin).
(Based on ref. 25)

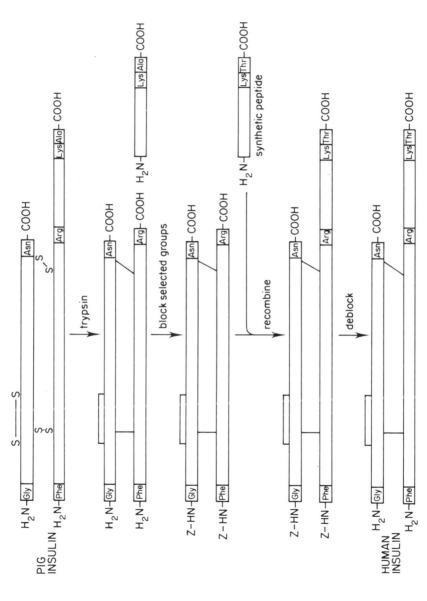

Figure 15.10 Scheme illustrating the way by which semisynthesis can be used to replace alanine at the C-terminus of the insulin B chain by threonine, and thus convert pig insulin to human insulin

human insulin. Several approaches are possible, one of which is illustrated in Figure 15.10. Residues B23–B30 are removed from pig insulin by trypsin. After introduction of appropriate blocking groups, the resulting des-octapeptide (B23–B30) insulin is then coupled with an octapeptide (synthetic or potentially also semisynthetic) in which the C-terminal residue is threonine.

6.4   Photoaffinity labelling of the insulin receptor[27–29]

The principles of the use of affinity labels for study of hormone-receptor interactions have been described in Section 2.6. Work with insulin provides an exciting

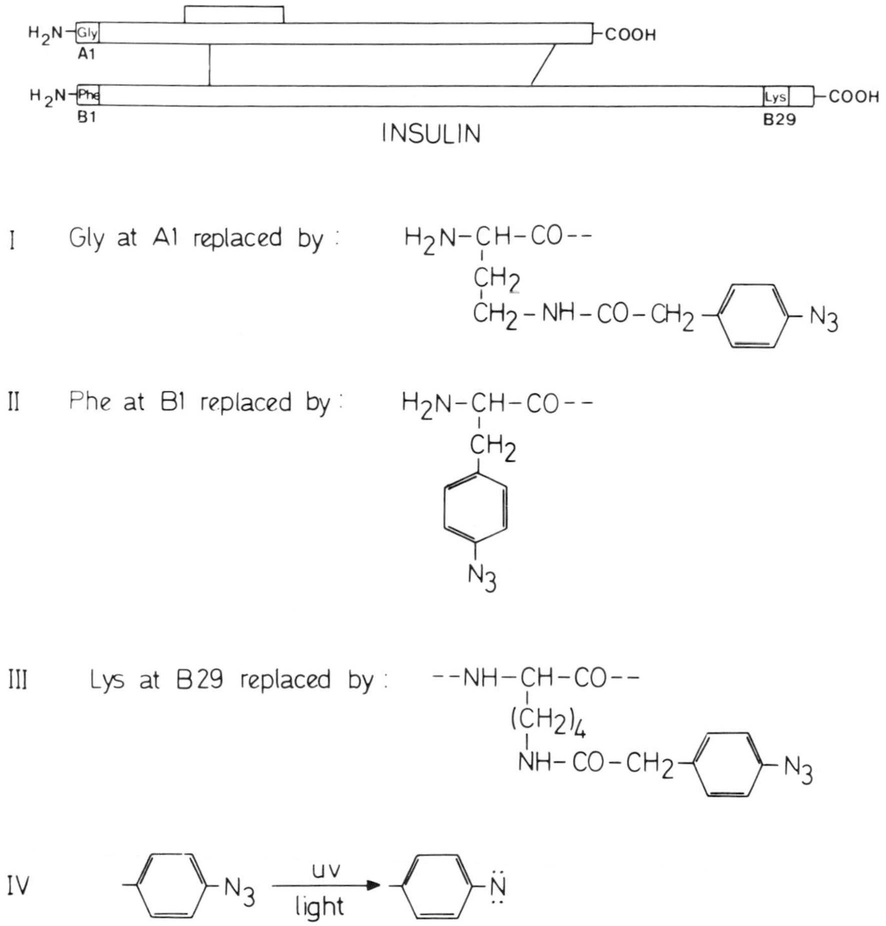

Figure 15.11   Diagram illustrating ways by which azido groups have been introduced on to each of the amino groups of insulin. Once introduced, irradiation with ultra-violet light can convert the azido group into a highly reactive nitrene (IV)

illustration of these principles. Various derivatives of insulin have been prepared and purified which incorporate an azido group on the amino groups of residues A-1, B-1 or B-29 (Figure 15.11). These derivatives retain reduced, but significant receptor-binding and biological activity and can be made highly radioactive using ^{125}I (Chapter 2). Such radiolabelled insulin affinity derivatives are stable in solution. When incubated with plasma membranes from liver

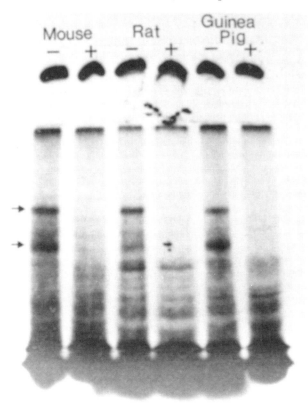

Figure 15.12 Use of photoaffinity labelling to identify subunits of the insulin receptor. Liver plasma membranes from mouse, rat and guinea pig were photo-affinity labelled by ^{125}I-MAB-insulin (insulin with an azidobenzoyl group on the lysine residue at B29—see Fig. 15.11 (III)) in the presence (+) or absence (−) of 50 μg of unlabelled insulin. They were then subjected to SDS-gel electrophoresis and autoradiography. The arrows indicate two well-defined bands that are specifically displaced by excess unlabelled insulin and which are thought to represent components of the insulin receptor. (Reprinted with permission from Yip, C. C., Yeung, C. W. T. and Moule, M. L. (1980) *Biochemistry* **19**, 70–76. Copyright (1980) the American Chemical Society)

containing insulin receptors, they bind to the receptors, tightly but non-covalently, as usual. If the receptor–hormone–derivative complex is then irradiated with ultraviolet light, the azido group is converted to a highly reactive nitrene $(R\text{-}N_3 \xrightarrow{hv} R\text{-}N: + N_2)$, which will react with any available group—potentially providing a covalent link to the receptor.

Several investigators have carried out studies of this kind.[27-29] In every case they have found that the ^{125}I-labelled insulin is covalently bound mainly to a polypeptide of molecular weight 90 000–130 000 (Figure 15.12). This appears to be a subunit of a membrane protein of molecular weight about 300 000, which is clearly a strong candidate for the insulin receptor. Such studies provide enormous potential for both identifying the insulin receptor and determining in detail the nature of the interactions between insulin and this receptor.

6.5 Mutant insulins

Although detailed analysis of mutant proteins provides a potentially powerful way of investigating structure–function relationships, few such mutants have been described for the polypeptide hormones. An exception is provided by mutant insulins.[30,31] A diabetic patient was found to have normal levels of immunoreactive insulin, and no evidence for defective target cell response or high levels of insulin antagonists. A portion of the pancreas of this individual was removed during exploratory abdominal surgery, and insulin was isolated from it. The insulin content of the pancreas was normal. The insulin was immunologically identical to normal human insulin, but possessed only about 50% of normal activity in receptor-binding assays, and 12% of normal activity in biological assays.

Chemical characterization of this 'defective' insulin suggested that it contained a (partial) substitution of phenylalanine by leucine at residue B-25. The patient was thus a heterozygote, producing approximately equal quantitites of this mutant insulin and the normal hormone. It also seems likely, in view of the much lower biological activity than receptor-binding activity, that the mutant insulin antagonizes the normal insulin of the patient.

The phenylalanyl residue at position B-25 lies within the hydrophobic region of the insulin molecule thought to be involved in receptor-binding. The reduced activity of the mutant thus accords well with the proposed role for this region of the molecule. Other mutant insulins with substitutions at the B-25 position have now been identified, although the detailed nature of the substitutions has not been defined.[31]

Another mutation of interest has been located in a patient who showed high levels of proinsulin in the circulation. A mutant proinsulin was identified in this patient, in which residue 65, normally arginine, is substituted by another amino acid. This prevents the normal conversion of proinsulin to insulin and confirms that the presence of a pair of basic amino acids at residues 64 and 65 is crucial for such conversion.[32]

7. STRUCTURE–FUNCTION RELATIONSHIPS IN OTHER PEPTIDE AND PROTEIN HORMONES

Studies of the kind described in Sections 4–6 have been carried out for most of the polypeptide hormones described in this book, but space does not allow a detailed consideration of them. Such studies are continuing, and have been given more impetus in the past few years both by the discovery of many new, fairly small, highly active peptides and by the development of improved methods of peptide synthesis and peptide chemistry. The area will increasingly become a meeting point between those interested in fundamental aspects of the mechanism of hormones and those primarily concerned with practical applications.

REFERENCES

1. Wallis, M. (1975) *Biol. Rev.,* **50**, 35–98.
2. Niall, H. D. (1982) *Ann. Rev. Physiol.,* **44**, 615–624.
3. Perutz, M. F. and Lehmann, H. (1968) *Nature,* **219**, 902–909.
4. Glazer, A. N., Delange, R. J. and Sigman, D. S. (1975) *Chemical Modifications of Proteins.* North-Holland, Amsterdam.
5. Glazer, A. N. (1976) In: *The Proteins,* Vol. II (Eds. Neurath, H., Hill, R. L. and Boeder, C.-L.). Academic Press, New York, pp. 1–103.
6. Finn, F. M. and Hofmann, K. (1976) In: *The Proteins,* Vol. II (Eds. Neurath, H., Hill, R. L. and Boeder, C.-L.). Academic Press, New York, pp. 105–253.
7. Erickson, B. W. and Merrifield, R. B. (1976) In: *The Proteins,* Vol. II (Eds. Neurath, H., Hill, R. L. and Boeder, C.-L.). Academic Press, New York, pp. 255–527.
8. Offord, R. E. (1980) *Semisynthetic Proteins.* Wiley, Chichester.
9. Chowdhry, V. and Westheimer, F. H. (1979) *Ann. Rev. Biochem.,* **48**, 293–325.
10. Milstein, C. (1980) *Scientific American,* **243**, (No. 4, October) 56–64.
11. Eisenbarth, G. S. and Jackson, R. A. (1982) *Endocrine Rev.,* **3**, 26–39.
12. Ivanyi, J. (1982) In: *Monoclonal Hybridoma Antibodies: Techniques and Applications* (Ed. Hurrell, J. G. R.). CRC Press, Boca Raton, pp. 59–79.
13. Cadman, H. F., Wallis, M. and Ivanyi, J. (1982) *FEBS Lett.,* **137**, 149–152.
14. Meyers, C. A., Coy, D. H., Murphy, W. A., Redding, T. W., Arimura, A. and Schally, A. V. (1980) *Proc. Nat. Acad. Sci. USA,* **77**, 577–579.
15. Rudinger, J., Pliška, V. and Krejči, I. (1972) *Recent Progr. Hormone Res.,* **28**, 131–172.
16. Manning, M., Gzronka, Z. and Sawyer, W. H. (1981) In: *The Pituitary* (Eds. Beardwell, C. and Robertson, G. L.). Butterworths, London, pp. 265–296.
17. Schwyzer, R. (1977) *Ann. N.Y. Acad. Sci.,* **297**, 3–26.
18. Ramachandran, J. (1973) In: *Hormonal Proteins and Peptides,* Vol. II (Ed. Li, C. H.), Academic Press, New York, pp. 1–28.
19. Eberle, A. N. (1981) *Ciba Symposium,* **81**, 13–31.
20. Seelig, S., Sayers, G., Schwyzer, R. and Schiller, P. (1971) *FEBS Lett.,* **19**, 232–234.

21. Ontjes, D. A., Ways, D. K., Mahaffee, D. D., Zimmerman, C. F. and Gwynne, J. T. (1977) *Ann. N.Y. Acad. Sci.*, **297**, 295–313.
22. Blundell, T. L., Pitts, J. E. and Wood, S. P. (1982) *CRC Critical Rev. Biochem.*, **13**, 141–213.
23. Pullen, R. A., Lindsay, D. G., Wood, S. P., Tickle, I. J., Blundell, T. L., Wollmer, A., Krail, G., Brandenburg, D., Zahn, H., Gliemann, J. and Gammeltoft, S. (1976) *Nature*, **259**, 369–373.
24. King, G. L. and Kahn, C. R. (1981) *Nature*, **292**, 644–646.
25. Geiger, R., Teetz, V., König, W. and Obermeier, R. (1978) In: *Semisynthetic Peptides and Proteins* (Eds. Offord, R. E. and Di Bello, C.). Academic Press, London, pp. 141–159.
26. Obermeier, R. (1978) In: *Semisynthetic Peptides and Proteins* (Eds. Offord, R. E. and Di Bello, C.). Academic Press, London, pp. 201–211.
27. Massague, J., Pilch, P. F. and Czech, M. P. (1980) *Proc. Nat. Acad. Sci. USA*, **77**, 7137–7141.
28. Wisher, M. H., Baron, M. D., Jones, R. H., Sönksen, P. H., Saunder, D. J., Thamm, P. and Brandenburg, D. (1980) *Biochem. Biophys. Res. Commun.*, **92**, 492–498.
29. Yip, C. C., Yeung, C. W. T. and Moule, M. L. (1980) *Biochemistry*, **19**, 70–76,
30. Tager, H., Given, B., Baldwin, D., Mako, M., Markese, J., Rubenstein, A., Olefsky, J., Kobayashi, M., Kolterman, O. and Poucher, R. (1979) *Nature*, **281**, 122–125.
31. Shoelson, S., Haneda, M., Blix, P., Nanjo, A., Sanke, T., Inouye, K., Steiner, D., Rubenstein, A. and Tager, H. (1983) *Nature*, **302**, 540–543.
32. Robbins, D. C., Blix, P. M., Rubenstein, A. H., Kanazawa, Y., Kosaka, K. and Tager, H. S. (1981) *Nature*, **291**, 679–681.

Chapter 16

The role of cyclic nucleotides in hormone action

W. MONTAGUE

Department of Biochemistry,
University of Leicester, Leicester, U.K.

1. INTRODUCTION

In order for a target cell to respond to a hormone there must be a specific part of the cell with which the hormone can interact to produce a response, i.e. the cell must have receptors for the hormone. These receptors must be able to distinguish the correct hormone from the jumble of hormones and other molecules impinging on the cell. In the case of those hormones which do not readily cross the plasma membrane of the target cell such receptors must be exposed on the outer surface of the plasma membrane of the cell. In addition, they must be linked to some form of intracellular signalling system which can relay the information concerning hormone–receptor interaction to the cellular components involved in the response. Many hormone receptors are closely related to or may even be part of an adenylate cyclase system (Figure 16.1). In this system the interaction of the hormone with a specific receptor localized on the outer surface of the cell leads to the activation of adenylate cyclase on the inner surface of

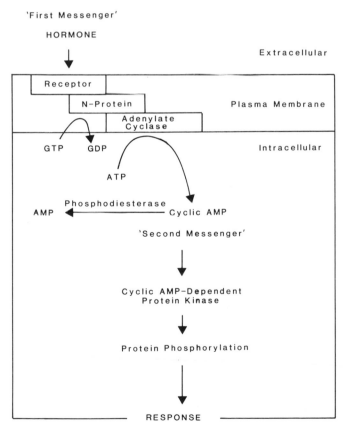

Figure 16.1 The adenylate cyclase system

the plasma membrane and a consequent increase in the intracellular concentration of adenosine 3'5' cyclic monophosphate. Adenosine 3'5' cyclic monophosphate (cyclic AMP) is then responsible for initiating the appropriate response of the cell. Hormones which are thought to act through cyclic AMP are listed in Table 16.1.

Table 16.1 Polypeptide hormone responses mediated by an increase in intracellular cyclic AMP

Hormone	Target tissue	Response
Glucagon	Liver	Glycogenolysis
	Liver	Gluconeogenesis
	Pancreas	Insulin secretion
Vasopressin	Kidney medulla	Water reabsorption
Thyrotropin (TSH)	Thyroid	Thyroglobulin hydrolysis
Adrenocorticotropic hormone (ACTH)	Adrenal cortex	Steroidogenesis
Luteinizing hormone (LH)	Corpus luteum	Steroidogenesis
Follicle stimulating hormone (FSH)	Ovary	Steroidogenesis
Chorionic gonadotropin	Ovary	Steroidogenesis
Parathyroid hormone	Renal cortex	Phosphaturia
	Bone	Calcium resorption ↑
Calcitonin	Bone	Calcium resorption ↓
Luteinizing hormone releasing hormone (LHRH)	Anterior pituitary	LH release
Thyrotropin releasing hormone (TRH)	Anterior pituitary	TSH release
Secretin	Exocrine pancreas	Bicarbonate secretion

In contrast, hormones such as the steroid and thyroid hormones readily enter their target cells and interact with intracellular receptors. In this situation the intracellular hormone–receptor complex acts as the signal and it is generally thought that cyclic AMP is not involved in the response mechanism.

2. CYCLIC AMP

Cyclic AMP was discovered by Earl Sutherland and co-workers in the late 1950s during the course of their investigation into the regulation of liver glycogenolysis by adrenaline.[1] They found that in cell-free extracts of liver, the breakdown of glycogen was increased by a small, heat-stable compound, later identified as cyclic AMP. Adrenaline was shown to stimulate the accumulation of cyclic AMP in the cell-free system by activating a plasma-membrane associated enzyme, adenylate cyclase. Sutherland's group and other investigators demonstrated that adenylate cyclase and cyclic AMP were present in almost all animal cells and that, in addition to adrenaline, many polypeptide hormones also increased the concentrations of cyclic AMP or the activity of adenylate cyclase in target tissues. It was also shown that there was an enzyme in cells, cyclic nucleotide phosphodiesterase, which inactivated cyclic AMP by hydrolysing it to 5'-adenosine monophosphate (AMP).

These observations led directly to the proposal by Sutherland of the 'second messenger' theory of hormone action. According to this proposal the hormone is the first messenger. It interacts with the plasma membrane of the target cell, an interaction which leads to the activation of adenylate cyclase and an increase in the intracellular concentration of cyclic AMP. Cyclic AMP then acts as a second messenger controlling the intracellular responses of the target cell. Soon after the discovery of cyclic AMP in cells, another cyclic nucleotide, guanosine 3'5' cyclic monophosphate (cyclic GMP) was demonstrated in a variety of tissues. The role of cyclic GMP in cellular regulation has not yet been defined in detail and the nucleotide will therefore only be considered in outline in this chapter.

Cyclic AMP is a versatile regulatory molecule which appears to be involved in mediating the effects of numerous hormones on a wide range of physiological processes in a great variety of tissues (Table 16.1). In order to establish that cyclic AMP is involved in mediating the effect of a hormone on a particular tissue, the following criteria need to be established.[1,2]

(1) The hormone should be capable of stimulating adenylate cyclase activity in broken cell preparations of the tissue.

(2) Physiological concentrations of the hormone should increase the intracellular concentration of cyclic AMP in the intact tissue.

(3) The increase in cyclic AMP should precede or at least not follow the physiological response.

(4) It should be possible to potentiate the response to the hormone by administration of a phosphodiesterase inhibitor such as theophylline, caffeine or isobutylmethylxanthine together with the hormone.

(5) It may be possible to mimic the physiological response to the hormone by addition of exogenous cyclic AMP or dibutyryl cyclic AMP.

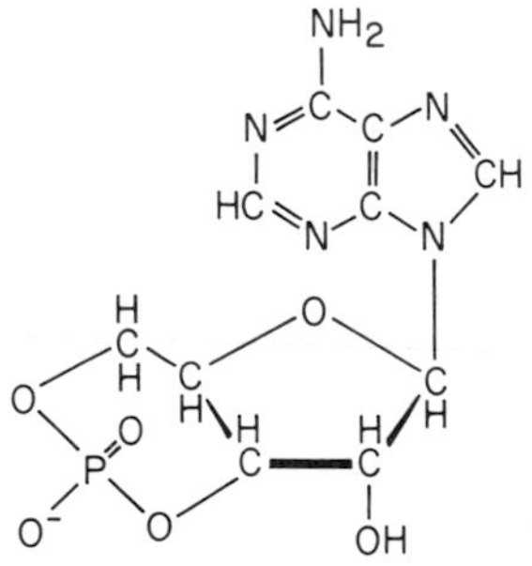

Figure 16.2 Structural formula of cyclic AMP

$$ATP \xrightarrow[\text{Adenylate cyclase}]{Mg^{2+}} \text{Cyclic AMP + Pyrophosphate}$$

$$\text{Cyclic AMP} \xrightarrow[\text{Phosphodiesterase}]{Mg^{2+}} 5'\text{AMP}$$

Figure 16.3 Adenylate cyclase and cyclic nucleotide
phosphodiesterase activities

For many of the hormones listed in Table 16.1 these criteria have mostly been satisfied and it is largely on this evidence that cyclic AMP has been implicated in their mechanism of action.

Cyclic AMP is a relatively stable molecule and it has the structure shown in Figure 16.2. The concentration of cyclic AMP in the cell is determined (Figure 16.3) largely by the relative rates of its synthesis from ATP by the enzyme adenylate cyclase and degradation to AMP by the enzyme cyclic nucleotide phosphodiesterase. In addition, the rate of its excretion from the cell may sometimes by important in controlling its intracellular concentration.

The basal amount of cyclic AMP in cells varies between 0.1 and 0.5 nmoles/g wet weight (0.5–2.5 pmoles/mg protein). This would give an intracellular concentration in the region of 10^{-6}–10^{-7} M assuming that cyclic AMP was distributed evenly in the intracellular water. It is likely, however, that the concentration of free cyclic AMP in the cell is much lower than 10^{-7} M, as much of the cyclic AMP is bound to a specific cyclic AMP receptor protein.

Since cyclic AMP is normally present in cells at relatively low concentrations, highly sensitive assays have had to be developed to determine the amount of cyclic AMP in tissues and body fluids. Cyclic AMP is most commonly assayed by saturation analysis either using a receptor-binding assay system or a radioimmunoassay system. In the receptor-binding assay, ^{3}H-labelled and unlabelled cyclic AMP compete for binding to cyclic AMP binding proteins extracted from tissues. In the radioimmunoassay procedure ^{125}I-labelled cyclic AMP derivatives and unlabelled cyclic AMP compete for binding to antibodies raised against cyclic AMP-protein conjugates.

Cyclic AMP has many of the features of an 'ideal' regulatory molecule. It functions only as an allosteric effector molecule to alter reaction rates, and never as a substrate or product, except in its own synthesis and degradation. It is small and is therefore able to diffuse rapidly from its site of synthesis to its site of action and since it is synthesized from ATP there is always adequate substrate available for its synthesis. In addition, the intracellular concentration of cyclic AMP is capable of undergoing rapid and dramatic changes since its synthesis and breakdown are both controlled by enzymes which can undergo rapid changes in activity.

3. ADENYLATE CYCLASE

The adenylate cyclase system appears to consist of three distinct structural components, the hormone receptor subunit, the regulatory subunit and the catalytic subunit.[3,4] The subunits are normally separate and free to diffuse laterally within the plane of the plasma membrane.[5] It is only when the hormone interacts with the receptor that the three components reversibly associate and this leads to the activation of adenylate cyclase.

3.1 Hormone-receptor unit

Hormone receptors are the means by which cells are able to recognize and discriminate between the various hormones present outside the cell. The receptors for hormones which do not enter the cell to exert their effects are localized on the outer surface of the plasma membrane of the target cell. In general, they are large glycoprotein molecules with a molecular weight in excess of 50 000 daltons. They are highly specific and will only bind the appropriate hormone or biologically active derivatives of the hormone, the extent of binding being proportional to the biological activity. The receptors have a high affinity for their specific hormone such that they will normally bind the hormone when it is present at physiological concentrations. Hormone binding is via non-covalent interactions and it is rapid and reversible and the hormone is released unaltered. The binding exhibits saturation kinetics since there is a finite number of receptors on the cell surface.

3.2 Catalytic unit

If the interaction of the hormone with its surface receptor is to lead to an appropriate response by the target cell then there must be some means of signalling the hormone–receptor interaction to the response mechanism. The interaction of the hormones listed in Table 16.1 with their respective receptors on target cells leads to the activation of the catalytic subunit of the adenylate cyclase system and signalling is achieved by means of an alteration in the intracellular concentration of cyclic AMP. Such activation was thought to be possible because of a close structural relationship between the receptor and catalytic subunits. However, it now appears likely that the receptor and inactive catalytic subunits are distinct structural entities which interact only after the hormone has bound to the receptor. As a result of this interaction the catalytic subunit is activated. In addition, the interaction of receptor and catalytic subunits depends on the presence of a third component the guanine nucleotide regulatory protein (see Figure 16.1).

3.3   Regulatory subunit

The regulatory subunit interacts with both receptor and catalytic subunits and serves to couple the two.[7] Activation of the catalytic subunit is dependent upon the presence of guanosine triphosphate (GTP) since the regulatory subunit has a specific guanine nucleotide binding site. When GTP is bound to this site the regulatory subunit is active but it is inactive if the site is occupied by GDP or GMP or is unoccupied. The binding of GTP to the regulatory subunit is facilitated when it interacts with the hormone–receptor complex and the active regulatory subunit can then interact with the catalytic subunit and activate it. Inactivation of the catalytic subunit can occur following the dissociation of hormone receptor, since this favours loss of GTP from the regulatory protein or it can occur as a consequence of a loss of activity of the regulatory protein consequent upon the hydrolysis of GTP to GDP. Thus, the regulatory protein plays an important role in modulating hormone-regulated adenylate cyclase activity (see also Chapter 11).

The extent to which hormones can activate adenylate cyclase and thereby influence their target cells by increasing intracellular cyclic AMP concentration depends on a number of factors. These include the hormone concentration, the affinity of the receptor for the hormone, the number of specific receptors and the degree of coupling between receptor and catalytic subunits, i.e. on the activity of the regulatory subunit.

3.4   Specificity

Since the effects of many hormones on a variety of tissues are mediated by the adenylate cyclase system, the question arises as to how a particular hormone can elicit a specific response in the correct target tissue. In the first instance such specificity is determined by the presence or absence of specific hormone receptors on the cell surface. Only those cells with the appropriate receptor can recognize a particular hormone. However, the possession of hormone receptors need not necessarily lead to a response since such receptors may not be able to couple with the adenylate cyclase enzyme, i.e. the receptor-coupling mechanism may not be present. If the interaction of the hormone with the appropriate receptor leads to a change in intracellular cyclic AMP, then the response that is elicited must clearly be an expression of the functional activities of that particular cell that are sensitive to cyclic AMP. Compartmentalization of such activities and their differential sensitivity to changes in cyclic AMP may provide additional means of ensuring the appropriate response.

4.   CYCLIC NUCLEOTIDE PHOSPHODIESTERASE

The intracellular concentration of cyclic AMP depends not only on the rate of formation of the nucleotide, but also on the rate of its degradation. The irreversible enzymatic hydrolysis of cyclic AMP to AMP by cyclic nucleotide phosphodiesterase is the only known route of cyclic AMP degradation. Many studies have established the existence, in various tissues, of multiple enzyme forms (isoenzymes) of cyclic nucleotide phosphodiesterase with characteristic substrate affinities, kinetic behaviour, molecular sizes and sensitivities to inhibitors and activators. Most tissues appear to contain at least three distinct phosphodiesterase activities: a low affinity cyclic nucleotide phosphodiesterase which will hydrolyse both cyclic AMP and cyclic GMP, a high affinity cyclic AMP phosphodiesterase and a high affinity cyclic GMP phosphodiesterase.

All cyclic nucleotide phosphodiesterases are sensitive to inhibition by methyl-xanthines such as theophylline, caffeine and isobutylmethylxanthine. This property has been of considerable use in studying the role of cyclic AMP in hormone action since addition of such methylxanthines to cells or tissues usually increases the intracellular concentration of cyclic AMP and potentiates cyclic AMP-mediated hormonal responses.

Physiological regulation of phosphodiesterase activity would provide a means of controlling intracellular cyclic AMP concentrations. However, it seems likely that most hormones do not directly affect the activity of this enzyme. The one major exception is insulin which has been shown to stimulate phosphodiesterase activity in a variety of tissues including liver and adipose tissue. The effect of insulin appears to be specific for the high affinity cyclic AMP phosphodiesterase. This effect is responsible, at least in part, for the ability of insulin to lower cyclic AMP concentrations in these tissues and it may represent part of the mechanism of action of the hormone.

5. MODE OF ACTION OF CYCLIC AMP

Most of the physiological responses of cells to hormones, which are mediated via the cyclic AMP system, involve changes in the activity of key regulatory enzymes or other functional proteins. This is immediately obvious in situations where the response directly involves a change in the activity of a metabolic pathway, although it is less obvious in situations where the response involves a change of membrane permeability or a change in a cellular activity such as protein synthesis or secretion. In these latter situations it has not yet been possible to define the response process in molecular terms and so we are unable as yet to determine exactly how cyclic AMP mediates the hormonal effect. This is in marked contrast to the situation in which the response involves a change in the activity of a metabolic pathway since the mechanism of action of cyclic AMP in these systems has been defined in detail.

5.1 Cyclic AMP-dependent protein kinases

The action of cyclic AMP in regulating physiological responses usually involves the activation of cyclic AMP-dependent protein kinase enzymes.[8,9] Such enzymes are tetrameric proteins composed of dimers of two types of subunits

(a) Cyclic AMP-dependent enzyme

$$R_2C_2 + 4\ cAMP \rightleftharpoons R_2(cAMP)_4 + 2C$$

186 K tetramer   110 K dimer   2 × 38 K monomer
(inactive)   (active)

(b) Cyclic GMP-dependent enzyme

$$E_2 + 2\ cGMP \rightleftharpoons E_2(cGMP)_2$$

165 K dimer   165 K dimer
(inactive)   (active)

Figure 16.4 Activation of cyclic nucleotide-dependent protein kinases

(Figure 16.4). The receptor subunits (R) have high-affinity binding sites for cyclic AMP and the catalytic subunits (C) are responsible for the catalytic activity of the enzyme. In the absence of cyclic AMP the R and C subunits associate as the tetrameric protein. The R and C subunits combine in such a way that the cyclic AMP binding sites on the R subunit are exposed, but the active centres of the C subunits are masked and are unable to interact with substrate. In this form the enzyme is therefore inactive. However, as the cyclic AMP concentration is increased the R subunit binds cyclic AMP and dissociates from the C subunit. The C subunit is then free to express its catalytic activity. The extent of the dissociation and hence the enzyme activity is largely dependent upon the concentration of cyclic AMP. When the cyclic AMP concentration falls, cyclic AMP dissociates from the R subunits which recombine with the C subunits and the enzyme activity ceases.

One of the classical criteria for establishing that cyclic AMP is involved in a particular hormonal response states that physiological concentrations of the hormone should increase the intracellular concentration of cyclic AMP in the intact tissue. However, cyclic AMP concentrations and biological activities do not always correlate, and in some instances there is a response before a detectable change in the concentration of cyclic AMP, even though cyclic AMP is involved in mediating the response. One reason for this may be that a small change in the cyclic AMP concentration in the vicinity of a cyclic AMP-dependent protein kinase, whilst sufficient to activate the kinase, is not sufficient to be detected when total cellular cyclic AMP is measured. In this situation it has been found that it is the amount of cyclic AMP bound to the receptor subunit that is the function which correlates best with the biological response.

The cyclic AMP-dependent protein kinase enzymes all function to transfer the γ-phosphate group of ATP to specific serine and/or threonine residues on their protein substrates. The substrates are generally proteins involved in the response mechanism, whose activity is rate limiting for the response. The covalent modification of such proteins either increases or decreases their functional activity. Since this change in activity arises as a result of covalent modification of the protein it would persist if there was not a mechanism for the removal of the phosphate groups. This is achieved by the activity of phosphoprotein phosphatase enzymes. The extent of phosphorylation and hence the degree of activity of the protein substrates is therefore a function of both the activities of the cyclic AMP-dependent protein kinases and the phosphoprotein phosphatases.

The phosphoprotein phosphatase activity of a number of tissues has been characterized and in general most tissues appear to contain one major multifunctional phosphoprotein phosphatase activity which acts to dephosphorylate a wide range of substrates, although there may also be smaller amounts of specific phosphatases.

In most tissues two major forms (type I and type II isoenzymes) of cyclic AMP-dependent protein kinase have been identified. These differ in their subcellular localization and in their sensitivity to cyclic AMP, although their substrate specificity appears to be similar. The type II protein kinase isoenzyme is capable of undergoing autophosphorylation. In this reaction the catalytic subunit phosphorylates two serine residues on the receptor subunit dimer. The phosphorylation favours dissociation and slows down the reassociation of catalytic and receptor subunits and it may represent a way of maintaining the biological effect of cyclic AMP at a time when tissue concentrations are returning to control levels.

5.2   Cyclic AMP and glycogen metabolism

The involvement of cyclic AMP in the regulation of glycogen metabolism was the first biological role of the nucleotide to be investigated and it is thus not surprising that this system is the best documented of all the actions of the cyclic nucleotide. It is worth considering in detail the involvement of cyclic AMP in this system since it is likely to provide information which may be of use in determining the action of cyclic AMP in other systems.

Cyclic AMP controls liver and muscle glycogen metabolism by coordinately regulating the activities of the rate-limiting enzyme of both the synthetic (glycogen synthase) and degradative (phosphorylase) pathways, such that when one is active the other is inactive. This coordinated regulation prevents the futile cycling of glucose between the two pathways. Reciprocal alterations in the activities of the two enzymes are achieved through a complex series of reactions which are outlined in a simplified form in Figure 16.5.

In essence both glycogen synthase and phosphorylase exist in active and inactive forms and changes in activity result from covalent modifications of the enzymes. Such covalent modifications involve the reversible phosphorylation of specific serine residues in the protein. In the case of phosphorylase, phosphorylation of the enzyme at specific sites increases the activity of the enzyme, while for glycogen synthase, phosphorylation at a number of sites inactivates

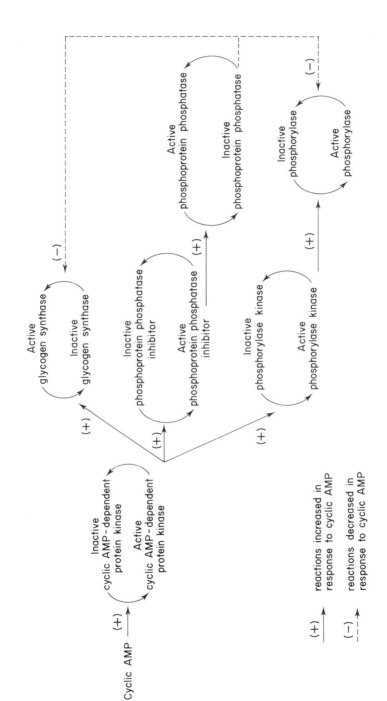

Figure 16.5 Cyclic AMP and glycogen metabolism

the enzyme. Cyclic AMP affects the level of phosphorylation and hence the activity of both enzymes by regulating the activity of the phosphorylating (protein kinases) and dephosphorylating (phosphoprotein phosphatases) enzymes.

The initial site of action of cyclic AMP is its interaction with a cyclic AMP-dependent protein kinase. Activation of this enzyme by cyclic AMP leads directly to the phosphorylation and inactivation of glycogen synthase, since glycogen synthase is a substrate of this protein kinase. Phosphorylase is not, however, a substrate for the protein kinase and phosphorylation of this enzyme involves the specific enzyme phosphorylase kinase. However, the activity of phosphorylase kinase is dependent upon the phosphorylation of specific sites in the protein by the cyclic AMP-dependent protein kinase.

In addition to phosphorylating glycogen synthase and phosphorylase kinase the cyclic AMP-dependent protein kinase also phosphorylates and activates an inhibitor of phosphoprotein phosphatase, thereby reducing the rate of dephosphorylation of the phosphorylated enzymes. Thus, when intracellular cyclic AMP levels are high, various enzymes of the system are in their phosphorylated form and glycogen breakdown is stimulated, while glycogen synthesis is inhibited.

In the absence of cyclic AMP or when cyclic AMP levels fall, the cyclic AMP-dependent protein kinase becomes inactive. Dephosphorylation of the phosphoprotein phosphatase inhibitor ensues and this leads to an increase in the activity of phosphoprotein phosphatase which dephosphorylates the various components of the system. This results in the inactivation of phosphorylase and the activation of glycogen synthase with the result that glycogen synthesis is stimulated and breakdown is inhibited.

5.3 Protein kinase substrates

Since the effect of cyclic AMP is mediated via the activation of cyclic AMP-dependent protein kinases the response elicited will depend upon the substrate specificity of the kinase. Where the response involves the activation of a metabolic pathway the substrates have been characterized and in general they are key regulatory enzymes of the pathway (Table 16.2).[10] The phosphorylation of the enzyme leads either to an increase in its activity (phosphorylase kinase, triglyceride lipase) or to a decrease in its activity (glycogen synthase). However, where the response involves a change in membrane permeability or a change in a cellular activity such as protein synthesis or secretion, the characterization of the substrates has been more difficult. To date cyclic AMP-dependent protein kinases have been shown to phosphorylate a variety of non-enzymatic proteins including histone H1 and troponin and proteins associated with membranes, ribosomes and microtubules. It seems likely that these proteins are involved in the response mechanism and that phosphorylation is associated with a change in their activity. Thus, the phosphorylation of histones and

Table 16.2   Cyclic AMP-dependent protein kinase substrates

Substrate	Effect of phosphorylation	Role
Triglyceride lipase	Activation	Regulation of lipolysis
Phosphorylase b kinase	Activation	Regulation of glyco-genolysis
Cholesterol ester hydrolase	Activation	Regulation of steroido-genesis
Fructose 1,6-diphosphatase	Activation	Regulation of gluconeo-genesis
Pyruvate kinase	Inactivation	Regulation of glycolysis and gluconeogenesis
Glycogen synthase	Inactivation	Regulation of glycogenesis
3-Hydroxy-3-methylglutaryl-CoA reductase	Inactivation	Regulation of cholesterol biosynthesis

ribosomal proteins may exert important regulatory effects on the rate or specificity of protein synthesis, while the phosphorylation of specific plasma membrane and mitochondrial membrane proteins may play an important role in the regulation of membrane permeability. In addition, the phosphorylation of microtubule-associated proteins may be related to the function of these organelles in secretory processes. In view of these findings it seems reasonable to conclude that all cyclic AMP-mediated responses involve the activation of cyclic AMP-dependent protein kinases which phosphorylate and thereby alter the activity of functional proteins involved in the response mechanism.

It is at present unclear what determines the substrate specificity of the cyclic nucleotide-dependent protein kinases, although there are similarities in the amino acid sequence of the phosphorylation site of the protein substrates so far identified. In general, hydrophobic side-chains usually occur on both sides of the phosphorylated serine and there is usually a basic amino acid on the amino terminal side of the phosphorylated residue. Clearly, the environment of a serine or threonine residue in a protein will play a major role in determining whether it can act as a substrate. However, it is uncertain whether this is the only factor responsible.

In a number of instances the phosphorylation of protein substrates by cyclic AMP-dependent protein kinase occurs at a number of different sites in the molecule.[11] This multisite phosphorylation has been observed with glycogen synthase and phosphorylase b kinase. The activity of these two enzymes is regulated predominantly by the phosphorylation–dephosphorylation of a few primary sites while secondary sites of phosphorylation appear to modify the rate of dephosphorylation of the primary sites. In the case of phosphorylase b kinase, multisite phosphorylation may play a role in delaying the inactivation of the enzyme following a decrease in intracellular cyclic AMP concentration.

5.4 Specificity

Since the wide range of effects of cyclic AMP in a variety of target tissues is mediated by an increase in cyclic AMP-dependent protein kinase activity, the question of how a specific response can be elicited has to be considered. Cells respond in different ways because they contain different substrates for the protein kinases. These substrates are normally proteins which are important components of the response mechanism and whose activity is modified by phosphorylation. It is the versatility of the kinase in being able to phosphorylate a wide range of protein substrates that is largely responsible for the fact that cyclic AMP can serve the role of a second messenger for various cellular responses.

6. AMPLIFICATION

The cyclic AMP system is an example of an amplification (or cascade) system which permits relatively few hormone molecules to trigger a substantial response in the target cell. Thus, it has been estimated that one molecule of glucagon can cause the release of 10^8 molecules of glucose from the liver. The amplification arises because some of the components of the system are enzymes which act catalytically to convert many substrate molecules to product. Amplification occurs initially at the level of adenylate cyclase since one molecule of activated cyclase catalyses the conversion of many molecules of ATP to cyclic AMP. The circulating concentration of many hormones is of the order of 10^{-10} M, whereas the concentration of cyclic AMP in a stimulated target cell is about 10^{-6} M. Thus, the synthesis of cyclic AMP by activated adenylate cyclase leads to an approximately 10^4-fold amplification of the hormonal signal. Amplification also occurs at the level of the cyclic AMP-dependent protein kinase since one molecule of activated kinase catalyses the phosphorylation of many molecules of substrate. If the substrate is itself an enzyme, then there is further amplification at this stage.

7. CYCLIC GMP

In addition to cyclic AMP, another cyclic nucleotide, the 3'5' phosphodiester of guanosine monophosphate (cyclic GMP) is also present in mammalian tissues.[12] The concentration of this nucleotide (10^{-8}–10^{-7} M) is normally 10–50 times lower than that of cyclic AMP. It is produced from GTP by guanylate cyclase, an enzyme that is found largely in the cytoplasmic compartment of the cell. Guanylate cyclase is stimulated by calcium ions, but it does not appear to respond to hormones, at least in broken cell preparations. Cyclic GMP is degraded by cyclic nucleotide phosphodiesterase and in addition most tissues appear to contain a relatively specific cyclic GMP phosphodiesterase. Thus, the concentration of cyclic GMP in cells is determined largely by the relative activities of guanylate cyclase and cyclic GMP phosphodiesterase.

Specific cyclic GMP-dependent protein kinases have been demonstrated in certain tissues.[13] They differ in a number of important respects from the cyclic AMP-dependent enzyme. The cyclic GMP-dependent enzyme is a dimer composed of two identical subunits, each subunit containing both cyclic GMP binding and catalytic sites (Figure 16.4). Since both sites are on a single protein there is no dissociation of the complex when it interacts with cyclic GMP. Activation must therefore occur as a result of a conformational change in the protein arising from its interaction with cyclic GMP. In spite of the structural differences of the cyclic AMP and cyclic GMP-dependent enzymes, their substrate specificities overlap and they will both phosphorylate pyruvate kinase, glycogen synthase and phosphorylase b kinase. The cyclic AMP-dependent enzyme is generally much more active with these substrates than the cyclic GMP-dependent enzyme.

It is clear therefore that there exists in cells an intracellular signalling system which utilizes cyclic GMP as the signal molecule rather than cyclic AMP. However, the role, if any, of this system in mediating the effects of hormones on their target tissues has not yet been defined in the same detail as the cyclic AMP system.

In some tissues the intracellular concentrations of the two cyclic nucleotides have been found to vary inversely to one another. These observations have led to the two cyclic nucleotides being likened to the Yin and Yang of Chinese philosophy, as opposing effectors of metabolism.[12] According to the Yin–Yang hypothesis, the dual involvement of cyclic AMP and cyclic GMP is restricted to cellular responses that require simultaneous control of two opposing processes. For example, in some rapidly dividing cells the cyclic AMP concentration is low and the cyclic GMP concentration is high. When proliferation ceases, the cyclic AMP concentration rises while the cyclic GMP concentration falls. Thus, cyclic GMP may mediate the effects of agents which stimulate cell division, whereas cyclic AMP may mediate the effect of inhibitory agents.

8. PROSTAGLANDINS AND CYCLIC NUCLEOTIDES

Prostaglandins have a number of effects on mammalian tissues, some of which involve modulating the responsiveness of the tissue to hormonal stimulation. Many such effects are thought to be mediated via changes in the intracellular concentrations of cyclic nucleotides.[14] Thus, the binding of prostaglandins, particularly those of the E series, to a variety of cells or tissues has been correlated with the activation of adenylate cyclase and the intracellular accumulation of cyclic AMP. In addition, prostaglandins of the F series may activate guanylate cyclase and elevate the intracellular concentration of cyclic GMP in certain tissues. These differential effects of prostaglandins E and F on cyclic AMP and cyclic GMP metabolism, correlate with the opposing effects of the prostaglandins on certain tissue and cell responses and they are compatible with the

'Yin–Yang' hypothesis of cyclic nucleotide action. The physiological significance of the actions of prostaglandins as hormonal modulators is unclear at the
present time, as is the molecular basis of their action on cyclic nucleotide
metabolism.

9. CALCIUM

Although cyclic AMP plays an important role in mediating the effects of many
hormones on their target tissues, it has become clear that it is not the only intracellular signal that mediates the cell's response to external stimuli. Another
system which is extremely important in regulating cellular activity uses calcium
as the signal molecule (Figure 16.6). It has been known for a long time that calcium ions affect a wide range of enzyme activities, although the role of calcium
as a signal molecule has only recently been appreciated. This system mediates
the effects of certain hormones on their target tissues (e.g. the α-adrenergic
effects of catecholamines on the liver) and in addition may be extremely important in modulating the activity of the cyclic AMP system.[15]

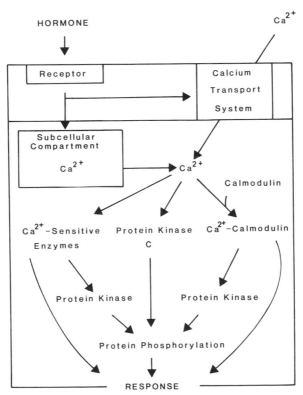

Figure 16.6 The calcium–calmodulin system

9.1 Regulation of intracellular calcium

The concentration of unbound calcium in the various cellular compartments is thought to play a major role in determining the metabolic activity within those compartments, and of the cell as a whole. The factors responsible for regulating such calcium concentrations include its rate of uptake and efflux from the cell, its rate of exchange between the various cellular compartments and the relative rates of its binding and dissociation from intracellular calcium-binding proteins.[16]

The concentration of unbound calcium within the cytoplasmic compartment of most cells is thought to be within the range of 10^{-8}–10^{-7} M. The extracellular calcium concentration is normally in the 10^{-3} M range and there is therefore a concentration gradient which favours calcium entry into the cell. The excessive accumulation of calcium within the cell is normally prevented by the activity of energy-dependent calcium transport systems in the plasma membrane which function to pump calcium ions from the cell and by the relative impermeability of the plasma membrane to calcium. These transport systems are capable of regulation and in this way the intracellular concentration of calcium may be controlled. In addition to the relatively high extracellular calcium concentration there exist within the cell various compartments (e.g. the mitochondrial matrix and the microsomal cisternal space) which contain relatively large amounts of calcium. The uptake and release of calcium from these compartments is subject to regulation and may also play an important role in determining cytosolic calcium concentrations. It is considered likely that some hormones exert their effects on target tissues by altering the concentration of calcium within specific compartments of the cell and that these alterations are mediated by changes in the activities of membrane-associated calcium translocases.

9.2 Mode of action of calcium

Various enzymes and effector proteins are known to be sensitive to changes in calcium concentration and it is thought that calcium interacts directly with these proteins to alter their activity. Indeed the recent demonstration of the calcium-activated protein kinase C (see section 9.4) in various tissues supports this hypothesis. In addition, however, it appears likely that some of the effects of calcium are mediated via the protein calmodulin.[17-20]

Calmodulin is a heat-stable globular protein with a molecular weight of 16 700. It is found in all mammalian tissues and its structure varies little from tissue to tissue or species to species. This high degree of structural conservation implies that the function of the molecule is highly dependent upon the correct three-dimensional structure of the protein. It is an acidic protein and is composed of a single chain of amino acids, 30% of which are aspartate or glutamate residues. This high proportion of acidic residues may be related to the calcium binding function of the molecule since calcium is bound to the carboxyl side chains of certain of the acidic amino acids. Each molecule of calmodulin has four high affinity ($\sim 10\mu M$) binding sites for calcium.

The binding of calcium to calmodulin is dependent upon the calcium concentration. The steady-state concentration of calcium in the cytosol of many cells ranges from 10^{-8}–10^{-7} M and is limiting for binding. Stimulation of the cell by a specific effector molecule may cause a transient increase of cytosolic calcium to 10^{-6} M or higher, a level sufficient to cause calmodulin to bind calcium. The binding of calcium to calmodulin produces an active calcium–calmodulin complex. The active complex in turn combines with the target enzyme or effector protein, inducing a change in the activity of the protein. The change in activity of the protein then leads to the appropriate physiological response.

9.3 Calcium and cyclic AMP

Several systems have been identified which appear to operate via a calcium–calmodulin mechanism (Table 16.3). In many instances, such as glycogen metabolism, muscle contraction and stimulus-secretion coupling, the effects of calcium and cyclic nucleotides are often interrelated since one may accentuate or attenuate the effect of the other.

Table 16.3 Calcium–calmodulin mediated processes

Cyclic nucleotide metabolism
  (a) Activation of adenylate cyclase
  (b) Activation of guanylate cyclase
  (c) Activation of cyclic nucleotide phosphodiesterase

Phosphorylation reactions
  (a) Activation of phosphorylase b kinase
  (b) Activation of myosin light chain kinase
  (c) Phosphorylation of specific membrane proteins

Stimulus-secretion coupling
  (a) Secretion of neurotransmitters
  (b) Secretion of insulin

The molecular basis of the link between these two signalling systems also appears to involve the calcium–calmodulin complex, since it acts as a regulator of adenylate cyclase and cyclic nucleotide phosphodiesterase. The regulation of both adenylate cyclase and phosphodiesterase by the calcium–calmodulin complex may allow a sequential stimulation of the synthesis and subsequent degradation of cyclic AMP, resulting in only a transient accumulation of cyclic AMP. This attenuation of cyclic AMP might follow the influx of calcium through the plasma membrane or the release of membrane-bound calcium in response to an extracellular stimulus. In this situation, therefore, calcium could clearly regulate cyclic AMP levels. However, it is also known that cyclic AMP can regulate intracellular calcium levels by stimulating its influx from the extracellular fluid or by stimulating its efflux from intracellular organelles. In addition, a calcium-dependent protein kinase, which is activated by the calcium-calmodulin complex, has been identified in several tissues. The substrate specificity of this enzyme has not been characterized in detail, although there is some overlap with the substrate specificity of the cyclic AMP-dependent enzyme, since both enzymes will phosphorylate phosphorylase b kinase. Thus, as with cyclic AMP, some of the effects of calcium may be mediated by protein phosphorylation and in this situation the effects of calcium would appear to accentuate those of cyclic AMP.

9.4   Calcium and protein kinase C

Recent studies have revealed the presence in a variety of tissues of protein kinase C, a phospholipid-dependent protein kinase with wide substrate specificity which is activated by calcium.[21] The affinity of this enzyme for calcium, and hence its activity, are greatly increased by diacylglycerol which is produced from phosphatidylinositol 4,5-bisphosphate by the activity of phospholipase C, as an early event in the hormone/response mechanism of several tissues.[22] The other product of this reaction, inositol 1,4,5-trisphosphate has also been suggested to play an important role in hormone action since it appears to increase cytoplasmic calcium concentrations by promoting the release of calcium from intracellular storage pools. These observations have been reviewed by Berridge[22] and they provide further insight into the mechanism of action of certain hormones on their target cells.

REFERENCES

1. Robison, G. A., Butcher, R. W. and Sutherland, E. W. (1971) *Cyclic AMP*. Academic Press, New York, pp. 17–47.
2. Pastan, I. and Perlman, R. L. (1971) *Nature New Biology*, **229**, 5–7.
3. Baxter, J. D. and Funder, J. W. (1979) *New Eng. J. Med.*, **301**, 1149–1161.
4. Ross, E. M. and Gilman, A. G. (1980) *Ann. Rev. Biochem.*, **49**, 533–564.
5. Jacobs, S. and Cuatrecasas, P. (1977) *Trends in Biochem. Sci.*, **2**, 280–282.
6. Schlessinger, J. (1980) *Trends in Biochem. Sci.*, **5**, 210–214.
7. Swillens, S. and Dumont, J. E. (1980) *Life Sciences*, **27**, 1013–1028.
8. Glass, D. B. and Krebs, E. G. (1980) *Ann. Rev. Pharmacol. Toxicol.*, **20**, 363–388.
9. Hoppe, J. and Wagner, K. G. (1979) *Trends in Biochem. Sci.*, **4**, 282–285.
10. Krebs, E. G. and Beavo, J. A. (1979) *Ann. Rev. Biochem.*, **48**, 923–959.
11. Soderling, T. R. (1979) *Mol. Cell. Endocrinol.*, **16**, 157–179.
12. Goldberg, N. D. and Haddox, M. K. (1977) *Ann. Rev. Biochem.*, **46**, 823–896.
13. Lincoln, T. M. and Corbin, J. D. (1978) *J. Cyclic Nucleotide Research*, **4**, 3–14.

14. Samuelsson, B., Goldyne, M., Granström, E., Hamberg, M., Hammarström, S. and Malmsten, C. (1978) *Ann. Rev. Biochem.,* **47**, 997–1029.
15. Berridge, M. J. (1975) In: *Advances in Cyclic Nucleotide Research,* Vol. 6 (Eds. Greengard, P. and Robison, G. A.). Raven Press, New York, pp. 1–98.
16. Bygrave, F. L. (1978) *Trends in Biochem. Sci.,* **3**, 175–178.
17. Cheung, W. Y. (1980) *Science,* **207**, 19–27.
18. Means, A. R. and Dedman, J. R. (1980) *Mol. Cell. Endocrinol.,* **19**, 215–227. 215–227.
19. Means, A. R. and Dedman, J. R. (1980) *Nature,* **285**, 73–77.
20. Klee, C. B., Crouch, T. H. and Richman, P. G. (1980) *Ann. Rev. Biochem.,* **49**, 489–515.
21. Nishizuka, Y. (1984) *Nature,* **308**, 693–698.
22. Berridge, M. J. (1984) *Biochem. J.,* **220**, 345–360.

Chapter 17

Polypeptide hormone receptors

ROBERT C. BAXTER and JOHN R. TURTLE

Department of Endocrinology,
Royal Prince Alfred Hospital and Department of Medicine,
University of Sydney, Sydney, N.S.W., Australia

1. INTRODUCTION

The first step in the action of a peptide hormone at its target tissue involves interaction between the peptide and a membrane-bound receptor. Peptide hormones span a 100-fold range of molecular weight, from approximately 400 for the tripeptide thyrotropin releasing hormone (TRH) to over 30 000 for the gonadotropins. Considering this wide range, it is perhaps surprising that receptors for peptide hormones show such uniformity of physical and chemical properties. In this chapter these properties, and the nature of the primary hormone–receptor interaction, will be examined. Much of the early work on receptors concerned insulin binding to adipocyte and other cell receptors, and insulin receptors are still among the most widely studied. For this reason many of the concepts discussed will be illustrated by reference to insulin binding studies.

1.1 Receptor or acceptor?

Since hormone action is observed as a metabolic effect on the target cell, it is clear that the role of receptors must involve more than the mere binding of hormones. The hormone–receptor complex must generate a signal which initiates metabolic changes leading to the cellular expression of the hormone action. However, following the introduction of simple methods for the study of receptor binding using radiolabelled ligands, many cell hormone binding sites have been defined without any knowledge of the metabolic changes resulting from the binding. Birnbaumer, Pohl and Kaumann[1] have emphasized the distinction between these putative receptors, termed 'acceptors', and true receptors, for which an end response has been clearly defined. Since this chapter is concerned with the primary interaction between hormone and cell, information will be drawn from studies of both proven and putative receptors.

2. CELL LOCATION OF RECEPTORS

2.1 Methods of study

A variety of accumulated evidence indicates that the principal location of peptide hormone receptors is the plasma membrane. This can be demonstrated in four types of experiment. In the first, agents are added to intact cell or tissue preparations to inhibit the effects of hormone–receptor interaction without themselves entering the cells. For example, hormone action can be reversed by the addition of an antibody directed against the hormone. Antibodies against receptors may have the same effect (see Section 9). Limited proteolysis can also be used to demonstrate the location of receptors on the plasma membrane. In particular, trypsin is capable of destroying the ability of fat cells to respond to insulin, without having any effect on cellular integrity or on responses to other hormones.

The second approach involves chemically coupling hormones to macromolecules and demonstrating that they retain biological and binding activity. Obviously, it is of critical importance in these experiments that the link between the hormone and the macromolecule is totally stable during the binding experiment. Uncertainty over this point has led to the validity of such experiments being questioned. Nevertheless, experiments with hormones linked to agarose, cellulose, ferritin and other substrates confirm that peptide hormones exert their membrane-mediated effects without entering cells—that is, the receptors must be located on the outer surface of the plasma membrane.

The location of receptors can also be determined by direct microscopic observation. Using fluorescent derivatives of insulin and epidermal growth factor, it has been demonstrated that these hormones initially bind to receptors diffusely distributed on the cell surface.[2] Within minutes of binding, the occupied receptors aggregate into clusters (see Section 6.3). It has been estimated that two receptors moving randomly across a cell surface can encounter each other within 50 msec.

The fourth method of localizing receptors depends on disruption and subcellular fractionation of cells. Such experiments show, for example, approximately 50-fold enrichment of insulin binding to mouse liver plasma membrane compared to crude liver homogenate.[3] The proven location of receptors for many peptides on plasma membranes has led to the wide use of this membrane preparation for receptor studies. In many other studies a crude microsomal preparation containing both plasma membrane and endoplasmic reticulum has been utilized. In such preparations the endoplasmic reticulum itself may bind hormones, in addition to the plasma membrane fragments present. Insulin binding sites on purified nuclei[4] and Golgi bodies[5] of rat liver have also been reported. While binding sites in Golgi membranes might be newly-synthesized receptors passing from their site of synthesis to the plasma membrane, it has been suggested that the presence of nuclear receptors could be evidence for an intracellular site of insulin action. In contrast, glucagon showed no binding to nuclear sites. Whatever the role of intracellular receptors, they appear to be immunologically different from those on the cell surface: an anti-plasma membrane antiserum which abolishes insulin binding to purified plasma membranes has been shown to have very little effect on binding to receptors on rough or smooth endoplasmic reticulum or on nuclear membranes.[6] Current views on internalization of plasma membrane receptors will be discussed in Section 6.

2.2 Receptor preparations

Suitable sources of receptor for structure and function studies range from metabolically active whole cells, through several stages of decreasing structural integrity, to solubilized and purified receptors. Some whole cell studies have used circulating cells, including monocytes and erythrocytes. Other studies involve a wide variety of cells in culture and cells such as hepatocytes, adipocytes and chondrocytes isolated from their tissue by mild digestion with collagenase or other proteolytic enzymes. Loss of surface receptors by proteolysis has sometimes been observed to result from enzymatic cell dispersion; thus a period of culture may be required for the regeneration of destroyed receptors.

Receptor studies with membrane preparations generally use purified plasma membranes or crude microsomal preparations, as mentioned above. The latter preparation, made by sedimentation of membrane at 100 000 g after vigorous disruption of tissue and removal of particles sedimenting at lower forces, has proved an excellent source of receptors for insulin, somatomedin-C (insulin-like growth factor-I), growth hormone and other peptides. Receptors may be purified from membranes by solubilization in detergents such as Triton X-100 or Lubrol. Solubility is normally defined as the inability to sediment at 200 000–300 000 g. Soluble receptors retain the binding properties of membrane-bound receptors; indeed, these properties are retained even on subsequent purification. Purification techniques applied cover the range of methods used in protein biochemistry, but the most powerful tool appears to be affinity chromatography. In this technique, a ligand specific for the required receptors, or a hormone closely related structurally, is covalently bound to an inert matrix such as agarose. The solubilized receptor binds to the agarose-linked hormone, and can be eluted subsequently with excess hormone or under dissociating conditions of altered ionic strength or pH. Purifications of several thousand-fold have thus been obtained for insulin,[7] prolactin,[8] gonadotropin[9] and other receptors.

3. PROPERTIES OF RECEPTORS

The majority of solubilized peptide hormone receptors have molecular weights in the range 200 000–400 000. Of the group of receptors surveyed by Kahn,[10] most consist of subunits. The insulin receptor is believed to consist of two α and two β glycoprotein subunits of approximate molecular weights 130 000 and 90 000, respectively, linked by disulphide bonds in a structure resembling an immunoglobulin G molecule.[11] The structurally-related growth factor, insulin-like growth factor-I, has a receptor of similar structure while, surprisingly, another closely related peptide, insulin-like growth factor-II, appears to bind preferentially to a receptor composed of a single 260 000 molecular weight polypeptide chain.[12] It might be expected that receptors like the heterotetrameric insulin receptor would have two hormone binding sites, although this is not proven. Other receptors also have heterologous subunits, such as the testicular gonadotropin receptor[9] which appears to combine subunits of hormone binding and adenylate cyclase activities.

Both lipid and carbohydrate are closely associated with receptors. Receptors embedded in lipid-protein membranes clearly have extensive hydrophobic regions. This is confirmed by the inability of many detergent-solubilized receptors to remain in solution when the detergent is dialysed away. Treatment of particulate or soluble receptors with phospholipases has a range of effects in different systems. For example, the action of both phospholipase A and C can inhibit gonadotropin binding to corpus luteum plasma membrane receptors, while the same enzymes increase insulin binding to fat cells, presumably by 'unmasking' receptors.[13] Similarly, lipid extraction of fat cell membranes with ethanol-ether results in higher insulin binding.[14]

The involvement of carbohydrate can be inferred from two types of experiment. First, treatment with enzymes such as neuraminidase and β-galactosidase frequently affects receptor binding.[14] Secondly, concanavalin A (a protein isolated from jack bean which specifically binds to α-D-glucosyl and related carbohydrate residues) can interact with many peptide hormone receptors to inhibit hormone binding. Indeed, this interaction has been used to advantage in the purification of glycoprotein receptors by affinity chromatography on agarose-bound concanavalin A. The carbohydrate residues on receptors are thought to play an important role in the processing of the precursors of receptors to their final form. A precursor form of the insulin receptor, rich in mannose, has been described with a molecular weight of about 200 000.[15] This may undergo both proteolytic cleavage and carbohydrate processing to yield the mature α- and β-subunits described above.

4.   THE HORMONE–RECEPTOR INTERACTION

4.1   Radioactively labelled hormones

Between the primary hormone–receptor interaction and the final expressed response are a number of intracellular events which differ for different systems and are in many cases poorly understood. Thus, examination of the biological response to hormone binding, while essential to establish the true receptor status of binding sites, can give only limited information on the nature of the primary interaction. The approach which has been most fruitful has involved the binding of radioactively labelled hormones to receptors.

The range of radioactive atoms used to label hormones includes ^{3}H, ^{14}C, ^{35}S, ^{125}I and ^{131}I. In general, peptide hormones labelled with ^{14}C have only limited value in receptor studies, while ^{3}H appears useful for some low molecular weight hormones. The main restrictions have been the long half-lives of these isotopes, resulting in labelled hormones of low specific radioactivity, and the difficulty of introducing the label, especially into peptides of high molecular weight. [35]S has a much higher specific radioactivity, but also presents problems with labelling. The most useful isotopes have proved to be ^{131}I and ^{125}I, following the introduction of a simple and mild chemical iodination method.[16] In this method, the mild oxidizing agent chloramine T is used to convert iodide to a form able to react with the aromatic ring of tyrosine. The reaction is carried out at room temperature and neutral pH. Subsequently a number of other satisfactory chemical and enzymatic iodination methods have been developed.[17] The 60-day half-life of ^{125}I, together with its relatively low energy gamma emission, makes it the isotope of choice for hormone labelling in most laboratories.

It is important that the introduction of iodine to a hormone should not change its binding or biological properties, although in practice this is often difficult to achieve. In particular, peptides iodinated with more than one atom per molecule frequently show reduced activity and stability. For this reason, most laboratories aim for monoiodination. Kahn[18] has reviewed the biological activity of a number of monoiodinated hormones. It appears that the smaller peptides, such as angiotensin and oxytocin, lose 25–90% of their activity, while intermediate and large hormones like insulin and chorionic gonadotropin retain 100% of activity.

4.2 Time-course of binding

Studies of biological responses to hormones have shown that many act within minutes of their exposure to cells. Thus, hormone interaction with receptors must be a very rapid process *in vivo*. While this is easily demonstrated *in vitro* for a small hormone such as angiotensin II,[19] many other hormones react very slowly with receptors *in vitro*, even when incubated at 37°C. For example, human growth hormone takes up to 24 hours to reach maximal binding to rat liver membranes. The reason for this slow association rate is not clear, but it may depend on an alteration in the conformation of the membrane during isolation, since receptors on intact cells usually bind hormones more rapidly. In practice, it is often impracticable to carry out binding studies at 37°C, especially when using crude membrane preparations, since proteolytic enzymes may damage the hormone. For this reason many studies are carried out at 15°C or 4°C. Clearly, the fluidity and conformation of cell membranes under these conditions will not be the same as at 37°C—therefore such studies should be interpreted with caution. A practical advantage of studying binding at low temperatures, apart from avoiding proteolysis, is the higher equilibrium binding sometimes observed. This may be due to the fact that dissociation rates show a greater increase with temperature than association rates, giving a higher affinity constant at lower temperatures.

Time-course studies can yield evidence of the coupling between hormone binding and biological activity. In one such study, Gliemann *et al.*[20] showed a clear coincidence in time between insulin binding to fat cells, and lipid synthesis resulting from the binding. This experiment indicated that hormone-stimulated events are intimately connected to receptor binding. In contrast, the loss of biological response following hormone withdrawal appears to be considerably slower that the dissociation of receptor-bound hormone,[21] suggesting that the sequence of events initiated by hormone binding does not terminate upon dissociation.

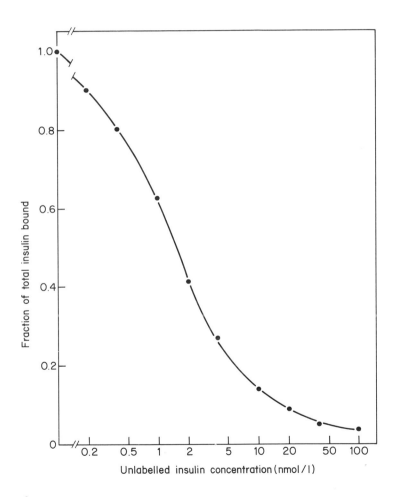

Figure 17.1 Competitive binding curve showing the displacement of labelled insulin from rat liver membrane by increasing concentrations of unlabelled insulin

4.3 Equilibrium studies

While the presence or absence of receptors may be inferred simply from the binding of radioactive tracer hormone, most information is obtained by the *competitive binding technique*. In this method, the principle of which was outlined by Berson and Yalow in 1958,[22] a fixed concentration of tracer hormone binds to equilibrium in competition with increasing concentrations of unlabelled hormone. The number of available binding sites must be limiting to ensure competition between labelled and unlabelled hormones. As the unlabelled hormone concentration increases, a decreasing proportion of tracer hormone is able to bind to receptor (Figure 17.1). Hormone bound to the receptor is quantitated by separating the bound and free fractions and counting the radioactivity in either fraction.

4.4 Graphical measurement of binding parameters

For a simple reversible bimolecular reaction at equilibrium, the association constant (or affinity constant)

$$K = \frac{[HR]}{[H][R]}$$

where

$[H]$ = free hormone concentration,
$[R]$ = free receptor concentration, and
$[HR]$ = receptor-bound hormone concentration.

If total receptor sites $[R_o] = [R] + [HR]$,
then

$$K = \frac{[HR]}{[H]([R_o] - [HR])}$$

This equation can be rearranged in various ways, for example;

(i)
$$\frac{1}{[HR]} = \frac{1}{[H](K[R_o])} + \frac{1}{[R_o]}$$

the 'double reciprocal' equation, analogous to the Lineweaver–Burk equation of enzyme kinetics;

(ii)
$$\frac{[HR]}{[H]} = K[R_o] - K[HR], \text{ the Scatchard equation.}[23]$$

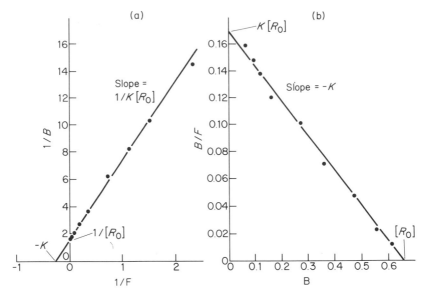

Figure 17.2 (a) Double reciprocal plot of human growth hormone binding to lactogenic sites in female rat liver membranes. (b) The same data as a Scatchard plot. These plots illustrate the graphical measurement of association constant (K) and binding site concentration (R_0) for interaction at homogeneous, non-cooperative sites

If bound hormone concentration, $[HR]$, is represented by B, and free hormone concentration, $[H]$, by F, then plotting either $1/B$ against $1/F$ (Figure 17.2a) or B/F against B (Figure 17.2b) allows simple graphical measurement of K and R_o. In terms of B and F, the Scatchard equation becomes:

(iii)
$$\frac{B}{F} = K(R_o) - K(B).$$

The Scatchard plot, a graph of B/F against B, has been used extensively for the estimation of binding parameters of cell receptors. The ordinate intercept is $K(R_o)$, the slope $-K$, and the abscissa intercept R_o.

Equation (iii) refers to a binding system having only one class of binding site (i.e. all receptors have equal affinity for ligand). When more than one class of binding site exists, the total amount of hormone bound, B, at a given free hormone concentration, F, is a function of the properties of each type of binding site, and may be represented, after rearrangement of equation (iii), as:

(iv)
$$B = \sum B_i = \sum \left(\frac{R_{oi} K_i F}{1 + K_i F} \right),$$

where K_i is the association constant, R_{oi} the number of binding sites and B_i the amount of ligand bound, for each of i independent classes of binding site. As explained in detail in Section 5, Scatchard plots may be linear or curved, depending on the number of classes of binding site present, and the extent of between-site interactions. It should be noted that, according to Scatchard's original definition,[23] the abscissa of a Scatchard plot represented the molar ratio of bound ligand to protein, or saturation fraction, and had no units, while the ordinate was expressed as litres per mole, being the saturation fraction divided by the free ligand concentration, F, in moles per litre. Then the slope, $-K$, was in litres per mole and the abscissa intercept represented moles of ligand binding sites per mole of protein. This differs somewhat from the usual treatment of the equation in hormone receptor studies, presented above in equations (ii)–(iv). In this treatment, B/F represents a ratio of concentrations and thus has no units, while the units of the abscissa are either concentration units such as moles per litre, or related specifically to the receptor preparation used, e.g. moles per mg membrane protein. Whichever way the binding site concentration is expressed, the affinity constant K is always in litres per mole.

While binding data are generally analysed using transformations such as the Scatchard or double reciprocal plot, it is important to note that the conditions implicit in such analyses are frequently not met. It has recently become clear that hormone–receptor interactions are not fully reversible (see Section 6.3). Furthermore, true equilibrium may often not be achieved owing to insufficient reaction time, constraints imposed by hormone degradation, etc. Thus, caution should be applied when interpreting Scatchard plots and similar analyses.

4.5 Non-specific binding

In addition to their interaction with receptors, most radioactively labelled hormone preparations have also been found to bind to other membrane structures as well as talc, glass and plastic tubes and various other non-receptor substances. This binding is termed 'non-specific', and should be subtracted from total binding to receptor to allow correct interpretation of results. The quantitation of non-specific binding frequently presents a problem. Hollenberg and Cuatrecasas[24] suggest that it should represent binding at hormone concentrations above those at which saturation of the biological dose–response curve can be demonstrated. For example, since insulin acts *in vivo* and in isolated cells over the approximate concentration range of 0.01 to 1 nmol/l, it could be assumed that insulin bound at concentrations above 10 nmol/l is bound non-specifically. Thus, non-specific binding is routinely measured by tracer remaining bound in the presence of excess unlabelled hormone. The definition of 'excess' is particularly difficult for hormone acceptors for which no end response has been demonstrated. In practice, excess hormone may range from 2-fold to 1000-fold greater than the highest standard concentration used.

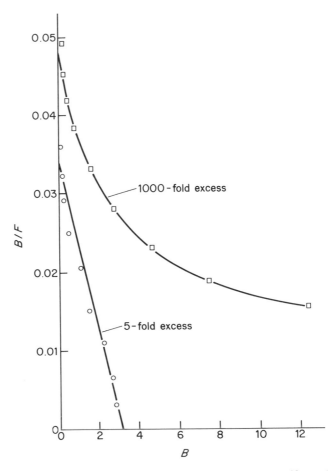

Figure 17.3 Scatchard plots illustrating the artifactual curvature that can be introduced when non-specific binding is measured in the presence of an excessively high unlabelled hormone concentration

Figure 17.3 compares Scatchard plots of human growth hormone binding to rat liver membrane when non-specific binding has been calculated in the presence of a 5-fold excess (4 µg/ml) or a 1000-fold excess (800 µg/ml) of unlabelled hormone over the highest standard concentration. Clearly, curvilinear Scatchard plots may result from the use of excessively high unlabelled hormone levels to calculate non-specific binding. The curvature arises from the contribution of a very low affinity component to hormone binding (see Section 5).

5. RECEPTOR AFFINITY

As explained above, the slope of a Scatchard plot gives a measure of the affinity of binding. Receptor systems showing a single class of binding site, without cooperative interactions, generally yield affinity constants in the range 10^8 to 10^{10} l/mol. If more than one class of binding site is present, the different sites having different affinities, a curvilinear Scatchard plot will be observed. The mathematics of resolution of Scatchard plots resulting from site heterogeneity are well established.[25] Multiple classes of binding site have been reported for a number of different hormone–receptor systems (Table 17.1). Generally, for systems having independent binding sites, the high affinity sites have association constants in the range 10^8 to 10^{10} l/mol, while low affinity sites are one or two orders of magnitude lower.

Table 17.1 Some hormone receptors reported to show heterogeneous binding sites

Hormone	Tissue
Angiotensin III	Rat adrenals
FSH	Rat testes
Glucagon	Rat liver
Human growth hormone	Female rat liver
Insulin	Hamster liver
Oxytocin	Rat mammary, uterus, etc.
Somatomedin C	Pig liver
Vasoactive intestinal peptide	Pancreatic acinar cells

An interesting example of a tissue with two independent types of binding site for a single hormone arises from the 'dual specificity' of human growth hormone binding to intact female rat hepatocytes.[26] These cells bind a variety of lactogenic hormones, including human growth hormone, at one site, and somatogenic hormones, which also include human growth hormone, at another. Thus, binding experiments with tracer and standard human growth hormone yield curved Scatchard plots. However, if the lactogenic sites are saturated by the addition of excess ovine prolactin, a single class of somatogenic sites, yielding a linear Scatchard plot, can be demonstrated by displacing the tracer with increasing concentration of bovine growth hormone. Similarly, if somatogenic sites are saturated with excess bovine growth hormone, a linear Scatchard plot for ovine prolactin is observed.

5.1 Negative cooperativity

Several years ago it was recognized that receptor affinity is not necessarily fixed, but may vary depending upon the relative occupancy of available sites. This important observation was first reported by De Meyts *et al.*[27] It was observed that the dissociation rate of labelled insulin from its receptors on cultured human lymphocytes, under conditions where rebinding of tracer did not occur, was accelerated in the presence of added unlabelled insulin. This was interpreted as an increase in the dissociation rate constant of sites containing bound insulin caused by the occupancy of other sites by unlabelled hormone—the phenomenon of negative cooperativity. In contrast, the dissociation of human growth hormone from the same cells was not affected by the addition of excess hormone.

Scatchard analysis of equilibrium hormone binding to a receptor exhibiting negative cooperativity yields a curvilinear plot, since the slope of the Scatchard plot (a measure of affinity) decreases with increasing addition of hormone and increasing site occupancy. Thus, receptor systems with cooperative binding cannot be distinguished by equilibrium binding studies from those with multiple classes of sites; this distinction can only be made by kinetic experiments. It is critically important to exclude the rebinding of dissociated hormone in such experiments, for if tracer were able to rebind to a receptor with multiple classes of binding site, the addition of unlabelled hormone would cause a redistribution of bound tracer from high affinity, slow dissociating sites, to low affinity, faster dissociating sites. This redistribution would have the effect of increasing the net dissociation rate, giving the superficial appearance of negative cooperativity.

Negative cooperativity has been demonstrated kinetically in insulin receptors from a variety of tissues, as well as several other peptide receptors (Table 17.2).

Table 17.2   Some hormone receptors reported to show negative cooperativity

Hormone	Tissue
Insulin	Human monocytes, rat liver, human placenta, etc.
Caerulein	Rat pancreas
Nerve growth factor	Chick dorsal root ganglia
Thyroid-stimulating hormone	Bovine thyroid membranes
Thyroliberin	Anterior pituitary

De Meyts[28] has provided a simple method of graphical analysis of curvilinear Scatchard plots in systems showing negative cooperativity. A typical application of this method is shown in Figure 17.4 for the insulin receptor of rat liver. Since the intercept on the ordinate represents the limiting B/F value when no ligand is bound, and the intercept on the abscissa represents the (fixed) total receptor concentration, the slope of the straight line joining these intercepts gives a measure of the limiting high affinity, when no binding sites are filled and cooperative interactions are absent (K_e in De Meyts' terminology). As the ligand concentration, and thus site occupancy, increases, the average affinity of receptors will decrease due to cooperative interactions, until a limiting low affinity (K_f) is reached. It must be emphasized that in curvilinear Scatchard plots due to negative cooperativity, while K_f is similar to the low affinity calculated from a 'two independent site' model, K_e is not the same as the high affinity for a two-site model (calculated from the slope of the tangent to the ordinate intercept).

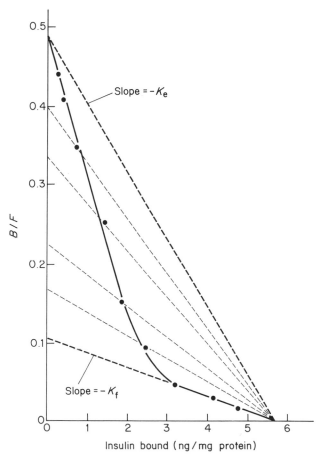

Figure 17.4 Scatchard plot of insulin binding to rat liver membrane, illustrating negative cooperativity in hormone–receptor interaction as described by De Meyts.[27] With increasing occupancy of insulin-binding sites, the receptor affinity (initially K_e) decreases (broken lines) until a limiting low affinity (K_f) is reached

Where negative cooperativity is demonstrated, the question arises whether the site–site interaction occurs between subunits of a single receptor or between different receptors. Microscopic studies show that hormone binding to receptors in intact cells may lead to clustering of the receptors, a possible mechanism for cooperative interaction. However, at least in the case of insulin receptors, the demonstration of negative cooperativity even in detergent-solubilized receptors, together with evidence that the receptor is tetrameric in structure, suggests that the interactions occur within a single receptor molecule.

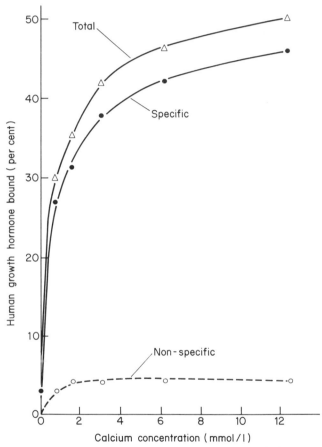

Figure 17.5 Effect of calcium on human growth hormone binding to the lactogenic receptor of female rat liver microsomal membranes

5.2 Other factors affecting affinity

A number of peptide receptors are sensitive to the presence of Ca^{2+} and Mg^{2+}. For example, the lactogenic receptor of female rat liver shows an absolute dependence on divalent cations for binding activity (Figure 17.5). Insulin receptor affinity is also increased by calcium, while, interestingly, the affinity of the structurally similar receptor for insulin-like growth factor-I is actually decreased by calcium.[29] Hydrogen ion concentration also affects receptor affinity: at pH 8, high affinity forms of the insulin receptor are favoured, while pH 5 favours lower affinity forms.

Insulin receptor affinity is also known to be altered in a number of disease states, although the mechanisms causing these alterations are not understood. For example, increased affinity has been reported in patients with insulinoma and acromegaly, and in fasted obese patients.[30] Normal subjects also show increased receptor affinity 5 hours after glucose ingestion.

6. RECEPTOR CONCENTRATION

6.1 Regulation of receptor concentration—down regulation

In addition to regulation by changes in receptor affinity, many cells have the ability to regulate hormone binding by changing the concentration of receptors. The best studied examples come once again from the area of insulin receptors.

Obesity in man and experimental animals is characterized by resistance to the normal actions of insulin, accompanied by elevated levels of the hormone. In the genetically obese mouse, liver membranes from obese animals bind only about 30% as much tracer insulin as membranes from thin litter mates. This decrease in binding is seen by Scatchard analysis to be due solely to a decrease in receptor number, with no change in affinity.[31] Subsequently, it has been shown, using cultured human lymphocytes, that receptors for both human growth hormone and insulin are regulated by the ambient concentrations of the respective hormones. These observations have led to the concept of 'down regulation' of receptor number by an increase in the concentration of homologous hormone.

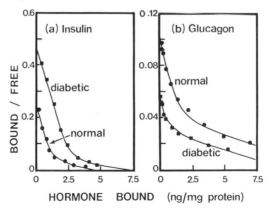

Figure 17.6 Specific binding of insulin (a) and glucagon (b) to liver membranes from normal (circles) and streptozotocin-diabetic (squares) rats

In the example shown in Figure 17.6b, rat liver glucagon receptors are decreased in experimental diabetes, concomitant with an increase in circulating serum glucagon. Conversely, under conditions of reduced hormone level, receptor numbers may increase. Figure 17.6a shows the increase in liver insulin receptors observed in diabetes, when serum insulin levels fall.

The concentration of receptors for a particular hormone may also be regulated by hormones which do not interact at the binding site. An example is the regulation by insulin of the somatogenic receptor of rat liver.[32] The receptor concentration decreases in diabetes, with no change in binding affinity, and is restored following insulin therapy. Lactogenic receptors are particularly sensitive to the sex steroid status of the animal. Thus, these receptors can be induced in male rats by injecting oestradiol or by castration, and in female rats are greatly increased in pregnancy. These changes appear to be mediated via the pituitary, since they are prevented by hypophysectomy. The observation that lactogenic receptors are greatly increased in animals bearing tumours which secrete growth hormone and prolactin suggests that these receptors are induced by receptor-active hormone—the opposite effect to 'down-regulation'.

6.2　Other effects on receptor number

The receptor concentration in various tissues is under a number of influences apart from hormone levels. Figure 17.7 shows how the concentration of angiotensin receptors in rat brain varies with the age of the animal, a phenomenon seen with a variety of receptors in different tissues. Studies with cells in synchronous culture also indicate that receptors for peptide hormones may change in number at various stages through the cell cycle.

6.3　Internalization

Experiments with cultured human lymphocytes have shown that the cytochalasins (drugs which modify the microfilament system of cells) are capable of reducing the number of receptors for both insulin and human growth hormone, without having any effect on binding affinity.[33] It was suggested that the receptor loss was due to redistribution of receptors within the cells. An increasing body of evidence indicates that receptors, once occupied, become internalized within the cell. This may merely be a reflection of the normal turnover of receptors which, like other membrane proteins, are in a continuous state of synthesis and degradation. Studies with cultured lymphocytes in which protein synthesis is blocked with cycloheximide indicate a half-life for growth hormone receptors of about 10 hours.

Other evidence suggests a more specific role for receptor internalization. One possibility is the transport of hormones to intracellular binding sites (Section 2.1). The degradation of a variety of hormones by target tissues appears to

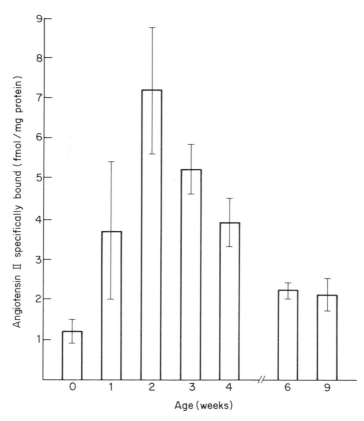

Figure 17.7 Specific binding of angiotensin II to receptors on brain membranes isolated from rats aged between 0 and 9 weeks (unpublished data of C. R. Baxter)

depend upon their interaction with membrane receptors. A specific endocytosis mechanism for receptor-bound hormone may then operate, comparable to systems described for lipoproteins. Summarizing observations made in a number of different laboratories,[2,34–37] the process appears to involve several steps. Hormone initially binds in a reversible manner to receptors distributed diffusely over the cell surface. Within one to two hours, up to 50% of bound hormone becomes irreversibly associated with the receptor, due at least in some cases to covalent bond formation.[38] During this period occupied receptors aggregate into clusters, perhaps in special regions or 'pits' in the plasma membrane which, with some exceptions, are able to enter the cell by endocytosis.[39] In this process, the receptor-rich regions invaginate and form vesicles which move into the cell where they fuse with lysosomes. Whether receptor fragments which might be formed by lysosomal proteolysis can be 're-cycled' into new receptors, and whether hormone fragments generate, or are themselves, second messengers, are intriguing questions. In one report[40] internalized insulin receptors in chick liver cells were not degraded even after 18 hours exposure to extracellular insulin; the number of binding sites measurable after extraction of whole cells with detergent was unchanged, despite a 60% decrease in cell surface receptors.

If receptor endocytosis only occurs for occupied receptors, there should be a correlation between receptor loss and the extent of occupancy. However, the loss of growth hormone receptors from cultured lymphocytes, and of LH receptors from rat testis, following exposure to hormone, has been shown not to relate directly to the extent of receptor occupancy. This discrepancy is currently unexplained. As noted previously, if internalization and proteolytic degradation are the normal fate of receptor-bound hormones, it may not always be possible to demonstrate reversibility of binding when studying hormone–receptor interaction *in vitro*, since extracellular dissociation of bound hormone may not occur.

Posner *et al.*[41] have compared Golgi and plasma membrane insulin receptors in hyperinsulinaemic obese mice and lean controls. While the plasma membrane insulin binding was much lower in the obese group, binding to Golgi cisternae was higher in obese animals than in the controls. On the basis of this observation, a hypothesis has been proposed which could explain both the inverse regulation of insulin receptors and the direct induction of lactogen receptors by their respective hormones. In this hypothesis the binding of hormone to its receptor results in two changes. There is an accelerated loss of plasma membrane receptors by internalization, and a concomitant increase in synthesis of new receptors. These new receptors could be detected in Golgi cisternae as they move from their site of synthesis to the plasma membrane. Whether the net change in plasma membrane receptors is an increase or decrease depends only upon the relative kinetics of the synthesis and endocytosis. However, in terms of metabolic regulation, the advantage to the cell of induction of receptors by some hormones, and repression by others, remains unknown.

7. RECEPTOR PHOSPHORYLATION

It has been known for over a decade that the action of certain hormones on their target cells results in the phosphorylation of membrane-associated proteins. In 1974 Cuatrecasas and co-workers postulated that cyclic AMP-dependent phosphorylation of fat cell membrane proteins might be involved in the regulation of insulin-stimulated glucose transport. However, they were unable to demonstrate any direct action of insulin on the phosphorylation reaction.[42] More recently, Cohen reported that epidermal growth factor (EGF) activates a membrane-associated protein kinase (cyclic AMP independent) which catalyses the phosphorylation of the EGF receptor.[43] The kinase activity also co-purified with the receptor, suggesting that the binding and kinase functions were present on the one protein. The major phosphorylated amino acid appeared to be phosphotyrosine.[44]

Insulin has also been shown to stimulate phosphorylation of its own receptor in a cyclic AMP-independent reaction which takes place selectively on the β-subunit.[45] The kinase activity, as judged by the presence of an ATP binding site, also occurs in this subunit, and the principal phosphorylated amino acid is, as in the case of EGF receptor kinase, phosphotyrosine.[46]

The significance of receptor phosphorylation is not yet clear. It is possible that the receptor is a coincidental substrate for its kinase activity, and that there are other phosphorylated substrates of major biological importance either in metabolic regulation or growth control. Alternatively, receptor phosphorylation may serve as a specific intracellular signal, be involved in receptor internalization, or initiate the generation of an intracellular second messenger.[45]

8. SPECIFICITY OF BINDING

Studies of specificity of binding at receptor sites can yield very important information about the hormone–receptor interaction, and its relationship to post-receptor hormone-mediated effects. The structural determinants of binding in a hormone can be examined by comparing the binding of hormone analogues to that of native hormone.

An important comparison of the specificity of receptor binding and post-receptor events was carried out by Freychet et al.[47] They compared the relative potency of a series of insulins and derivatives in displacing labelled porcine insulin from rat liver plasma membranes, and in stimulating glucose oxidation in isolated fat cells. The insulins tested ranged from bonito insulin, with 50% of the potency of porcine insulin, to guinea-pig insulin, with only 1% of the potency. Over the entire range tested, the relative ability to compete with insulin binding to liver membranes paralleled the ability to stimulate fat cell metabolism. This study provided evidence that expression of the biological response to insulins and analogues was directly proportional to the binding of the peptides to their receptor.

The same does not always appear to be the case for peptides such as angiotensin II and ACTH. Thus, some synthetic angiotensin analogues displace labelled angiotensin II from adrenal membranes just as insulin analogues displace bound insulin, but unlike the insulin analogues, the angiotensin derivatives have no biological activity and thus block the activity of angiotensin II in stimulating aldosterone synthesis.[19]

Studies of structure–function relationships may yield different results depending on the method used to measure the biological response to a hormone. As an example, stimulation of adrenocortical plasma membrane adenylate cyclase has commonly been used as a measure of potency of ACTH. Finn *et al.*[48] tested several analogues of ACTH containing amino acid substitutions, in this system, and found that they antagonized the ACTH-stimulated membrane cyclase activity. However, the same analogues were found to stimulate cyclic AMP accumulation and steroidogenesis in intact adrenocortical cells. These studies emphasize that the relative specificities of some hormone analogues may differ in different biological systems, so that analogues acting as receptor agonists in one system may appear to be antagonists in another.

9. ANTIRECEPTOR ANTIBODIES

One of the aims of purifying receptors to homogeneity has been the generation of antireceptor antibodies. Shiu and Friesen[49] were the first to achieve this goal for a peptide hormone receptor. Their antibody against the purified rabbit mammary prolactin receptor was shown to inhibit binding of [^{125}I] prolactin to membrane receptors. The antibody also blocked specific prolactin-sensitive functions such as leucine incorporation into casein, without any effect on general cell metabolism. This important study provided clear evidence for an obligatory role of the prolactin receptor in mediating the action of prolactin.

Antibodies to insulin receptors did not initially yield such clear-cut results. Jacobs, Chang and Cuatrecasas[50] found that antibodies against rat liver insulin receptors did not block insulin binding but did precipitate solubilized receptors labelled with [^{125}I] insulin. They were also insulin-like in activity, stimulating glucose oxidation and inhibiting lipolysis. Thus, while clearly binding at a site different from the hormone binding site, these antibodies were able to initiate a biological response. In contrast, spontaneous antibodies against insulin receptors, observed in a small group of insulin-resistant patients with the skin condition, acanthosis nigricans, severely inhibited insulin binding to receptors from a variety of sources, and were capable themselves of stimulating muscle glucose metabolism.[51] Some of these antibodies react both with the insulin receptor and with the receptor for insulin-like growth factor-I, indicating that some common antigenic determinants are shared by the two receptor types.[52] The recent development of monoclonal antibodies against insulin receptors[53,54] should yield more information on structural similarities and differences between receptors for insulin and related peptides.

A novel approach to the problem of raising antireceptor antibodies is the use of anti-idiotypic antibodies raised against insulin antibodies.[55] The rationale of this approach is that a proportion of anti-insulin antibodies will have ligand binding sites similar to the binding sites of insulin receptors. Then antibodies formed against these receptor-like antibodies will themselves act as antireceptor antibodies. The demonstration that such antibodies do inhibit insulin binding to fat cell receptors suggests that this approach might have a wide application in hormone–receptor studies, particularly if combined with the recent monoclonal antibody technology which allows selection and amplification of particular antibody types from the mixture normally present in a polyclonal antiserum.

10. CONCLUSION

The past decade has seen some remarkable advances in our knowledge of hormone–receptor interactions. Among the milestones achieved over this period are the discovery that receptor concentration can be regulated by changes in hormone levels and the ensuing observations of receptor internalization, the discovery of cooperative interactions within some receptors and the demonstration of nuclear and other intracellular receptors for some peptide hormones. What progress can we expect in the next decade?

The use of receptor antibodies, particularly monoclonal antibodies, to probe receptor structure and function is one area of rapid advancement. The development of radioimmunoassays for receptors should provide a useful tool for these studies. Detailed mapping of functionally important regions in hormone molecules, as already attempted for insulin,[56] will also yield important knowledge about the nature of hormone binding. Finally, when the detailed mechanism of hormone and receptor internalization following binding to surface receptors has been elucidated, we should be much closer to a complete understanding of the hormone–receptor interaction.

REFERENCES

1. Birnbaumer, L., Pohl, S. L. and Kaumann, A. J. (1974) *Adv. Cyclic Nucleotide Res.,* **4**, 239.
2. Schlessinger, J., Schechter, Y., Willingham, M. C. and Pastan, I, (1978) *Proc. Nat. Acad. Sci. USA,* **75**, 2659.
3. Kahn, C. R., Neville, D. M. and Roth, J. (1973) *J. Biol. Chem.,* **248**, 244.
4. Goldfine, I. D. and Smith, G. J. (1976) *Proc. Nat. Acad. Sci. USA,* **73**, 1427.
5. Bergeron, J. J. M., Evans, W. H. and Geschwind, I. I. (1973) *J. Cell Biol.,* **59**, 771.
6. Goldfine, I. D., Vigneri, R., Cohen, D., Pliam, N. B. and Kahn, C. R. (1977) *Nature,* **269**, 698.
7. Cuatrecasas, P. (1972) *Proc. Nat. Acad. Sci. USA,* **69**, 1277.
8. Shiu, R. P. C. and Friesen, H. G. (1974) *J. Biol. Chem.* **249**, 7902.
9. Dufau, M. L. and Catt, K. J. (1976) *Antigens and Antibodies, Polypeptide Hormones and Small Molecules* (Eds. Beers, R. F. and Bassett, E. G.). Raven Press, N.Y., pp. 135–163.
10. Kahn, C. R. (1976) *J. Cell Biol.* **70**, 261.
11. Czech, M. P., Massague, J. and Pilch, P. F. (1981) *Trends Biochem. Sci.,* **6**, 222.
12. Massague, J. and Czech, M. P. (1982) *J. Biol. Chem.,* **257**, 5038.
13. Cuatrecasas, P. (1971) *J. Biol. Chem.,* **246**, 6532.
14. Cuatrecasas, P. (1973) *Federation Proc.,* **32**, 1836.
15. Hedo, J. A., Kahn, C. R., Hayashi, M., Yamada, K. M. and Kasuga, M. (1983) *J. Biol. Chem.,* **258**, 10020.
16. Hunter, W. M. and Greenwood, F. C. (1962) *Nature,* **194**, 495.

17. Roth, J. (1975) *Methods Enzymol.*, **37**, 223.
18. Kahn, C. R. (1975) *Methods Membrane Biol.*, **3**, 81.
19. Devynck, M. A., Koreve, V., Matthews, P. G., Meyer, P. and Pernollet, M. G. (1977) *Adv. Nephrol.*, **7**, 121.
20. Gliemann, J., Gammeltoft, S. and Vinten, J. (1975) *J. Biol. Chem.*, **250**, 3368.
21. Ciaraldi, T. P. and Olefsky, J. M. (1980) *J. Biol. Chem.*, **255**, 327.
22. Berson, S. A. and Yalow, R. S. (1958) *Adv. Biol. Med. Phys.*, **6**, 349.
23. Scatchard, G. (1949) *Ann. N.Y. Acad. Sci.*, **51**, 660.
24. Hollenberg, M. D. and Cuatrecasas, P. (1976) *Methods in Receptor Research,* Part II, (*Methods Mol. Biol.*, Vol. 9, Ed. Blecher, M.). Marcel Dekker Inc, N.Y., pp. 429–477.
25. Klotz, I. M. and Hunston, D. L. (1971) *Biochemistry*, **10**, 3065.
26. Ranke, M. B., Stanley, A., Tenore, A., Rodbard, D., Bongiovanni, A. M. and Parks, J. S. (1976) *Endocrinology*, **99**, 1033.
27. De Meyts, P., Roth, J., Neville, D., Gavin, J. and Lesniak, M. A. (1973) *Biochem. Biophys. Res. Commun.*, **55**, 154.
28. De Meyts, P. and Roth, J. (1975) *Biochem. Biophys. Res. Commun.*, **66**, 1118.
29. Baxter, R. C. and Williams, P. F. (1983) *Biochem. Biophys. Res. Commun.*, **116**, 62.
30. Bar, R. S., Harrison, L. C., Muggeo, M., Gorden, P., Kahn, C. R. and Roth, J. (1979) *Adv. Intern. Med.*, **24**, 23.
31. Kahn, C. R. and Roth, J. (1976) *Mod. Pharmacol. Toxicol.* Vol. 9 (Ed. Levey, G. S.). Marcel Dekker Inc, N.Y., p. 1.
32. Baxter, R. C., Bryson, J. M. and Turtle, J. R. (1980) *Endocrinology*, **107**, 1176.
33. Van Obberghen, E., De Meyts, P. and Roth, J. (1976) *J. Biol. Chem.*, **21**, 6844.
34. Das, M. and Fox, C. F. (1978) *Proc. Nat. Acad. Sci. USA*, **75**, 2644.
35. Baldwin, D., Prince, M., Marshall, S., Davies, P. and Olefsky, J. M. (1980) *Proc. Nat. Acad. Sci. USA*, **77**, 5975.
36. Amsterdam, A., Berkowitz, A., Nimrod, A. and Kohen, F. (1980) *Proc. Nat. Acad. Sci. USA*, **77**, 3440.
37. Gorden, P., Carpentier, J. L., Freychet, P. and Orci, L. (1980) *Diabetologia*, **18**, 263.
38. Saviolakis, G. A., Harrison, L. C. and Roth, J. (1981) *J. Biol. Chem.*, **256**, 4924.
39. Hazum, E., Chang, K. J. and Cuatrecasas, P. (1980) *Proc. Nat. Acad. Sci. USA*, **77**, 3038.
40. Krupp, M. and Lane, M. D. (1981) *J. Biol. Chem.*, **256**, 1689.
41. Posner, B. I., Raquidan, D., Josefsberg, Z. and Bergeron, J. J. M. (1978) *Proc. Nat. Acad. Sci. USA*, **75**, 3302.
42. Chang, K. J., Marcus, N. A. and Cuatrecasas, P. (1974) *J. Biol. Chem.*, **249**, 6854.
43. Cohen, S., Carpenter, G. and King, L. (1980) *J. Biol. Chem.*, **255**, 4834.
44. Ushiro, H. and Cohen, S. (1980) *J. Biol. Chem.*, **255**, 8363.
45. Kasuga, M., Karlsson, F. A. and Kahn, C. R. (1982) *Science*, **215**, 185.
46. Shia, M. A. and Pilch, P. F. (1983) *Biochemistry*, **22**, 717.
47. Freychet, P., Roth, J. and Neville, D. M. (1971) *Proc. Nat. Acad. Sci. USA*, **68**, 1833.
48. Finn, F. M., Johns, P. A., Nishi, N. and Hofmann, K. (1976) *J. Biol. Chem.*, **251**, 3576.
49. Shiu, R. P. C. and Friesen, H. G. (1976) *Science*, **192**, 259.
50. Jacobs, S., Chang, K-J. and Cuatrecasas, P. (1978) *Science*, **200**, 1283.
51. Le Marchand-Brustel, Y., Gorden, P., Flier, J. S., Kahn, C. R. and Freychet, P. (1978) *Diabetologia*, **14**, 311.
52. Jonas, H. A., Baxter, R. C. and Harrison, L. C. (1982) *Biochem. Biophys. Res. Commun.*, **109**, 463.

53. Roth, R. A., Maddux, B., Wong, K. Y., Styne, D. M., Van Vliet, G., Humbel, R. E. and Goldfine, I. D. (1983) *Endocrinology, 112*, 1865.
54. Krull, F. C., Jacobs, S., Su, Y. F., Svoboda, M. E., Van Wyk, J. J. and Cuatrecasas, P. (1983) *J. Biol. Chem., 258*, 6561.
55. Sege, K. and Peterson, P. A. (1978) *Proc. Nat. Acad. Sci. USA, 75*, 2443.
56. De Meyts, P., Van Obberghen, E., Roth, J., Wollmer, A. and Brandenburg, D. (1978) *Nature, 273*, 504.

Chapter 18

Genetic manipulation and the study of polypeptide hormones

1. INTRODUCTION

During the past few years techniques have become available whereby 'foreign' DNA can be inserted experimentally into microorganisms, particularly the bacterium *Escherichia coli*. Once inserted, this foreign DNA (in many cases equivalent to a 'gene' for a foreign protein) can be replicated, and in some cases transcribed into RNA and translated as protein. The development of these techniques has given rise to a powerful new methodology in biochemistry, variously known as 'genetic manipulation', 'genetic engineering' or 'recombinant DNA technology', application of which is having a revolutionary effect on our understanding of the eukaryote genome and on many areas of biochemistry and biology. Hormone biochemistry has already felt the impact of these new techniques, as has been mentioned several times in this book. They open up possibilities of:

(a) understanding in detail how hormones work on gene expression;

(b) studying the structures of the genes producing individual hormones (including alterations which may occur when hormone production is defective); and

(c) preparing substantial quantities of protein and polypeptide hormones (and their analogues) from bacteria as an alternative to preparing them from endocrine organs or by chemical synthesis.

The techniques of genetic engineering have already made a dramatic impact in some areas of hormone biochemistry and their use is likely to increase over the next few years. The topic is therefore given specific coverage in this chapter. References 1–3 may be consulted for further details.

2.   INSERTION AND MANIPULATION OF EUKARYOTIC DNA INTO PROKARYOTES: GENERAL PRINCIPLES

The general principle of the genetic engineering approach is that DNA derived from one organism is inserted into a different organism, usually a bacterium, in which it is replicated and sometimes expressed. Several specific steps can be identified:

(a) preparation of suitable DNA for insertion;

(b) insertion of the DNA in a form in which it will be replicated in the host bacterium;

(c) selection/cloning of the bacteria into which DNA has been inserted;

(d) recovery and identification of the cloned DNA; and

(e) in some cases, expression of the DNA as mRNA and protein.

2.1   Preparation of suitable DNA

DNA corresponding to all or part of a 'gene' for a polypeptide or protein hormone has been prepared in three ways:

(a) by reverse transcription of the corresponding mRNA (cDNA);

(b) by extensive cleavage of the total genome DNA using specific (restriction) endonucleases; and

(c) by chemical synthesis of a DNA sequence corresponding to a specific amino acid sequence (i.e. based on the genetic code).

2.1.1   Preparation of cDNA

During the past few years, good methods have been developed for the purification or partial purification of mRNAs for specific proteins.[4] Total RNA is readily purified from a tissue using phenol extraction (or other procedures) and total mRNA can then be purified from this by chromatography on oligo-dT cellulose, which specifically binds mRNA by base-pairing with the poly A tails of the latter. The total mRNA can then be fractionated by methods such as gel electrophoresis, gel filtration or density gradient centrifugation, and the distribution of a specific mRNA can be followed after such a fractionation by translation in a cell-free, protein-synthesizing system. An alternative method which has been devised for preparing specific mRNAs involves immunoprecipitation of polysomes. If a fresh tissue is homogenized and subjected to differential centrifugation, polysomes can be prepared from it which consist of several ribosomes attached to a mRNA molecule, together with corresponding nascent polypeptide chains. If such a polysome preparation, containing mRNAs for several different proteins, is immunoprecipitated with antiserum to a specific protein, the nascent chains of that particular protein, together with ribosomes and the corresponding mRNA, will be precipitated specifically. The mRNA can then be readily extracted from the precipitate. Obviously, whatever the method used for purification, it is advantageous to start with a tissue which synthesizes a considerable quantity of the protein (and mRNA) of interest. Thus, the anterior pituitary gland contains a high concentration of mRNA for growth hormone.

Once a pure or partially purified mRNA is available, this can be used as a template for the preparation of complementary (or copy) DNA (cDNA).[5] The enzyme reverse transcriptase is usually used for this purpose, together with an oligo-dT primer (a polynucleotide sequence comprising 10–20 deoxythymidines linked together). The primer base-pairs with the poly A tail of the mRNA (a sequence of adenosines is usually, though not invariably, linked to the 3′ terminus of an eukaryote mRNA molecule) and reverse transcriptase attaches nucleotides to this primer using the mRNA as a template. The result is a complementary, double-stranded polynucleotide in which one strand is the mRNA and the second strand is cDNA. The mRNA can now be removed from this duplex by alkaline or enzymic hydrolysis, leaving a single-stranded cDNA. This can in some circumstances be used directly for insertion into a bacterium, but is normally used as a template for production of double-stranded cDNA, using reverse transcriptase or DNA polymerase to synthesize the second strand of DNA. Again, a primer may be used in the conversion of single-stranded to double-stranded cDNA, but often the template acts as its own primer for synthesis of the second strand, in which case the cDNA duplex contains a hairpin loop at one end, which must subsequently be cleaved by a nuclease specific for single-stranded nucleic acid. The procedure used for preparation of cDNA is summarized in Figure 18.1. It should be noted that various features of the procedure used for cDNA preparation tend to lead to heterogeneity and incomplete molecules. A way of reducing the heterogeneity, which has been used frequently, is to cleave the cDNA with a specific restriction endonuclease (see Section 3.1) and then to purify the fragments so produced. This *ensures* that the DNA produced is only an incomplete copy of the mRNA, but eliminates much of the heterogeneity introduced during copying.

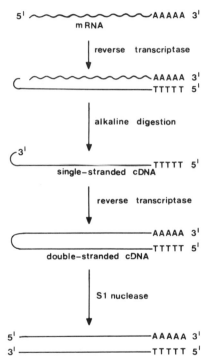

Figure 18.1 Preparation of double-stranded cDNA from a mRNA template, using reverse transcriptase

2.1.2   Genomic DNA fragments

A more direct source of DNA for insertion into microorganisms would at first sight appear to be the genomic (nuclear) DNA. However, two important factors have led to this being a less satisfactory source than cDNA.

First is the problem of identifying the individual region(s) of DNA of interest. There is sufficient DNA in the mammalian cell nucleus to code for about one million proteins of average size (although it is unlikely that this number of individual proteins is in fact made). Even though restriction endonucleases can cleave the genomic DNA into fragments which have an average size of about one gene, identification and purification of a fragment which contains all or part of the gene for any one polypeptide hormone, from the hundreds of thousands which do not, is an extremely difficult task. The only approach which has been used at all generally for such identification involves use of highly labelled, pure, cDNA or mRNA for the particular protein of interest, as a probe which can hybridize (i.e. form a specifically based-paired duplex) with the corresponding sequence in the genomic DNA, and thereby allow the identification of this. Since this procedure requires purer mRNA than is normally needed for preparation of cDNA, it is clear that in the first instance, anyway, direct cloning of cDNA is likely to be more convenient than use of genomic DNA.

The other problem encountered in using fragments of genomic DNA for insertion into bacteria is that it is now clear that many eukaryotic genes include 'intervening sequences' (introns) which are transcribed into RNA, but excised specifically in the processing of mRNA.[6] If fragments of genomic DNA are used directly for cloning experiments, these introns will be retained. Since bacteria are probably unable to excise these, they will be retained in 'mRNA transcripts' and will make expression of the protein of interest impossible.

Despite these drawbacks, many cloning experiments have been carried out using genomic DNA fragments. Such studies are usually performed after a suitable cDNA has been cloned, which can provide a probe for identification of an appropriate genomic DNA fragment. Cloning of genomic DNA has led to extremely valuable information about the organization of the eukaryote genome (including, for example, the organization of the human growth hormone gene—see below) and will eventually lead to a much greater understanding of how gene expression is controlled. It is unlikely, however, to be a very useful method where the aim of cloning experiments is the production of polypeptide hormones in bacteria.

2.1.3   Preparation of DNA by chemical synthesis

The third way of producing DNA for genetic manipulation experiments is by chemical synthesis. Until a few years ago such an approach would have been quite unrealistic, but rapid advances in the techniques available for such chemical synthesis have made feasible the synthesis of specific DNAs up to a hundred or so nucleotides long.[7] As a consequence, if the amino acid sequence of a peptide or small protein is available it is now possible to synthesize a corresponding nucleotide sequence (based on the genetic code) which can then be inserted into a bacterium and cloned. An advantage of this approach is that the codons chosen for the various amino acids in the protein can be those most readily utilized by the bacterial transcription and translation system rather than those actually used in the eukaryote from which the peptide is originally obtained. Disadvantages are that the method can give no information about the nature of the *real* nucleotide sequence of the mRNA or gene for the peptide, and that DNA probes produced will not match exactly the natural nucleotide sequence and may therefore be of limited use. Genetic engineering using synthetic DNA has been successfully carried out for somatostatin, the A and B chain of insulin and the first 24 residues of human growth hormone (see Sections 4–6). Synthetic oligodeoxynucleotides have also proved important as aids in the preparation and identification of specific cDNA sequences.

2.2 Physical insertion of DNA into the bacterium

Two main approaches have been explored for the insertion of DNA into bacteria in a form in which it can subsequently be replicated. The first uses a lysogenic bacteriophage as a cloning vector (transduction) and the second uses a plasmid (transformation). Lysogenic bacteriophages are bacterial viruses which can infect a bacterium and, under some circumstances, insert their DNA into that of the host. If foreign DNA is inserted (*in vitro*) into the bacteriophage DNA strand, this will also be inserted into the host's chromosome and will be replicated with it as the bacterium divides. Plasmids are small circular pieces of DNA which occur in many bacteria, replicate, at least partly, independently of the main bacterial chromosome, carry some important bacterial genes (for example, genes conferring resistance to many antibiotics) and are readily transferred from one bacterium to another by the process of transformation.[8] If a piece of 'foreign' DNA is inserted, experimentally, into a plasmid and the modified plasmid is then used to transform a bacterial colony, subsequent replication of that plasmid will then also involve replication of the foreign DNA, and in some cases will allow its expression. Of the two types of cloning vehicle, plasmids have become the most popular, for various technical reasons, and the subsequent discussion will be confined exclusively to these.

The development of suitable plasmids as cloning vehicles and of specific techniques for inserting DNA into such plasmids has been one of the main reasons for the rapid advances made in genetic engineering during the past few years. It is discussed in detail in Section 3. Suffice it to say here that once a suitable recombinant plasmid has been prepared (or, in practice, a mixture of recombinant plasmids and unmodified plasmids), its insertion into the host bacterium, in which no plasmid is initially present, is a simple procedure which involves simply mixing the two under suitable conditions. Such procedures resemble naturally-occurring transformation occurring in wild populations of bacteria.

2.3   Selection and cloning of transformed bacteria

The plasmids used as cloning vehicles usually possess at least two genes confer-ring antibiotic resistance, for example resistance to ampicillin, a derivative of penicillin, and tetracycline. The foreign DNA is usually inserted into one or other of these genes, and this insertion normally inactivates that gene. The host bacterium is usually sensitive to both antibiotics. After a transformation exper-iment three types of bacteria will be present, with different antibiotic sensitivi-ties. Those bacteria, which have not received a plasmid, will remain sensitive to both antibiotics, those which have received a plasmid without inserted foreign DNA will be resistant to both antibiotics, while those which have received a recombinant plasmid will be resistant to only one antibiotic, resistance to the other having been eliminated by the insertion of foreign DNA (Figure 18.2). Long-established microbiological techniques are available for selecting just the third class of bacteria, resistant to only one antibiotic, even when this comprises only a small proportion (one in a thousand or less) of the whole population.

Once the bacteria containing recombinant plasmids have been selected, they are cloned; i.e. colonies are grown, all bacteria in which are derived from a sin-gle individual, and all plasmids in which are therefore, usually, identical. This crucial step circumvents the problem introduced by the use of a heterogeneous DNA preparation for insertion into the plasmid—such heterogeneity now becomes resolved into a large number of bacterial clones, each of which con-tains a homogeneous recombinant DNA which is different from that in most of the other clones.

2.4   Identification and recovery of cloned DNA

Once a series of bacterial clones has been obtained, each containing a recom-binant-DNA-containing plasmid, there remains the identification of those clones which contain DNA fragments of interest, and then the extraction of this DNA. If the DNA used for preparing the recombinant plasmid was far from pure, as is often the case, then only a small proportion of the bacterial clones produced will in fact contain DNA fragments of interest. These can be identi-fied in several ways, including:

(a) specific identification of expressed protein, for example using specific antibodies against the protein of interest; but this approach is somewhat restric-ted since, as explained, the DNA fragment in a clone may not be expressed;

(b) hybridization with a suitable, usually labelled, cDNA or mRNA probe; (a quick and convenient method but often limited by lack of availability of a suitable probe; the Grunstein–Hogness technique is a specific application which is described in Section 3.3);

(c) the hybrid arrest technique (which allows identification of the specific mRNA, out of a complex mixture, with which recombinant DNA will hybridize); and

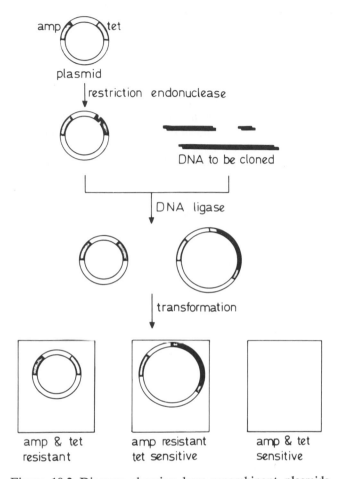

Figure 18.2 Diagram showing how recombinant plasmids containing a DNA insert in one of two antibiotic resistance genes can be selected for. Following insertion of DNA and transformation of a suitable bacterial host, bacterial colonies that are sensitive to just one of the two antibiotics, tetracycline (tet) or ampicillin (amp), are selected

(d) physicochemical characterization of the recombinant DNA (determination of nucleotide sequence, restriction endonuclease digestion patterns, etc.).

By use of one or several of these techniques those clones containing the recombinant DNA of interest can be identified. They can then be grown up in moderate amount, plasmid can be prepared from them, by lysing the bactcrial cells followed by differential or gradient centrifugation, and the recombinant DNA can be separated from the plasmid by cleaving the latter with a suitable restriction endonuclease and fractionating the digest.

2.5 Expression of recombinant DNA[9]

Where the aim of genetic manipulation is to produce a eukaryotic protein (such as insulin) in a bacterium, it is necessary to insert the foreign DNA into the plasmid in such a way as to facilitate its transcription as mRNA and the translation of the mRNA by the bacterial protein-synthetic machinery. It was originally thought that this would be a major problem, but it is fast becoming clear that prokaryotes are able to express eukaryotic genes more easily than was expected. In order to maximize the chances of expression, however, and particularly to maximize yields of expressed eukaryotic protein, various further manipulations of the recombinant DNA have been used. A successful approach has been to insert the eukaryotic DNA into the gene for β-galactosidase, tryptophan synthetase or β-lactamase, fairly close to the 5′ end of the gene. The consequence is that the bacterial clone containing eukaryotic DNA in such a position expresses a 'fusion protein' consisting of the N-terminal sequence of β-galactosidase (etc.), followed by the amino acid sequence of the eukaryotic protein. In some cases (e.g. somatostatin, see Section 6) the latter can be conveniently cleaved from the bacterial protein; in other cases such cleavage cannot be carried out so readily, but the eukaryotic protein can fold giving a conformation close enough to the natural one to be recognizable by antibodies. In order for satisfactory expression of the eukaryotic DNA, it must, of course, be in phase, and in the right orientation, with the bacterial gene into which it has been inserted.

Techniques for achieving controlled expression of eukaryotic DNA in prokaryotes are rapidly improving, and it is likely that with appropriate manipulation expression of almost any suitable fragment of DNA is achievable.

3. THE TECHNIQUES OF GENETIC MANIPULATION

A proper understanding of the potentialities and limitations of the new method of genetic manipulation requires knowledge of the remarkable new techniques on which it is based. It is not possible to detail all of these in a book of this kind, but a few of the more important experimental methods will be explained.

3.1 Restriction endonucleases[10]

Discovery of the highly specific restriction endonucleases provided a tool whereby DNA can be cleaved selectively into pieces of a useful size for genetic manipulation. The importance of these enzymes was recognized in 1978 by the award of the Nobel prize for Physiology and Medicine to W. Arber, H. O. Smith and D. Nathans who first discovered and applied them. About 100 restriction endonucleases have now been described, each with a different specificity. They are usually named after the initials of the bacterium from which they are obtained (e.g. *Eco RI*: *Escherichia coli* restriction endonuclease I). Most restriction endonucleases are specific for a sequence of 4, 5, or 6 bases in DNA, although sometimes the specificity is more complex; since there are 4 different bases, this means that cleavage points for a given enzyme will occur on average once in 4^4 (256), 4^5 (1024) or 4^6 (4096) bases. Restriction endonucleases normally cleave both strands of the DNA chain; many will cleave both single and double stranded DNA. As examples we will consider two restriction endonucleases, *Eco RI* and *Hae III*, with rather different specificities.

Eco RI cleaves a *symmetrical* sequence of 6 bases at the two (identical) points marked by arrows:

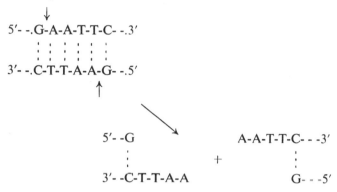

In order to separate the products, hydrogen (- - -) bonds have to be broken between the paired bases. An important consequence of this type of cleavage is that the products have complementary ('sticky') ends. They can be reannealed to each other, or to other products of *Eco RI* digestion, to give recombinant DNA molecules.

Hae III cleaves a symmetrical sequence of 4 bases:

$$5'- - -G-G-C-C- - -3'$$
$$3'- - -C-C-G-G- - -5'$$

$$5'- - -G-G \quad C-C- - -3'$$
$$3'- - -C-C \quad + \quad G-G- - -5'$$

In this case the products are blunt ended.

3.2 Insertion of foreign DNA into a plasmid

As already indicated, insertion of a foreign DNA into a plasmid forms a recombinant DNA which can then be cloned in a bacterium. Plasmid pBR 322[11] is frequently chosen for such purposes (see Figure 18.3). DNA can be inserted into this in several ways, three of the most important of which will be discussed here.

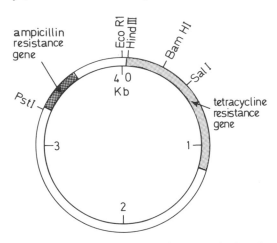

Figure 18.3 A simplified diagram of plasmid pBR322, showing the position of genes for antibiotic resistance and a few sites susceptible to restriction enzyme cleavage. The numbers indicate the length of DNA around the plasmid circle, in kilobases of nucleotide pairs

If the DNA to be inserted has 'sticky ends' (e.g. a fragment produced by *Eco RI* digestion), and if the plasmid contains a single site for the corresponding restriction endonuclease, complementary base pairing between these sticky ends can be used to insert the fragment. Thus, if the single *Eco RI* site of pBR 322 is cleaved, a linear DNA duplex is produced. This can combine with an *Eco RI* fragment, as indicated in Figure 18.4—the cleaved plasmid and the fragment are simply incubated together for a short while. Once base pairing has occurred, formation of covalent bonds can be achieved using the enzyme DNA ligase. Note that such a procedure will allow many plasmid DNA circles to reform without including foreign DNA (although procedures are available for reducing this) as well as the formation of some recombinants including more than one fragment of foreign DNA.

A second procedure for producing recombinant plasmids is known as 'tailing'. The enzyme deoxynucleotidyl terminal transferase will attach nucleotides to the 3' end of DNA. The plasmid is cleaved with a restriction endonuclease and a 'tail' of (say) thymidine is attached to the 3' ends of the linear DNA duplex, by incubating with thymidine triphosphate and terminal transferase.

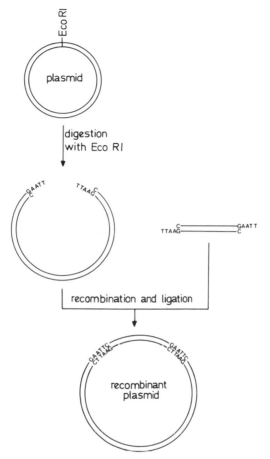

Figure 18.4 Insertion of DNA into a plasmid
using DNA-pairing at 'sticky ends'

The foreign DNA to be inserted is 'tailed' in a similar fashion, but with deoxy-adenosine. When plasmid and foreign DNA are now mixed, deoxyadenosine and thymidine tails base-pair (Figure 18.5), and covalent circles are completed using DNA ligase. In this case it is not possible for plasmids to recircularize without including a foreign DNA insert, although complexes including two or more plasmid molecules and two or more inserts are possible.

The third method used to construct recombinant DNA uses 'linkers'. These enable one kind of sticky end to be linked to a different, non-complementary kind. A short linear DNA duplex, made by chemical synthesis, is used, which includes a DNA sequence cleavable by a specific restriction endonuclease. Its use is illustrated in Figure 18.6. It should be noted that use of linkers in this way, or direct recombination of 'sticky ends', has the important advantage that the

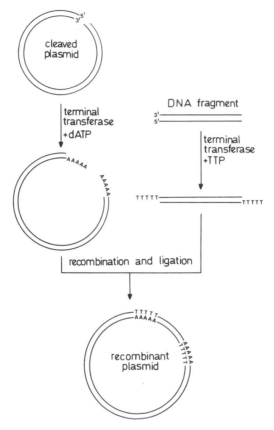

Figure 18.5 Insertion of DNA into a plasmid
following addition of poly A and poly T 'tails'
to plasmid and insert respectively

inserted fragment is readily excised from the recombinant plasmid by use of the appropriate restriction endonuclease(s).

3.3   Use of DNA (or RNA) probes; the Grunstein–Hogness technique[12]

A specific, highly-labelled fragment of DNA or RNA can be used as a probe to identify a complementary fragment of nucleic acid by virtue of its ability to hybridize with it. An example is the Grunstein–Hogness technique for identifying clones of bacteria producing a specific nucleic acid.[12] When a rather impure DNA fragment is cloned in a bacterium, only a small fraction of the recombinant clones will contain the DNA sequence of interest—the others contain the 'impurities', The clones of interest can be identified if a labelled probe is available. Several clones, growing on a solid medium, are 'blotted' onto a sheet of nitrocellulose, lysed and heated (to allow extrusion of DNA from the cells and strand separation). They are then exposed to the labelled probe, and those clones containing DNA sequences complementary to the probe form hybrids (duplexes) with it. All clones are then exposed to a nuclease (S1 nuclease)

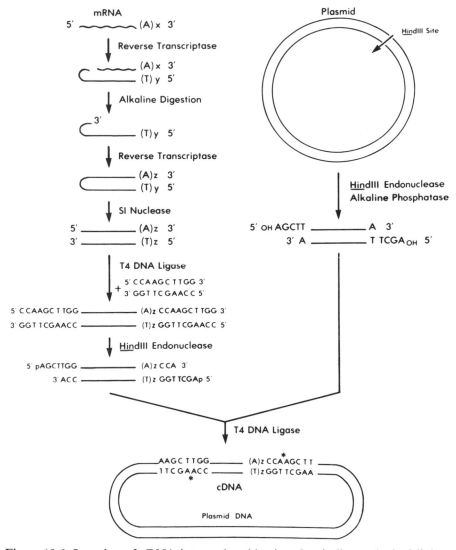

Figure 18.6 Insertion of cDNA into a plasmid using chemically synthesized linkers containing a specific restriction site (*Hind III* in this case). (Reproduced with permission from Ullrich, A. Shine, J., Chirgwin, J., Pictet, R., Tischer, E., Rutter, W. J. and Goodman, H. M. (1977) *Science* **196**, 1313–1319. Copyright (1977) the American Association for the Advancement of Science)

which digests only single-stranded DNA, when the probe is completely destroyed except where it has formed a double-stranded nucleic acid (i.e. in those clones containing the complementary DNA). After washing, the whole preparation is then subjected to autoradiography, when the clones of interest can be identified because they are radioactive. The original clones from which the blot was made can then be turned to for live bacteria. A related, powerful technique, 'Southern blotting', uses labelled DNA probes to identify specific DNA sequences in a complex mixture, after electrophoresis.

3.4　Determination of DNA sequences

The powerful new techniques of genetic manipulation have been accompanied by striking developments in the last few years in the techniques available for determining sequences of DNAs. These were devised originally by Sanger,[13,14] and further developed by Maxam and Gilbert[15] and others. They will not be described in detail here; suffice it to say that with these new techniques a DNA sequence of several hundred nucleotides can be determined in a single experiment taking only a few days. This means that the sequences of cloned DNAs can be determined readily. From such a sequence the amino acid sequence of the corresponding protein can be deduced, using the genetic code, and this both allows a very complete characterization of cloned DNA fragments and provides a rapid method for determining amino acid sequences.

4.　SOMATOSTATIN[16]

The 14-residue peptide somatostatin (Chapter 3) was the first peptide hormone produced in bacteria by genetic engineering. DNA for cloning this peptide was produced entirely by chemical synthesis. The amino acid sequence of the hormone is known (Chapter 3) and from this a corresponding nucleotide sequence was deduced. For each amino acid a codon was selected which was thought to be one favoured for expression in *E. coli*. An important feature of the peptide is that it does not contain methionine, since a specific chemical procedure is available for cleavage of a polypeptide chain after this amino acid (cleavage with cyanogen bromide). The DNA sequence synthesized is shown in Figure 18.7. In addition to nucleotides coding for the 14 amino acids of somatostatin, a triplet coding for methionine was placed at the start of the sequence, corresponding to the N-terminus of the peptide, and nucleotides specifying stop signals were placed at the end, corresponding to the C-terminus. Furthermore, a sequence corresponding to the *Eco RI* restriction enzyme site was constructed at the beginning of the sequence and one corresponding to the *Bam HI* restriction enzyme site was placed at the other end.

　　This synthetic DNA was inserted into the pBR 322 plasmid, together with a large DNA fragment containing the gene for the *E. coli* enzyme β-galactosidase

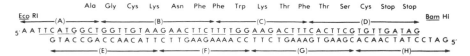

Figure 18.7 A synthetic deoxyoligonucleotide coding for somatostatin. Eight deoxyoligonucleotides (A–H) were synthesized chemically. These were designed to have at least five nucleotide complementary overlaps. When the oligonucleotides were mixed they combined, as a consequence of complementary base pairing, to give the structure shown. Treatment with DNA ligase then linked adjacent oligonucleotides covalently, to give a double-stranded DNA molecule that would code for somatostatin (amino acid sequence shown). This coding sequence is preceded by the codon for methionine and an Eco R1 restriction site, and followed by two 'stop signals' and a *Bam HI* restriction site. (Reproduced with permission from Itakura, K., Hirose, T., Crea, R., Riggs, A. D., Heyneker, H. L., Bolivar, F. and Boyer, H. W. (1977) *Science* **198**, 1056–1063. Copyright (1977) the American Association for the Advancement of Science)

and the control sequence for this gene (lac) (Figure 18.8). The insertion was such that the somatostatin gene was at the C-terminal end of the β-galactosidase gene and in phase with it. When this recombinant plasmid was inserted into *E. coli* and clones were grown, several clones were found which produced somatostatin-like activity (detectable by radioimmunoassay) after cleavage of the cellular protein with cyanogen bromide. Production of somatostatin by a clone was closely correlated with production of β-galactosidase. No immunoassayable somatostatin could be detected without cyanogen bromide cleavage.

Further investigation proved that a fusion product is produced, in which the 14 residues of somatostatin are attached via a methionine residue to the C-terminus of β-galactosidase. Cleavage with cyanogen bromide liberated somatostatin which was chemically, immunologically and biologically identical to natural material. (The cleavage also destroys the β-galactosidase, which contains several methionine residues.) The advantage of inserting the somatostatin gene adjacent to the β-galactosidase gene in this way is that it then becomes subject to the transcriptional and translational controls operating on the bacterial protein; the inserted gene for a eukaryotic peptide is thus made as if it were part of a bacterial protein. DNA-sequence analysis of a somatostatin-producing clone showed that the somatostatin gene was indeed inserted into the plasmid in the way expected (Figure 18.8).

A somatostatin-producing clone of *E. coli* has been grown up on a fairly large scale, and substantial quantities of the peptide have been produced from it. It is calculated that the cost of producing the peptide in this way is a small fraction of the cost of chemical peptide synthesis. A further advantage of the approach used—preparation of somatostatin as a fusion protein—is that the stability of the product is increased. Somatostatin itself, like many other small peptides, is rather unstable in the bacterial environment.

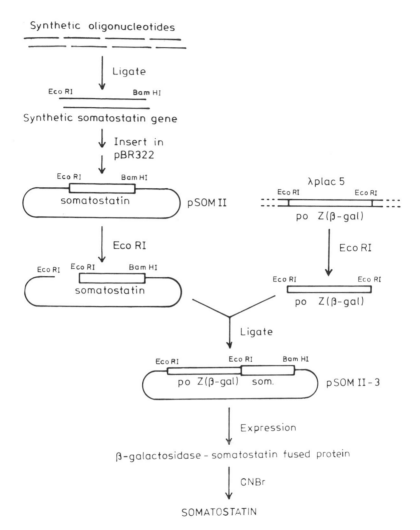

Figure 18.8 Bacterial expression of a cloned synthetic somatostatin gene. A recombinant plasmid pSOMII-3 was constructed, inserting the synthetic somatostatin gene (Figure 18.7) adjacent to a cloned fragment of the *E. coli* lac operon (po Z(β-gal)). Subsequent expression produced a β-galactosidase-somatostatin fused protein which could be cleaved with cyanogen bromide to release the somatostatin. (Reproduced with permission from Craig, R. K. and Hall, L. (1983). In *Genetic Engineering* **4** (Ed. Williamson, R.), pp. 57–125. Copyright (1983) Academic Press Inc. (London) Ltd)

5. INSULIN[17]

Cloning of the genes for insulin, and ultimately production of insulin in bacteria, has been seen as a major goal of genetic engineering ever since the new technology was conceived. An increasing shortage of insulin from conventional sources has added urgency and commercial incentive to the project.

The first successful cloning was reported in 1977, when Ullrich *et al.*[18] described the construction and cloning of plasmids containing cDNA made from rat insulin mRNA. They extracted mRNA from isolated rat islets of Langerhans, and from this prepared single- and then double-stranded cDNA (Section 2.1). They inserted such cDNA, and restriction endonuclease (*Hae III*) fragments prepared from it, which were thought to correspond to insulin-related sequences, into the plasmid pMB9, using linkers (which contained a site cleavable by another restriction endonuclease—*Hind III*) and then cloned recombinant plasmids. The clones so produced were screened for plasmids containing insulin-related DNA sequences. One such recombinant was found among the plasmids prepared from whole cDNA, and three were obtained from the cloned *Hae III* fragments. The insulin-related DNAs were cleaved from these recombinants (at the *Hind III* sensitive sites in the linkers) and their nucleotide sequences were determined. The sequences overlapped (Figure 18.9). It was deduced that three of the cloned DNAs corresponded to rat preproinsulin I, while the fourth corresponded to rat preproinsulin II (the rat has two, slightly different insulins, produced by non-allelic genes). From the DNA sequences of the fragments, an almost complete nucleotide sequence for the mRNA for rat preproinsulin I could be deduced, and from this the almost complete sequence of the protein itself could be derived. This study thus provided important detailed information about the structure of rat preproinsulin and its mRNA. However, no evidence for expression of the inserted DNA for insulin was obtained.

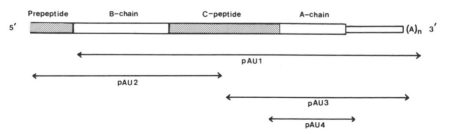

Figure 18.9 Figure illustrating the structure derived for the mRNA for rat preproinsulin, and the 4 cloned cDNA fragments (pAU1 to pAU4), the overlapping DNA sequences of which were used to deduce the mRNA structure

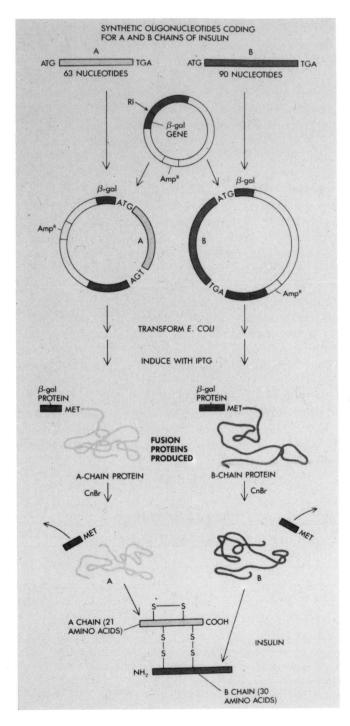

cDNA for human preproinsulin, made from mRNA prepared from a human pancreatic tumour, has been cloned.[19] Availability of the corresponding cloned DNA for rat preproinsulin was of great help in identifying the recombinant clone containing cDNA for human preproinsulin, using the Grunstein–Hogness technique (Section 3.3). Again this study has provided details of the mRNA for human preproinsulin, and of the protein for which it codes, but no expression was obtained. Expression of the rat preproinsulin gene has now been achieved by transferring the cloned cDNA to an 'expression plasmid' of the kind used to obtain expression of somatostatin (Section 4). The product of such expression is a fusion protein consisting of preproinsulin fused to β-lactamase. However, some expression plasmids have now been constructed which allow synthesis of an insulin-like product, carrying a prokaryotic and/or eukaryotic signal peptide, which is secreted from the bacterial cell (*E. coli*), and converted to pro-insulin.[20] Secretion has also been demonstrated in *Bacillus subtilis*. In these cases the bacterial cell is capable of recognizing the signal peptide as a determinant of secretion, and of subsequently removing this signal peptide, but conversion of proinsulin to insulin does not occur.

An alternative approach to expressing insulin genes in bacteria has been to synthesize the corresponding DNA chemically, as was done for somatostatin. This approach has now been successfully carried out.[21] DNAs which should code for the A and B chains of human insulin (neither of which contains methionine, unlike the preproinsulin precursor) were synthesized separately. Additional nucleotides were attached to the ends of the synthetic insulin A and B chain genes (as for somatostatin) to provide an extra N-terminal methionine, C-terminal stop signals and 'sticky ends'. These DNAs were then inserted, separately, into 'expression' plasmids in the same manner used for the somatostatin 'gene', so that they were at the end of, and in phase with, a β-galactosidase gene (Figure 18.10). The recombinant plasmids were cloned, and clones containing them were shown to yield a fusion protein containing β-galactosidase linked to the appropriate insulin chain. The insulin chains could be released from the fusion protein by treatment with cyanogen bromide, and purified from the reaction mixture. They were identical to natural A and B chains of human insulin in various physicochemical tests. In order to form native insulin, it is necessary for the A and B chains to associate. Procedures for carrying out such association *in vitro* in moderate yield are well established, using chains in which the cysteine residues have been sulphonated (converted to $-SSO_3^-$ groups). Association of the A and B chains so-produced gave rise to immunoreactive insulin.

Figure 18.10 Scheme showing how synthetic DNA sequences coding for the A and B chains were used to produce human insulin in *E. coli*. (Reproduced with permission from Watson, J. D., Tooze, J. and Kurtz, D. T. (1983) *Recombinant DNA: A Short Course*, Copyright (1983) the authors)

Human insulin obtained in this way is obtained in good yield. The material is highly active in animals and man, clinical trials in diabetic patients have been completed, and this genetically engineered insulin has been approved for human use. To what extent it will replace insulin produced from conventional sources over the next few years will depend on various factors, including production costs and the importance of side-effects experienced with insulin from the different sources.[21a]

The genomic DNA containing insulin genes has also been cloned, from rat and man,[22] and characterized in considerable detail. The insulin gene in most species, including man, contains two introns (Figure 18.11), but one of the two rat insulin genes contains only one. Putative regulatory DNA sequences close to the 5' end of the insulin gene have been recognized (see also Chapter 9).

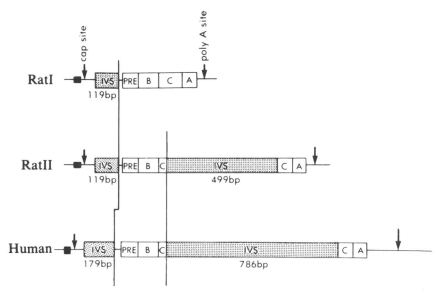

Figure 18.11 Schematic comparison of human and rat insulin genes. The overall structure of the two rat insulin genes (I and II) and the single human insulin gene is shown. Coding sequences for the peptide chains (pre, B, C and A) of preproinsulin are represented by clear boxes, while intervening sequences (IVS; introns) are distinguished by the shaded areas; the positions of homologous introns are marked by the vertical lines. Arrows indicate sites of the beginning and end of the transcribed sequences, and the black box indicates a potential site for the initiation of transcription. Note that IVS 1 occurs within the 5' untranslated region of the gene. (Reprinted by permission from Bell, G. I., Pictet, R. L., Rutter, W. J., Cordell, B., Tischer, E. and Goodman, H. M. (1980) *Nature* **284**, 26–32. Copyright © (1980) Macmillan Journals Ltd.)

cDNA or genomic DNA for preproinsulins has also been transferred to animal cells, after recombining with the DNA of a virus such as SV40, which allows replication in the animal cells and expression under the control of the viral promoter. When such recombinant DNA was transferred to a monkey kidney cell line, cells that secreted proinsulin into the medium were obtained. When a mouse pituitary cell line (AtT-20) was used (which normally produces, stores and secretes ACTH) the cell line obtained secreted mature insulin, and a substantial proportion of the material produced was stored in granules in the cells, along with ACTH.[23] Secretion of insulin was under similar controls to secretion of ACTH. Thus, the preproinsulin cDNA/SV40 recombinant introduced into this secretory cell type was able to feed insulin-like material into the secretory pathway in a fashion similar to that occurring in a normal pancreatic cell.

6. GROWTH HORMONE

Cloning of a cDNA made from mRNA for rat growth hormone was achieved in 1977.[24] mRNA was prepared from membrane-bound polysomes from cultured rat pituitary tumour cells, in which growth hormone production had been stimulated by dexamethasone and triiodothyronine. cDNA was prepared from this. The cDNA so prepared was cleaved with various restriction endonucleases and the fragments so produced were fractionated by gel electrophoresis. Major fragments were tentatively identified as corresponding to growth hormone, and DNA-sequence determination confirmed this identification. The intact cDNA was then fractionated by gel electrophoresis and an 800 base pair component was identified as containing mainly growth-hormone-related sequences. This was converted to double-stranded cDNA and inserted into plasmid pBR 322 using *Hind III* linkers. Several clones containing inserted DNA were obtained, of which 40% were shown to contain cDNA for growth hormone (identified by removing the inserted DNA by *Hind III* digestion and determining restriction endonuclease digestion patterns and in one case the complete DNA sequence). The determination of the DNA sequence of one of the cloned inserts allowed deduction of the complete sequence of rat pre-growth hormone, for which only a partial sequence was previously available.

Subsequently, expression of the cloned cDNA for rat growth hormone was achieved,[25] by transferring it to a plasmid in which it was linked to (and under the control of) the β-lactamase gene. The expression plasmid so produced (Figure 18.12), when reinserted into *E. coli*, directed the production of a fusion protein of molecular weight about 44 000, containing the N-terminal 181 residues of the β-lactamase precursor linked covalently to the 214 residues of rat pre-growth hormone. This was identified by gel electrophoresis and shown to cross-react immunologically with antibodies to rat growth hormone.

Production of growth hormone by genetic engineering clearly has considerable clinical potential since there is a marked shortage of growth hormone for treatment of human pituitary dwarfism. However, human growth hormone is needed for this purpose (see Chapter 7). Cloning of cDNA complementary to growth hormone mRNA made from human pituitary tumours has now been achieved,[26] using procedures similar to those used to clone rat growth hormone cDNA. The cloned human growth hormone cDNA has been sequenced, allowing deduction of the complete sequence of human pre-growth hormone. The cloned human growth hormone cDNA was also transferred to an expression plasmid containing part of the tryptophan operon (by which it is controlled). Again, expression results in production of a hybrid protein consisting of the N-terminal part of the trp D protein linked to human pre-growth hormone. This hybrid is recognized by antibody to human growth hormone.

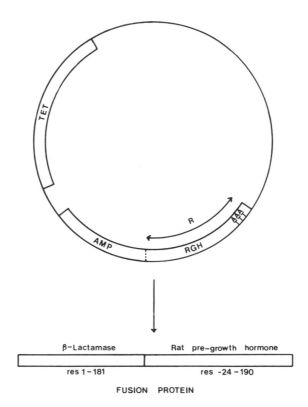

Figure 18.12 Diagram illustrating the expression plasmid used to produce a β-lactamase-rat pre-growth hormone fusion protein in *E. coli*. A cloned sequence (R) derived from rat pre-growth hormone cDNA was inserted into a plasmid derived from pMB9 and pBR322, in such an orientation that it could be expressed as part of a fusion protein containing the N-terminal part of β-lactamase (corresponding to the ampicillin resistance gene–AMP) followed by the pre-growth hormone sequence. (Based partly on ref. 25)

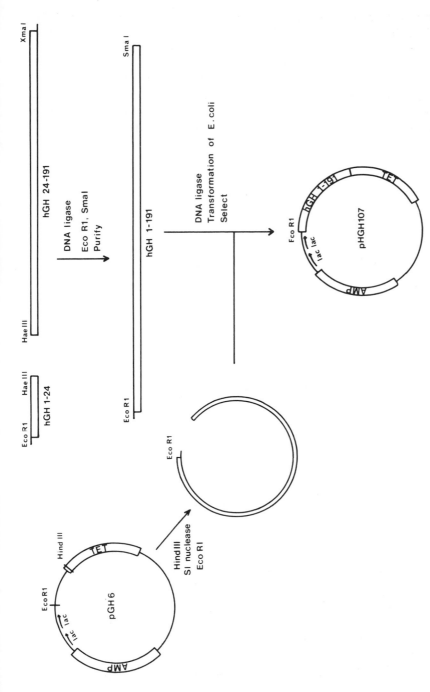

Figure 18.13 Scheme showing the procedure used to produce a plasmid that would give rise to expression of human growth hormone (hGH) in *E. coli*. A synthetic deoxynucleotide corresponding to the first 24 residues of hGH was combined with a restriction fragment of cDNA corresponding to the remainder of the molecule. The complete sequence was inserted into an expression plasmid (pGH 6) so as to be under the control of tandem lac promoters, leading to efficient expression of hGH. (Based in part on ref. 27)

A rather different approach to the cloning and expression of human growth hormone cDNA has also been described.[27] In this case a restriction endonuclease fragment was made from cDNA (derived from pituitary mRNA) and then cloned. This fragment was known, from previous studies, to include the coding sequence for residues 24–191 of human growth hormone (plus part of the 3' non-coding region). It was then linked, covalently, to a DNA sequence made by chemical synthesis which should code for residues 1–23 of human growth hormone, with an additional bacterial initiation codon (ATG). A 'hybrid gene' was thus constructed which should be expressible as growth hormone (*not* pre-growth hormone), when inserted in a plasmid 'down stream' of a suitable bacterial promoter. Such a plasmid was constructed, putting the growth hormone gene under the control of the lac operon (Figure 18.13). When this plasmid, plus several variants of it, were grown in suitable strains of *E. coli*, human growth hormone (detected by radioimmunoassay and polyacrylamide gel electrophoresis) was produced in good yield.

Such material has now been produced on a large scale and used in clinical trials for treatment of human hypopituitary dwarfism. It differs from pituitary-derived growth hormone by the presence of an additional N-terminal methionine, and some doubts remain as to whether it should be used therapeutically instead of material derived from human pituitaries. Various animal growth hormones have been produced by similar means, and offer potential as growth promotants in farm animals.

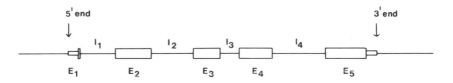

Figure 18.14 The structure of the growth hormone gene. The coding sequence is divided into 5 exons (E_1–E_5) by 4 introns (I_1–I_4)

The availability of cloned cDNA sequences for growth hormones from several species has enabled extensive studies to be carried out on the cloning and characterization of genomic DNA encoding growth hormone genes (see ref. 28). Genes for growth hormone from several species have been shown to contain four introns (Figure 18.14). In man a cluster of growth-hormone-related genes has been shown to occur in the long arm of chromosome 17 (Figure 18.15), including the hGH 1 gene encoding 'normal' human growth hormone, the hGH 2 gene encoding a variant of human growth hormone, the physiological significance of which is not yet clear, and several genes for the closely-related human placental lactogen. Studies on the genes in this cluster, as well as genes for prolactin, have provided new insight into the molecular evolution of the growth

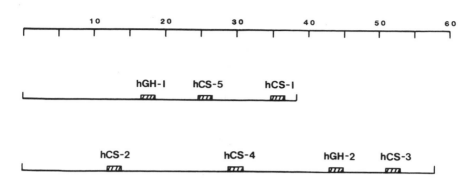

Figure 18.15 The overall organization of the growth hormone gene cluster in man. Two growth hormone-like genes (hGH-1, hGH-2) and five placental lactogen genes (hCS-1–hCS-5) are located on the long arm of chromosome 17. The organization of two sub-groups within this cluster is shown; the relationship between these two sub-groups is not clear. The scale at the top of the figure gives distances along the gene cluster in kilobase pairs, (Based on ref. 28)

hormone–prolactin family of hormones[28] (Chapter 8). In the rat there is no evidence for more than one gene for growth hormone.

Several studies have shown that when growth hormone cDNA or genomic DNA is linked with an appropriate viral vector, such as SV40 or bovine papilloma virus, and used to transform mammalian cells, cell lines that produce and secrete growth hormone can be obtained.[29,30] In some cases expression of growth hormone 'genes' introduced in this way can be stimulated by glucocorticoids, in the same way as the normal growth hormone genes, and this type of system thus holds great promise for the exploration of the DNA sequences that are involved in such regulation. It is also possible for growth hormone genes, linked to an appropriate promoter (to enable expression), to be incorporated directly into the genome of mammalian cells. The most spectacular application of this approach has been the production of giant mice by microinjection of fused rat metallothionein–growth hormone genes into fertilized mouse eggs followed by reimplantation of the eggs.[31] Mice were obtained in which there was a massive production of rat growth hormone by liver and other tissues, extremely high serum growth hormone levels, and giantism (Figure 18.16). The implications for increasing body weight in farm animals, studies on gene regulation, production of peptides and gene therapy are considerable. Similar studies have now been carried out involving introduction of metallothionein–human growth hormone fusion genes into mice.[32]

A final application of cDNA corresponding to growth hormone is in the study of the molecular basis of growth hormone deficiency, as has been discussed already in Chapter 7.

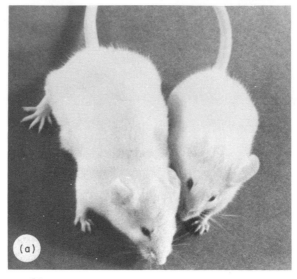

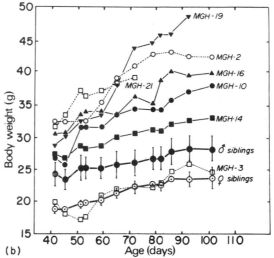

Figure 18.16 Giantism in transgenic mice. Mice in which rat GH genes had been introduced by injection of eggs showed elevated circulating growth hormone levels and increased growth. (a) A transgenic mouse (left) is compared with a normal littermate. (b) Growth of transgenic mice (MGH-2,3,10,14,16,19,21) is compared with that of male and female littermates. (Reprinted by permission from Palmiter, R. D., Brinster, R. L., Hammer, R. E., Trumbauer, M. E., Rosenfeld, M. G., Birnberg, N. C. and Evans, R. M. (1982) *Nature* **300**, 611–615. Copyright © (1982) Macmillan Journals Ltd.)

7. OTHER APPLICATIONS OF GENETIC ENGINEERING IN THE STUDY OF POLYPEPTIDE HORMONES

There have been many other applications of the recombinant DNA technology of the kind already described. For example, cDNAs have been cloned corresponding to all or part of human placental lactogen,[33] rat prolactin,[34] bovine pro-opiocortin (see Chapter 5),[35] pre-proparathyroid hormone[36] and the α and β chains of gonadotropins.[37,38] In all of these studies the cloned cDNAs have been subjected to DNA sequence determination which has provided extensive information about the structures of the corresponding mRNAs and the proteins for which they code. In the case of pro-opiocortin cDNA, the information obtained about the precursor of the ACTH/MSH/endorphin family provided amino acid sequence information which would have taken years to obtain by conventional methods; this study even allowed the prediction of the existence of a new hormone—γ-MSH (see Chapter 5).

Expression of polypeptide hormone 'genes' in bacteria has been most extensively studied in the cases of growth hormone, insulin and somatostatin, but similar studies on several other peptide hormones have now been reported,[9] as well as work on peptide growth factors and interferons. Similarly, the type of study of gene structure at the genomic level that has been mentioned above with regard to the growth hormone genes is being extended to other protein and polypeptide hormones.

The use of DNA probes produced by recombinant DNA studies for studying the expression of polypeptide hormone genes in the hormone-producing cells is also being applied to many different hormones. Growth hormone has already been mentioned; prolactin is another example.

Application of recombinant DNA techniques promises to provide extensive information on the nature of receptors for polypeptide hormones. A striking example is the recent elucidation of the complete sequences of the α and β subunits of the insulin receptor, including the demonstration that the two are formed as parts of the same precursor.[39]

Finally, it must also be stressed that recombinant DNA technology is being energetically applied to study of the actions of hormones on gene expression. Studies on the mechanism whereby oestrogens induce production of ovalbumin and other specific proteins in the chicken oviduct provide a spectacular application in this area. The structure of the ovalbumin gene has now been delineated in very great detail, and knowledge of this, with its adjacent and intervening sequences, may well provide the vital clues eventually to the mechanism whereby steroid hormones can 'switch on' the expression of this gene. Similarly, detailed studies of the structure of genes for milk proteins should provide a valuable basis for understanding the induction of such proteins by prolactin and other hormones, as discussed in Chapter 8.

The techniques of genetic engineering have only been available for a few years. They have already made a big impact on the polypeptide hormone field and it seems likely that they will soon be applied to all protein and polypeptide hormones. The information that they can provide about the structure and expression of genes and mRNAs, and indirectly protein precursors, for homones is unique, and this plus the great scope for commercial exploitation should ensure that their impact on the field is an increasing one for many years to come.

REFERENCES

1. Glover, D. M. (1980) *Genetic Engineering—Cloning DNA*. Chapman & Hall, London.
2. Craig, R. K. and Hall, L. (1983) In: *Genetic Engineering*, Vol. 4 (Ed. Williamson, R.). Academic Press, London, pp. 57–125.
3. Watson, J. D., Tooze, J. and Kurtz, D. T. (1983) *Recombinant DNA*. Scientific American Books, New York.
4. Taylor, J. M. (1979) *Ann. Rev. Biochem.*, **48**, 681–717.
5. Williams, J. G. (1981) In: *Genetic Engineering*, Vol. 1 (Ed. Williamson, R.). Academic Press, London, pp. 1–59.
6. Breathnach, R. and Chambon, P. (1981) *Ann. Rev. Biochem.*, **50**, 349–383.
7. Gassen, H. G. and Lang, A. (Eds.) (1982) *Chemical and Enzymatic Synthesis of Gene Fragments*. Verlag Chemie, Weinheim.
8. Thompson, R. (1982) In: Genetic Engineering, Vol. 3 (Ed. Williamson, R.). Academic Press, London, pp. 1–52.
9. Harris, T. J. R. (1983) In: *Genetic Engineering*, Vol. 4 (Ed. Williamson, R.). Academic Press, London, pp. 127–185.
10. Malcolm, A. D. B. (1981) In: *Genetic Engineering*, Vol. 2 (Ed. Williamson, R.). Academic Press, London, pp. 129–173.
11. Bolivar, F. (1979) *Life Sciences*, **25**, 807–817.
12. Grunstein, M. and Hogness, D. S. (1975) *Proc. Nat. Acad. Sci. USA*, **72**, 3961–3965.
13. Sanger, F. and Coulson, A. R. (1975) *J. Mol. Biol.*, **94**, 441–448.
14. Sanger, F. and Coulson, A. R. (1978) *FEBS Lett.*, **87**, 107–110.
15. Maxam, A. M. and Gilbert, W. (1977) *Proc. Nat. Acad. Sci. USA*, **74**, 560–564.
16. Itakura, K., Hirose, T., Crea, R., Riggs, A. D., Heyneker, H. L., Bolivar, F. and Boyer, H. W. (1977) *Science*, **198**, 1056–1063.
17. Riggs, A. D., Itakura, K., Crea, R., Hirose, T., Kraszewski, A., Goeddel, D., Kleid, D., Yansura, D. G., Bolivar, F. and Heyneker, H. L. (1980) *Recent Progr. Hormone Res.*, **36**, 261–276.
18. Ullrich, A., Shine, J., Chirgwin, J., Pictet, R., Tischer, E., Rutter, W. J. and Goodman, H. M. (1977) *Science*, **196**, 1313–1319.
19. Bell, G. I., Swain, W. F., Pictet, R., Cordell, B., Goodman, H. M. and Rutter, W. J. (1979) *Nature*, **282**, 525–527.
20. Talmadge, K., Brosius, J. and Gilbert, W. (1981) *Nature*, **294**, 176–178.
21. Goeddel, D. V., Kleid, D. G., Bolivar, F., Heyneker, H. L., Yansura, D. G., Crea, R., Hirose, T., Kraszewski, A., Itakura, K. and Riggs, A. D. (1979) *Proc. Nat. Acad. Sci. USA*, **76**, 106–110.
21a. Johnson, I. S. (1983) *Science*, **219**, 632–637.
22. Bell, G. I., Pictet, R. L., Rutter, W. J., Cordell, B., Tischer, E and Goodman, H. M. (1980) *Nature*, **284**, 26–32.
23. Moore, H. P. H., Walker, M. D., Lee, F. and Kelly, R. B. (1983) *Cell*, **35**, 531–538.

24. Seeburg, P. H., Shine, J., Martial, J. A., Baxter, J. D. and Goodman, H. M. (1977) *Nature,* **270**, 486–494.
25. Seeburg, P. H., Shine, J., Martial, J. A., Ivarie, R. D., Morris, J. A., Ullrich, A., Baxter, J. D. and Goodman, H. M. (1978) *Nature,* **276**, 795–798.
26. Martial, J. A., Hallewell, R. A., Baxter, J. D. and Goodman, H. M. (1979) *Science,* **205**, 602–607.
27. Goeddel, D. V., Heyneker, H. L., Hozumi, T., Arentzen, R., Itakura, K., Yansura, D. G., Ross, M. J., Miozzari, G., Crea, R. and Seeburg, P. H. (1979) *Nature,* **281**, 544–548.
28. Miller, W. L. and Eberhardt, N. L. (1983) *Endocrine Rev.,* **4**, 97–130.
29. Rosenfeld, M. G., Amara, S. G., Birnberg, N. C., Mermod, J. J., Murdoch, G. H. and Evans, R. M. (1983) *Recent Progr. Hormone Res.,* **39**, 305–351.
30. Pavlakis, G. N. and Hamer, D. H. (1983) *Recent Progr. Hormone Res.,* **39**, 353–385.
31. Palmiter, R. D., Brinster, R. L., Hammer, R. E., Trumbauer, M. E., Rosenfeld, M. G., Birnberg, N. C. and Evans, R. M. (1982) *Nature,* **300**, 611–615.
32. Palmiter, R. D., Norstedt, G., Gelinas, R. E., Hammer, R. E. and Brinster, R. L. (1983) *Science,* **222**, 809–814.
33. Shine, J., Seeburg, P. H., Martial, J. A., Baxter, J. D. and Goodman, H. M. (1977) *Nature,* **270**, 494–499.
34. Gubbins, E. J., Maurer, R. A., Hartley, J. L. and Donelson, J. E. (1979) *Nucleic Acids Res.,* **6**, 915–930.
35. Nakanishi, S., Inoue, A., Kita, T., Nakamura, M., Chang, A. C. Y., Cohen, S. N. and Numa, S. (1979) *Nature,* **278**, 423–427.
36. Kronenberg, H. M., McDevitt, B. E., Majzoub, J. A., Nathans, J., Sharp, P. A. Potts, J. T. and Rich, A. (1979) *Proc. Nat. Acad. Sci. USA,* **76**, 4981–4985.
37. Fiddes, J. C. and Goodman, H. M. (1979) *Nature,* **281**, 351–356.
38. Talmadge, K., Vamvakopoulos, N. C. and Fiddes, J. C. (1984) *Nature,* **307**, 37–40.
39. Ullrich, A., Bell, J. R., Chen, E. Y., Herrera, R., Petruzzelli, L. M., Dull, T.J., Gray, A., Coussens, L., Liao, Y-C., Tsubokawa, M., Mason, A., Seeburg, P. H., Grunfeld, C., Rosen, O. M. and Ramachandran, J. (1985) *Nature,* **313**, 756–761.

Index